World Soils Book Series

Series editor

Prof. Alfred E. Hartemink
Department of Soil Science, FD Hole Soils Laboratory
University of Wisconsin–Madison
Madison
USA

Aims and Scope

The *World Soils Book Series* brings together soil information and soil knowledge of a particular country in a concise and reader-friendly way. The books include sections on soil research history, geomorphology, major soil types, soil maps, soil properties, soil classification, soil fertility, land use and vegetation, soil management, and soils and humans.

International Union of Soil Sciences

More information about this series at http://www.springer.com/series/8915

Pavle Pavlović · Nikola Kostić
Branko Karadžić · Miroslava Mitrović

The Soils of Serbia

Springer

Pavle Pavlović
Department of Ecology, Institute for Biological Research "Siniša Stanković"
University of Belgrade
Belgrade
Serbia

Nikola Kostić
Department of Ecology, Institute for Biological Research "Siniša Stanković"
University of Belgrade
Belgrade
Serbia

Branko Karadžić
Department of Ecology, Institute for Biological Research "Siniša Stanković"
University of Belgrade
Belgrade
Serbia

Miroslava Mitrović
Department of Ecology, Institute for Biological Research "Siniša Stanković"
University of Belgrade
Belgrade
Serbia

ISSN 2211-1255 ISSN 2211-1263 (electronic)
World Soils Book Series
ISBN 978-94-024-1404-2 ISBN 978-94-017-8660-7 (eBook)
DOI 10.1007/978-94-017-8660-7

Printed on acid-free paper

This Springer imprint is published by Springer Nature
The registered company is Springer Science+Business Media B.V.
The registered company address is: Van Godewijckstraat 30, 3311 GX Dordrecht, The Netherlands

Preface

'The Soils of Serbia' is part of the Springer international series on World Soils and brings together all the information that is available on soils in Serbia. It is the first publication to consolidate all the efforts of the many generations of Serbian soil scientists over 165 years of soil survey and also the first monograph on Serbian soils to be published in a foreign language. We believe that this remarkable, informative and timely work will provide future generations of soil scientists with great impetus.

In particular, this book is in homage to all the scientists who excelled in their research into soils in the former Yugoslavia, including Serbia, and is also a tribute to soil science on the 200th anniversary of the birth of the great Serbian botanist, Josif Pančić (1814–1888), who was the first professor of Natural History and Agriculture, the first rector at the Great School in Belgrade, and the founder of the Institutes of Mineralogy and Geology, Zoological and Botanical Departments, and the Experimental Botanical Gardens in Belgrade. Coincidentally, work on this book started in 2015—a year declared the International Year of Soils by the 68th UN General Assembly.

Integrated soil research began in the mid-nineteenth century (1849) when Prince Aleksandar Karađorđević founded the agricultural school, which was allocated agricultural and forest land in Topčider in Belgrade. Over time, it grew into the Institute for Soil Sciences. The period after World War I was important for soil research with the Faculty of Agriculture being founded in Zemun and soil research gaining new impetus, thanks to Aleksandar Stebut, professor of pedology. He contributed to the formation of a well-established soil science research team at the faculty in the first decade of the twentieth century. As a result, the first soil maps for the Kingdom of Yugoslavia (scale: 1:3,500,000) were published and later, in 1931, at the scale of 1:200,000. In the same period, Prof. Stebut proposed the first national Serbian soil classification system, and since then several soil classifications have been published. The last version of the Classification of Yugoslav Soils (Škorić et al. 1985) is still in use in all the republics of the former Yugoslavia. Antonović and Protić (1997) proposed a modified classification for soils in Serbia, suggesting the need for harmonization of the national classification with the FAO, WRB and ESD systems, based on the results of satellite and aerial images and additional investigations in some areas.

The most intensive period of development in the field of soil sciences in Yugoslavia was witnessed post-World War II, from 1945 onwards. Over the 60 years that followed, in spite of difficulties during the 1990s, the whole territory of Serbia was mapped at a scale of 1:50,000. The use of contemporary geographic information systems in soil surveying resulted in a new digitized soil map of Serbia, which was published by Mrvić et al. (2013). The current Soil Map of Serbia, with 20 mapping units, was created on the basis of 1:50,000 pedological maps, produced by the Institute of Soil Science (Belgrade), the Institute of Field and Vegetable Crops (Novi Sad), the 'Jaroslav Černi' Institute for the Development of Water Resources (Belgrade) and the Environmental Protection Agency of Ministry of Agriculture and Environmental Protection of Republic of Serbia.

Serbian soils have been investigated and presented in numerous monographs, with contributions by authors from the former Yugoslavia, as well as those from Serbia. In the beginning, soil research focused on soil fertility and plant production, but in the 1970s, there was gradually a more holistic approach and research branched out into more specialized fields of soil survey. Advances in instrumental methods made a critical review of soil pollution possible, as well as the genesis of Serbian soils through pedogeochemical and pedomineralogical studies.

Soil is a limited but very valuable resource that attracts many human activities, such as agriculture, forestry, architecture and civil engineering. Hence, it is essential to understand the processes that lead to soil degradation, pollution and devastation. This will allow for the better management, protection and eventual remediation of this important resource in terms of modern land use and current soil utilization.

Today, soil degradation is a multifaceted problem of mining and industrial activity, and in some places even uncontrolled municipal and industrial waste disposal. It is intensified by water and aeolian erosion due to climate change and has caught researchers' attention more and more. More intensive land use and the expansion of urban areas and transport systems in Serbia have placed increased pressure on soil resources. Changes in agricultural soil management have also led to the depletion of organic matter in soil, as well as soil acidification and in some places even soil salinization. In short, there is an urgent need for better soil protection to safeguard this important resource and its capacity to support food production in future times of climate change.

The book contains 11 chapters that cover a variety of topics related to the comprehensive presentation of all the soils of Serbia. However, bearing in mind the factors that influence the occurrence and formation of soils, such as relief, climate, geological substrate and vegetation, considerable attention has been paid to the climatic, geological, mineralogical and vegetation characteristics of Serbia. Part of the book is also dedicated to the evermore marked problem of soil degradation due to various anthropogenic activities.

The authors hope that this book's contents will attract the interest of soil scientists, geologists, ecologists, students, soil managers and stakeholders, and that it will be of interest to the general public as well. We believe that this extensive book will contribute to the better understanding of the soil resources of Serbia, but will also be of use to an international professional audience as it is published in English.

Belgrade, Serbia

Pavle Pavlović
Nikola Kostić
Branko Karadžić
Miroslava Mitrović

Acknowledgements

The data presented in this book include present-day research and also that collected during the second half of the twentieth century, which led to the publication of the Soil Map of Serbia. Therefore, cordial thanks goes to all those scientists from Serbia and the former Yugoslavia who have worked in this field over the last two centuries and contributed greatly to knowledge on processes in soils, soil classification and soil mapping in Serbia, beginning with the first 1:3,500,000 Soil Map of the Kingdom of Yugoslavia and the first Serbian national soil classification system by Prof. A. Stebut in 1926.

Now the book has been completed, we should express our gratitude to all those who have contributed at any point, from the moment the idea to write this book was born right up to the final phases of this demanding project, including publication.

We would like to thank the series editor, Dr. Alfred E. Hartemink, who has initiated an excellent series of books entitled the *World Soils Book Series*. Likewise, our thanks go to Dr. Elodie Tronche, assistant editor of Springer Verlag (Dordrecht, the Netherlands), for her great patience and welcome support during this project. We would also like to thank the Springer project coordinators for their help during the final stage of editing this book.

We owe a debt of gratitude to our colleagues from those institutions concerned with various aspects of soil science: the Serbian Academy of Sciences and Arts (SANU), the Institute for Soil Science in Belgrade, the Faculty of Agriculture (the University of Belgrade), the Faculty of Forestry (the University of Belgrade), the Faculty of Agriculture (the University of Novi Sad), the Institute of Field and Vegetable Crops in Novi Sad, the Institute for Biological Research 'Siniša Stanković' (the University of Belgrade), the Faculty of Mining and Geology (the University of Belgrade), the Faculty of Biology (the University of Belgrade), the Institute of Forestry in Belgrade and the Institute of Lowland Forestry and Environment in Novi Sad. The results of their research contributed significantly to the writing of certain chapters in this book.

Furthermore, we are grateful that the data used in this book were made available to us by state institutions of the Republic of Serbia: the Government of the Republic of Serbia, the Ministry of Education, Science and Technological Development, the Ministry of Agriculture and Environmental Protection, the Serbian Environmental Protection Agency, the Statistical Office of the Republic of Serbia and the Privatization Agency of the Republic of Serbia.

Special gratitude goes to our colleagues: Dr. V. Mrvić, Dr. G. Antonović, Dr. D. Čakmak, Dr. V. Perović, Dr. S. Maksimović, Dr. E. Saljnikov and M. Nikoloski, who provided data for the map of Serbian soils. The Soil Map was presented by Dr. Mrvić et al. (2013) at the 1st International Congress and XIII Congress of Soil Science, held in Belgrade in 2013.

We express our thanks to Dr. Snežana Belanović, Dr. Olga Kostić and Dr. Jovica Vasin, who were of great assistance in the compilation of certain sections of this publication.

We are also very grateful to all our colleagues from the Institute of Biological Research 'Siniša Stanković', in particular Dr. M. Paunović, who provided us with invaluable help with the graphical maintenance used in this book and useful assistance by providing the databases we needed during this project.

Belgrade, Serbia
December 2016

Pavle Pavlović
Nikola Kostić
Branko Karadžić
Miroslava Mitrović

Contents

1 Introduction . . . 1
- 1.1 General Information Regarding Serbia . . . 1
  - 1.1.1 Geographic Position . . . 1
  - 1.1.2 Topography . . . 1
  - 1.1.3 Hydrology . . . 4
  - 1.1.4 Soil . . . 11
- 1.2 A Short History of Soil Science and Soil Survey . . . 14
- References . . . 23

2 Soils as Natural Resources . . . 25
- 2.1 Soil Types and Their Production Value . . . 27
- 2.2 Soil Under Agriculture . . . 29
- References . . . 29

3 Climate . . . 31
- 3.1 Temperature . . . 31
- 3.2 Precipitation . . . 33
- References . . . 39

4 Vegetation . . . 41
- 4.1 Forest Vegetation . . . 42
- 4.2 Scrub Vegetation . . . 48
- 4.3 Herbaceous Vegetation . . . 49
- References . . . 52

5 Geology . . . 55
- 5.1 Geomorphology . . . 55
  - 5.1.1 Dinarides . . . 55
  - 5.1.2 Vardar Zone . . . 57
  - 5.1.3 Serbian Massif . . . 58
  - 5.1.4 Carpatho-Balkanides . . . 58
  - 5.1.5 Pannonian Basin . . . 59
- 5.2 Lithology . . . 59
  - 5.2.1 The Pannonian Basin and Peripannonian Rim Region . . . 62
  - 5.2.2 The Mountainous Region . . . 66
- 5.3 Soil Mineralogy . . . 78
- References . . . 85

6 Soil Classification . . . 87
- 6.1 A Historical Overview of Soil Classification . . . 87
- 6.2 Soil Map Research . . . 88
- 6.3 Actual Soil Classification . . . 88

6.3.1 Definitions and Labelling of Genetic Soil Horizons 89
6.3.2 Soil Classification 90
References 99

7 Order of Automorphic Soils 101
7.1 Class I: Undeveloped Soils with (A)-C or (A)-R Soil Horizon Sequence 101
7.1.1 Lithosol 101
7.1.2 Regosol 102
7.1.3 Arenosol 104
7.1.4 Colluvial Soils (Colluvium) 105
7.2 Class II: Humus-Accumulative Soils with A-C, A-R or A-C-R Soil Horizon Sequence 107
7.2.1 Ranker 107
7.2.2 Rendzina 113
7.2.3 Chernozem 117
7.2.4 Vertisol 122
7.2.5 Kalkomelanosol (Limestone-Dolomite Black Soil) 125
7.3 Class III: Cambic Soils with A-(B)-C or A-(B)-R Soil Horizon Sequence 128
7.3.1 Eutric Cambisol 128
7.3.2 Dystric Cambisol 131
7.3.3 Kalkocambisol (Brown Soil on Limestone and Dolomite) 135
7.3.4 Red Soil (*Terra rossa*) 136
7.4 Class IV: Eluvial-Illuvial Soils of A-E-Bt-C or A-E-Bt-R Soil Horizon Sequence 139
7.4.1 Luvisol (Luvic Soil) 139
7.4.2 Podzol 142
7.4.3 Brown Podzolic Soil (Brunipodzol) 143
7.5 Class V: Anthropogenic Soils with P-C Soil Horizon Sequence 145
7.5.1 Rigosol 145
7.5.2 Hortisol 145
7.6 Class VI: Technogenic Soils (Technosols) 145
7.6.1 Soil of Disposal Sites (Deposol) 145
References 154

8 Order of Hydromorphic Soils 157
8.1 Class I: Soils with Stagnation of Rainwater with A-Eg-Bg-C Soil Horizon Sequence 157
8.1.1 Pseudogley 157
8.2 Class II: Initial Hydromorphic Soils with I-II-III-IV..., or (A)-G, or (A)-C Horizon Sequence 160
8.2.1 Fluvial or Alluvial (Fluvisol) 160
8.3 Class III: Semi-gleyic Soils with A-C-G Soil Horizon Sequence 163
8.3.1 Fluvial Meadow (Humofluvisol) 163
8.4 Class IV: Hypogleyic Soils with A-G Soil Horizon Sequence 164
8.4.1 Wetland Black Soil (Humogley) 164
8.4.2 Wetland gley (Eugley) 165
8.5 Class V: Peat Soils with T-G Soil Horizon Sequence 168
8.5.1 Low Moor Peat (Planohistosol) 169
8.5.2 High Moor Peat (Acrohistosol) 171
8.5.3 Transient Peat (Plano-Acrohistosol) 172

8.6 Class VI: Anthropogenic Soils with P-G Horizon Soil Sequence 172
8.6.1 Rigosol 172
8.6.2 Hydroameliorated Soil 173
References 173

9 Order of Halomorphic and Subaquatic Soils 175
9.1 Order of Halomorphic Soils 175
9.2 Class I: Salty Soils with Asa-G or Asa-Cg Soil Horizon Sequence 175
9.2.1 Solonchak 175
9.3 Class II: Alkalized Soil with A/E-Btna-C Soil Horizon Sequence 176
9.3.1 Solonetz 176
9.4 Order of Subaquatic Soils 178
9.5 Class I: Initial Subaquatic Soils with (A)-CG Soil Horizon Sequence 179
9.5.1 Protopedon 179
9.6 Class II: Humic Subaquatic Soils with A-C or A-G Soil Horizon Sequence 180
9.6.1 Gyttja 180
9.6.2 Dy 180
9.6.3 Sapropel 180
9.7 Class III: Anthropogenic Subaquatic Soils with P-C or P-G Soil Horizon Sequence 180
9.7.1 Drained Gyttja 180
9.7.2 Drained Sapropel 180
References 180

10 Land Use 181
10.1 Current State and Trends of Land Use 181
10.2 Changes in the Use of Agricultural Soil 184
References 189

11 Environmental Issues 191
11.1 Soil Degradation 191
11.1.1 Water and Eolian Erosion 191
11.1.2 Soil Acidification 192
11.1.3 Depletion of Organic Matter in Soils 195
11.1.4 Soil Salinization 196
11.1.5 Mining Activity and Industry Effects on Soils 197
11.1.6 Disposal of Industrial and Municipal Waste 198
11.2 Soil Pollution 201
11.2.1 Heavy Metals 201
11.2.2 Pesticides 214
11.2.3 Radionuclides 215
11.3 Climate Change 217
References 219

Index 223

About the Authors

Pavle Pavlović graduated from the University of Belgrade's Faculty of Forestry, Serbia, in 1986. He went on to complete his M.Sc. thesis at the Faculty of Natural Sciences (1992) and his Ph.D. thesis at the Faculty of Biology (the University of Belgrade) (1998). Since 1988, he has been at the Institute for Biological Research 'Siniša Stanković' (IBISS) (the University of Belgrade), becoming a full research professor in 2007. He is currently the director of this institute, a member of the University Senate, the President of the Council of Institutes (the University of Belgrade) and a member of the Presiding Board of the Association of Institutes of Serbia. He lectures on the subject 'Phytoremediation' as part of the 'Ecological engineering for the protection of soil and water resources' doctoral study programme at the Faculty of Forestry (the University of Belgrade). He is an accredited mentor on the 'Ecology' doctoral study programme at Belgrade University's Faculty of Biology and to date, he has mentored nine Ph.D. and M.Sc. theses. He is the leader of several projects in the area of basic and applied ecology. His research has resulted in the publication of over 150 papers and eight chapters, while he was also the editor of one book. His research work includes soil science and ecological phenomena related to soil evolution and classification in Serbia, the metabolism of the soil component of ecosystems, soil degradation and soil pollution, the ecophysiology of plants in natural and urban ecosystems, including plants on technogenic substrates and allelopathy. Current research by his Terrestrial Ecosystem Ecology study group is focused on ecophysiological adaptive strategies of plants as a response to the effects of multiple stress in degraded habitats. He has performed various duties as a member of expert committees of the Serbian Ministry of Education, Science and Technological Development, the Serbian Ministry of Agriculture and Environmental Protection and the University of Belgrade. He is a member of the 'Man and Environment' Departmental Committee of the Serbian Academy of Sciences and Arts. He was a member of the Governing Board of the Soil Science in Belgrade and the Institute of Forestry. He also participated as a member of the Scientific Committees of the 4th International Congress EUROSOIL 2012—Soil Science for the Benefit of Mankind and Environment; the 6th and 7th Symposium Chemistry and Environmental Protection—EnviroChem (2014, 2015); the 2nd International Conference on Plant Biology, the 21st

Symposium of the Serbian Plant Physiology Society, and COST ACTION FA1106 QUALITYFRUIT Workshop (2015); the 'Energy and Environment' Symposium of the Serbian Academy of Sciences and Arts (2013), and the Editorial Board of the 9th Congress of the Yugoslav Society of Soil Sciences (2005).

Professor Emeritus Nikola Kostić was born in 1940, in Jagodina, Serbia. In 1960, he enroled at the Faculty of Mining and Geology, University of Belgrade. After graduation, he worked as an assistant in the laboratory for mineralogy and attended post-graduate studies in applied mineralogy. He obtained M.Sc. and Ph.D. degrees in mineralogy at the University of Belgrade in 1975 and 1981, respectively. In 1972, he was appointed senior research mineralogist, at the R&D Department of the Anglo-American Corporation, in Kitwe, Zambia. In 1977, he became the Head of the Mineralogy Department, leading the team of 22 mineralogists/geologists responsible for providing services to the mining-geology departments in the evaluation of mineral raw materials of Cu–Co and Pb–Zn–V ores for reserve calculations. He was also involved in the development of the 'Leitz—TAS" image analyser for the control of mineral beneficiation processes at metallurgical plants. In 1985, he was invited to teach at the Faculty of Agriculture, Belgrade, where he taught for 20 years basic subjects to graduate students and advanced courses to M.Sc. and Ph.D. candidates in Soil sciences and Environmental protection programs. Research with Ph.D. candidates was in the areas of soil/clay mineralogy and geochemistry of heavy metals in soils and their availability to plants and indirectly to animals, and perhaps as threats to human health. He was involved on the professional level as the director of the Soil Irrigation Department (1987/96), the chair of Soil Science Division (1996–2004) and the vice dean, Faculty of Agriculture, University of Belgrade (2004/06). In 1985 and 1986, under the auspice of the Ministry of Science grants he was a visiting researcher at the Liege and Cologne Universities. He has also made working connections with the faculties in England, Belgium, Austria, Hungary and Germany. On several occasions, from 1990 to 1996, the author had conducted research in the MacAulay Land use Research Institute, Aberdeen, Scotland, on advanced studies of clay minerals and heavy metals in Serbian soils under the auspices of the British Council ALIS Program and the Royal Society of London awards, respectively. The data from his research have resulted in the publication of over 100 papers, several chapters and two books. In 2000, he published the book 'Agrogeology', providing an advanced knowledge about rock weathering, soil mineralogy and applied geology that students of agriculture should acquire. In 2002, he was the visiting professor at the Faculty of Natural Sciences, Salzburg under the auspices of the ÖAD award and CEEPUS Organisation grant. In 2003/2004, he has given lectures at the Environmental Sciences Department, University Bicocca-Milano, on Milanković's Canon of insolation and its relation to the rhythm of the paleosoils of the Batajnica loess plateau. He is the member of the Serbian Geological, Mineralogical and Soil Sciences Societies, the chairman of the Soil Mineralogy Committee and the chairman of the Serbian Mineralogical Association. Present days he dedicates his time to scientific writing, studying, arts and enjoying the company of his extended family.

Branko Karadžić was born in 1956, in Peć, Serbia. Since 1968, he lives in Belgrade, where he completed primary and secondary education. He graduated in 1982, at Department of Biology of the Faculty of Natural Sciences and Mathematics, University of Belgrade. In 1984, he started as an assistant at the Institute for Biological Research 'Siniša Stanković', at Belgrade University. He defended his M. Sc thesis at the Faculty of Natural Sciences and Mathematics, University of Belgrade, in 1987. In 1989, he received Ph.D. stipend and went to Sweden, at Department of plant ecology (Vaxtbio) of Uppsala University, where he cooperated with the most prominent vegetation scientists. He defended his Ph.D. thesis at Faculty of Biology, University of Belgrade in 1994. The author was the head of the Department of Ecology at Institute for Biological Research 'Siniša Stanković', during the period from 1999 to 2002. The Government of Republic of Serbia appointed him to be the first director of the Serbian Environmental Protection Agency (SEPA). In period from 2004 to 2007, he established conceptual framework of SEPA and environmental information system in Serbia. In 2007, he was offered teaching position at the Faculty of Biology, University of Belgrade, where he organized the lectures of quantitative ecology and ecological modelling for postgraduate students. In 2014, he was invited to be the visiting professor at the Faculty of Biology, University of Niš, where he organized advanced course of quantitative biology for postgraduate students. The author was a member of the Serbian Biological Society Managing Board twice (from 1994 to 2000 and from 2008 to 2014). He was Serbian focal point for clearing-house mechanism issues, within the UN Convention on Biological Diversity (2003–2004). Since 2004, the author is a member of the section for environmental protection, within the Serbian Academy of Science and Art. The author published five textbooks, chapters for six books and more than 100 articles in scientific journals. His main scientific interest covers community ecology, vegetation science and ecological modelling. He investigated diversity and ecology of endemo-relic forest ecosystems in canyons and gorges in central Balkan. Also, he studied serpentine vegetation in Serbia. The author published syntaxonomic review of vegetation in Serbia. With his colleagues from Slovenia, Croatia, Bosnia and Herzegovina and FYR Macedonia, he revised syntaxonomy of xeric oak forests in south-eastern Europe. The author also investigates the response of different species to variation in environmental factors. The author published numerous articles on ecological modelling and multivariate statistical analyses. After publishing the book

Quantitative ecology (Karadžić and Marinković, 2009), the author received an award from the Serbian Ministry of Science for the most prominent contribution in development of natural sciences and mathematics in Serbia. He developed the powerful software package 'FLORA' and databank of vegetation in Serbia. The databank is compatible with other pan-European databases.

Miroslava Mitrović graduated from the Faculty of Natural Sciences at the University of Belgrade, Serbia in 1987. She then completed her M.Sc. thesis at the Faculty of Natural Sciences (1992) and her Ph.D. thesis at the Faculty of Biology (the University of Belgrade) (1998). She has been at Belgrade University's Institute for Biological Research 'Siniša Stanković' (IBISS) since 1988, becoming a full research professor in 2007. She is currently a member of the Scientific Committee for Biology at the Serbian Ministry of Education, Science and Technological Development (since 2011), a member of the Scientific Board of IBISS (the University of Belgrade) and a member of its governing board; and a member of the governing board of the Institute for Soil Research, Belgrade. She is currently a member of the Editorial Board of the Archives of Biological Science, Belgrade. She is an accredited mentor on the 'Ecology' doctoral study programme and has been mentor to several Ph.D. and M.Sc. candidates. Over 100 papers and eight chapters have been published thanks to her research. Her main fields of interest are plant ecophysiology and interdisciplinary research within ecological, environmental and soil sciences. Currently, her research is focused on the ecophysiological adaptive strategies of plants as a response to the effects of multiple stress in degraded habitats. She was on the Scientific Committee of the 7th Symposium Chemistry and Environmental Protection—EnviroChem (2014, 2015), the 2nd International Conference on Plant Biology, 21st Symposium of the Serbian Plant Physiology Society and COST ACTION FA1106 QUALITYFRUIT Workshop (2015).

1 Introduction

The Republic of Serbia is geographically situated in south-eastern Europe (80% of the territory), in the north-western part of the Balkan Peninsula. Because Serbia covers part of the Pannonian Plain in the north it belongs to the Central-European, Pannonian and Danube regions, since this the second longest European river flow through and drains about 90% of its territory, while due to its southern region, in terms of geography and climate, Serbia is also considered a sub-Mediterranean country.

Serbia stands at the crossroads of Eastern and Western Europe, its routes leading via the Morava–Vardar and Nišava–Marica valleys to the Aegean coast, or to Asia Minor and to the Middle East, respectively. European transport corridors 7 (the Danube) and 10 (road and rail) pass through Serbia and meet in the capital Belgrade city.

In respect of the administration system, the Republic comprises of central Serbia and two autonomous regions: Vojvodina (21,506 km^2) and Kosovo and Metohija (10,887 km^2). Since June 1999, AP Kosovo and Metohija have been under the jurisdiction of UN Interim Administration Mission in Kosovo (UNMIK).

1.1 General Information Regarding Serbia

1.1.1 Geographic Position

The Republic of Serbia is located in the south-eastern part of Europe, within the geographic coordinates: 41° 51′ and 46° 11′ N and 18° 47′ and 23° 01′ E, and covers the area of 88,361 km^2. Serbia has borders of 2361.7 km in length, with several countries, on the north with Hungary (175 km), on the northeast with Romania (548 km), on the east with Bulgaria (360 km), on the south with Former Yugoslav Republic of Macedonia (283 km), on the south-west with Albania (114 km) and Montenegro (250 km) and on the west with Republic Srpska (370 km) and Croatia (262 km) (Statistical Office of the Republic of Serbia 2014), Fig. 1.1.

1.1.2 Topography

From the geomorphological point of view, the territory of Serbia belongs to the great Euro-Asian system of younger chain mountains. It is actually part of the western, Mediterranean zone. The relief of Serbia could be divided into the following parts:

- the Pannonian area with the Peripannonian rim and
- the mountains and valleys regions

Both regions are about the same size 43,705 km^2 of the Pannonian and 44,656 km^2 of the mountains and valleys, in total 88,361 km^2 or 49.46 and 50.54%, respectively (Fig. 1.2).

The Pannonian area comprises lowland and hilly parts of northern and eastern Serbia and is an expansive plain (between 100 m and 200 m a.s.l.) with two low mountains, Fruška Gora (538 m) and Vršačke Mts. (639 m), rising from it. It also extends over Mačva, lower Posavina and Podunavlje, as well as rolling hilly terrains of the Pannonian basin southern rim represented with: Posavina and Pocerina of Šabac, Serbian Podrinje, Podgorina of Valjevo, Kolubara, Šumadija, western and greater Pomoravlje, Stig and Braničevo.

Mountain and valleys regions extend over the southern Pomoravlje, Carpathian and Balkan Serbia, Vlasina and Krajište; as well as Toplica and Jablanica; Stari Vlah with Raška; Ibar-Kopaonik area, Metohija and Kosovo with Drenica basin.

Morphometric classification of the relief based on the heights has shown that Serbia is generally hilly and mountainous country with an average altitude of 565.3 m.s.l.

P. Pavlović et al., *The Soils of Serbia*, World Soils Book Series,
DOI 10.1007/978-94-017-8660-7_1

Fig. 1.1 Geographic position

(Table 1.1). In terms of altitude, hilly mountain terrains (altitudes greater than 500 m) cover 33,992 km^2, or 38.47% of the Serbian territory; mountain terrains (exceeding 1 000 m) cover 9887 km^2 (11.19%); medium high mountains with altitudes of 1000–1500 m are spread over 9681 km^2; the high mountains (over 1500 m) cover just 206 km^2, or 0.23% of the territory. The large river valleys with wide alluvial planes, loess plateaus and sand dune plains cover 31.6% of the territory. Over 68.4% of the country belongs to the altitudes above 200 m a.s.l. Hills and rolling lands, mainly of the Peripannonian region, as well as dissected lake and smaller river terraces extend over 24%, while the low mountains and Pešter Plateau and Stari Vlah highlands extend over 38%, while to the medium mountains, higher than 1500 m, belongs only about 6% of the territory.

The Pannonian part consists of all Vojvodina as well as the Mačva, Posavina, Podunavlje and Pomoravlje regions around the Sava, Velika Morava and Danube rivers. Pannonian depression is not entirely flat between these mountains. Across the lowest-lying land flow the big rivers: Danube, Sava, Tisa and Tamiš. On both sides of each of the rivers expand alluvial plains and between the rivers lie smaller and larger loess plateaus of Banat, Bačka, Titel and Srem and sand plains of Banat, Bačka and Ram—Golubac with specific aeolian morphology.

The Peripannonian Serbia includes the southern rim of the Pannonian basin, in other words, the regions of Pocerina and Posavina, between the towns of Šabac and Obrenovac, Serbian Podrinje along the river Drina, the Podgorina around Valjevo, then Kolubara, Šumadija and the area along the river Zapadna Morava, as well as Pomoravlje, Stig and Braničevo, along the Velika Morava and Mlava rivers. The Peripannonian part comprises of predominantly rolling hills, with the altitudes between 200 and 500 m. Besides, in the eastern Serbia, only Danube Ključ and Negotin Krajina of the expansive Dacian or Valachian-Pontian basin are on Serbian territory. Like the Pannonian basin, it is depression, with the lowest-lying land in Serbia, only 28 m.s.l. Alluvial plains lie along the Danube and Timok; further off the river there are terraces and hills, locally covered by the loess and sand plains.

The area south from the Sava and Danube rivers, the Peripannonian margin, where hills and low mountains dominate, corresponds to the Balkan Peninsula. The southern, Balkan part of Serbia has more complex geomorphological relationships; western part of the country belongs to the Dinaride, eastern to the Carpatho-Balkanide, and between these two extends the Serbian composite mountain terrain.

The mountain–valley region, which is higher than 500 m, consists of the rest of the Republic: south Pomoravlje, Carpathian and Balkan Serbia, Vlasina and Krajište, Toplica and Jablanica, then the region of mountain Kopaonik and highlands of Stari Vlah and Pešter plateau, as well as the Raška area around the river Ibar and Kosovo and Metohija depression. On the south, there is the prominent mountainous region of Serbia, which consists of several mountain systems: a part of Dinaride, the Serbian massif and the Carpatho-Balkanide. It has more varied relief with medium mountains surrounded by hilly sloping terrain and with steep peaks above 1500 m, and undulating lowland of the Kosovo and Metohija.

The Serbian massif of old mountains and depressions extends between the Dinaride and Carpatho-Balkanide is noted for its particular features. In the south, high Dukat (1.881 m) and Besna Kobila (1.922 m) mountains of the Serbian massif gradually descend towards north over Vlajna (1.441 m), Jastrebac (1.492 m), Juhor (773 m) and Crni vrh (707 m) to the periphery of the Pannonian plain. The descending impression is increased by the large Morava graben in the north of the Serbian massif, the biggest southern Peripannonian area.

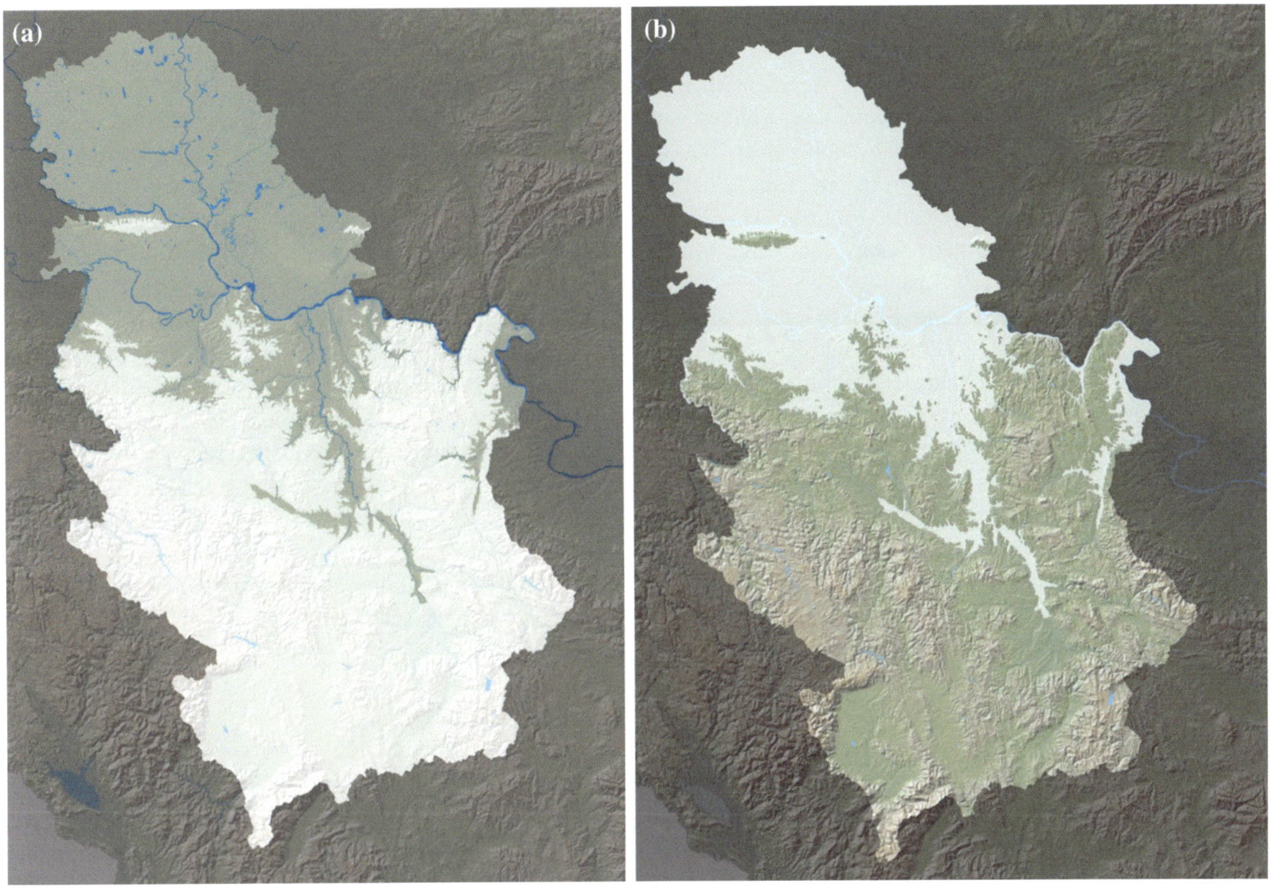

Fig. 1.2 Topography of Serbia: Pannonian area with the Peripannonian rim, the Morava and Danube valleys (**a**); mountains and valleys region (**b**)

Table 1.1 Distribution of the altitudes (m)

Altitude (m)	Altitude (%)	Statistical data	Altitude (m)
0–100	21.12	Minimum	28
100–200	10.48	Percentile 10%	94
200–300	11.17	Quartile 25%	217
300–500	12.90	Median (Md) 50%	625
500–750	14.30	Quartile 75%	1489
750–1000	12.70	Percentile 90%	2318
1000–1500	11.28	Maximum	2656
1500–2000	4.43	–	–
>2000	1.62	Average (Ma)	565.50
Total	100.00		

The eastern part of mountainous Serbia belongs to the Carpatho-Balkanide. This system of mountains extends between the Serbian massif in the west and the Valachian-Pontian depression in the east. The mountains are dissected by several large depressions: Zvižd, Homolje, Sokobanja, Zaplanje, Crna reka, Svrljig, Bela Palanka, Zaječar, Knjaževac and Pirot.

On the south from Danube and Sava rivers, the set of lake planes is developed as cascades from which the hills and low mountains Avala, Kosmaj, Bukulja, Cer, Boranja and Jagodnja are rising as islands from a lake. The South and Great Morava rivers flow throughout the central part of Serbia, which is dissected by their valleys surrounded with old mountains of the Serbian massif, e.g. Crni vrh and Juhor in the north over Jastrebac to Vidojevica (1.155 m) and Kukavica (1.440 m) in the south.

The south-western part of Serbia is represented with medium mountain chains dissected with deep valleys and canyons. The most important mountains higher than 1000 m are as follows: Maljen (1103 m), Povlen, Tara, Zlatibor (1.496 m), Javor, Radan (1.408 m), Rogozna (1.479 m), Čemerno, Radočelo (1.643 m), Zlatar, Jadovnik, Golija (1.833 m) and Kopaonik (2.017 m).

On the east of the Great and South Morava with its heights dominate Beljanica (1.339 m), Veliki Krš (1.148 m), Deli Jovan (1.141 m), Rtanj (1.565 m), Tupižnica (1.160 m), Ozren (1.174 m), Svrljiške Mts. (1.334 m), Ruj (1.706 m), Čemernik (1.638 m), Vardenik (1.874 m) and Stara planina (Midžor 2.169 m) (Fig. 1.3).

There are about 30 high peaks, over 2000 m, in southern Serbia, the most of them on Šar-planina (2.498 m), Mokra Gora (2.155 m), Hajla, Žljeb, Ošljak (2.212 m), Koritnik (2.395 m) and Paštrik (1987 m). The highest peak of Serbia, Djeravica (2.656 m), occurs in the Prokletije Mts., at the border with Albania.

The landforms in soil mapping are described primarily by their morphology and neither by their genetic origin nor by the processes responsible for their formation and shape. The regional slope class is the main differentiating criterion, followed by relief intensity. At the highest level of separation, three groups of landforms are distinguished: level (plains and flats), sloping (hills) and steep (mountains) lands.

The plains, flats and low gently sloping lands with a low relief intensity of less than 10% slope gradient or 50 m/slope unit relief intensity and altitudes below the 300 m.s.l. cover about 80% of the Vojvodina (Table 1.2).

The other 20% of the area belongs to the slopes of Fruška Gora and Vršačke Planina as well asloess plateau escarpments with the gradients of more than 10%. The central Serbia and Kosovo and Metohija have similar distribution of gradients with 13 and 15% of the low gradients (<9%), as well as more than 10% gradients with 87 and 85%, respectively.

The part of Serbia with medium gradient hills and low mountains of the sloping and rolling lands of the Peripannonian rim has medium slopes with 10–30% gradient or 50–150 m/km relief intensity with the altitudes up to 600 m above the level base. The medium level Pešter plateau, Stari Vlah highlands and low mountains with the altitudes between 750 and 1500 m of moderately sloping and steep lands with relief intensity greater then 600 m per km^2 and over 30% slope gradient extend over the largest area in Serbian mountains.

The mountain ranges of Dinaride and Carpatho-Balkan arch have shown the highest vertical division (300–800 m/km), and their peaks area (Prokletije Mts., Koritnik, Šara and Suva Planina, Rtanj, etc.), which were modelled either by the Pleistocene glaciation or by strong tectonic activity (faulting, thrusting), have all characteristics of extremely rugged relief with the relief intensity over 500 m/km.

The plains, flats and low gently sloping lands with a low relief intensity of less than 10% slope gradient and altitudes up to 200 m, extend over 37% of the Serbian territory. Hilly (200–500 m) and low mountain (500–1000 m) zones are of the same areas of about 26% each, while high mountains (over 1000 m) occur over 11% of the territory.

It should be noted that there is about 43% of steep and very steep terrains (slopes over 30%), with shallow and thin soils developed which are susceptible to erosion and so unsuitable for agriculture. The plains and flats with slopes of less then 5% and low gently sloping lands with less than 10% slopes extend over one-third of the soil cover. The remaining 24% of soil cover, with the slopes between 10 and 30%, could be used for agricultural production but with corresponding agrotechnical measures for soil protection.

1.1.3 Hydrology

From the hydrographic point, Serbia has relatively well-developed river network. Density of the drainage network is in close relationship with the lithological composition of the terrain. In silicate-rich terrains, the drainage is more dense compared to the areas with carbonate rocks, which is under the influence of the karstic processes reduced only to the alogene rivers.

The rivers in Serbia belong to the watersheds of the Black, Adriatic and Aegean seas (Fig. 1.4). On more than 90% of the Serbian territories, there are rivers that join the Danube, thus going to the Black Sea. The entire territory of the Vojvodina Province and most parts of central Serbia drain by the Danube into the Black Sea. Only 2.17% of the territory of central Serbia (or 1215 km^2) is drained by the Pčinja and Dragovištica rivers into the Aegean Sea. Due to the specific geographic position and characteristics of its topography, the territory of the Kosovo Province drains into three seas: 50.74% (or 5524.61 km^2) drains into the Black Sea, 42.85% (or 4665 km^2) into the Adriatic Sea and 6.41% (or 697.4 km^2) into the Aegean Sea.

The most important rivers are Danube, Sava, Velika, Zapadna and Južna Morava, Ibar, Drina, Tisa, Nišava, Tamiš, Begej, Kolubara, etc. (Table 1.3). The longest navigable river in Serbia is the Danube with more than one-fourth of this river's bed stretched through Serbia (Fig. 1.5).

Sava and Tisa, as well as part of the Morava, which is the largest Serbian river, are also navigable rivers. The Morava flows through the most fertile area of the central Serbia, called Pomoravlje. Other freshwater rivers are as follows:

Fig. 1.3 Stara planina Mt., eastern Serbia (*photo* Mitrović, M., Paunović, M.)

Table 1.2 Slope gradient

Slope gradient	<9% (%)	>10% (%)
Vojvodina	80	20
Central Serbia	13	87
Kosovo and Metohija	15	85
Serbia (average)	29.4	70.6

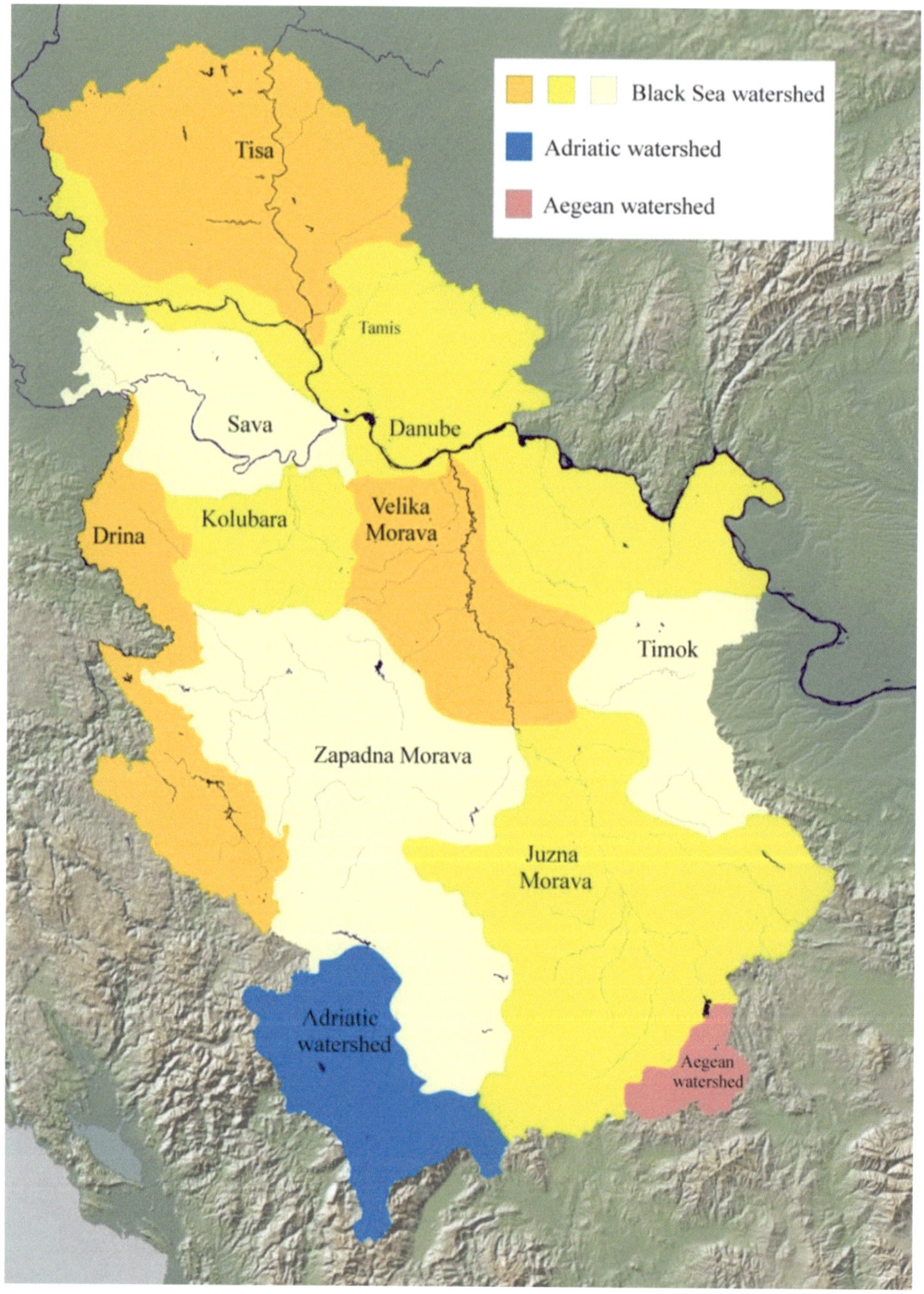

Fig. 1.4 Most important rivers in Serbia

Table 1.3 Most important rivers in Serbia

River	Length in Serbia (km)	Total length (km)
Danube	588	2783
Zapadna Morava	308	308
Južna Morava	295	295
Ibar	272	272
Drina	220	346
Sava	206	945
Timok	202	202
Velika Morava	185	185
Tisa	168	966
Nišava	151	218
Tamiš	118	359
Begej	75	244

Fig. 1.5 Danube river near Donji Milanovac, eastern Serbia (*photo* Pavlović, P.)

Fig. 1.6 Sava river near Kalemegdan Fortress, Belgrade (*photo* https://www.google.rs/search?q=sava+river)

Sava, Zapadna and Južna Morava, Drina, Ibar, Tisa, Nišava, Timok, Begej, etc. (Figs. 1.6, 1.7, 1.8 and 1.9).

The Velika, Južna and Zapadna Morava are the most important rivers, with their numerous tributaries which drain the most of the Serbian territory. Along with them are also Timok, which drains the eastern parts of Serbia, Kolubara drains Valjevo Podgorina and western Šumadija and Drina right side tributaries drain western parts of Serbia. All those rivers belong to the Black Sea drainage area. The international rivers Danube, Sava and Tisa are navigable in the whole length flow throughout Serbia, also belong to the Black Sea drainage area.

Beli Drim is the largest river in Metohija, which flows to the Adriatic Sea. Its tributaries drain north-western parts of

Fig. 1.7 Drina river, western Serbia (*photo* Pavlović, P.)

Šara and Jezerske Mts., as well as eastern parts of Prokletije Mts. Small area of south-eastern Serbia is drained to the Aegean Sea by Nerodimka, Lepenac and Pčinja as well as Božička rivers, which are the tributaries of Vardar and Strumica rivers, respectively.

There are many canyons and gorges in Serbia that are protected nature reserves. The Uvac river canyon is a veritable jungle inhabited by griffons that are protected by forest rangers; the Drina river canyon is the third deepest canyon in the world; Lazar's Canyon is the least accessible; the Vratna river canyon is home to a world phenomenon, the Gates of Vratna; the Miruša and Rosomača Canyon is strikingly beautiful (Figs. 1.10, 1.11, 1.12 and 1.13).

The largest natural lake is Belo Jezero in Vojvodina, covering 4.8 km^2. The largest artificial reservoir, Đerdap, covers a total area of 253 km^2 of which 163 km^2 is on the

Fig. 1.8 Velika Morava river, central Serbia (*photo* Paunović, M.)

Fig. 1.9 Ibar river, central Serbia (*photo* Paunović, M.)

Serbian side and the rest belongs to the Romanian side (Table 1.4; Fig. 1.14). The largest waterfall, Jelovarnik, located in Kopaonik, is 71 m high. Vlasina Lake is part of Vlasina river, Perućica and Zvornik lakes are located on Drina river, Zlatar lake on river Uvac and Potpeć lake is located on river Lim (Figs. 1.15, 1.16 and 1.17).

1.1.4 Soil

According to the results of the Census of Agriculture 2012, agricultural land in Serbia includes soil under agriculture (73.1%), meadows and pastures (20.7%), fruit plantations (4.8%), vineyards (0.6%), gardens (0.7%) and other permanent crops (0.1%). The area under cereals is the largest and covers 68% of the total soil under agriculture, with cultures of maize (39%) and wheat (24%) predominating. In terms of the kinds of fruit, plums are the principal fruit, covering 45% of the total area under fruit trees (Statistical Office of the Republic of Serbia 2013). The average utilized agricultural area per holding covers 5.4 ha. However, not all of the agricultural land area is utilized. For example, results from a survey conducted in 2007 indicate that farm holdings in Serbia's mountainous regions do not use a

Fig. 1.10 Rosomača river canyon, Stara planina Mt., eastern Serbia (*photo* Mitrović, M.)

quarter to a third of their land due to poor soil quality, high water levels and inaccessibility by road (Bogdanov 2007) (Fig. 1.18).

In 2011 and 2012, 12% of the total number of holdings applied irrigation to 3% of the utilized agricultural area. The following facts are notable: the crops that are most irrigated are vegetables, water melons and strawberries (grown in the open), the irrigated area of which accounts for 64% of the total area under these crops; 61% of agricultural holdings declared ground water at site as the main source of water for irrigation; 65,303 holdings apply surface irrigation, 13,174 holdings apply sprinkler irrigation, and 29,323 holdings use the drop irrigation method, while the municipality with the largest share of holdings that irrigate utilized agricultural areas is Bosilegrad (73%) (Statistical Office of the Republic of Serbia 2013).

The area to which mineral fertilizers are applied covers 67% of the total utilized agricultural area. Solid dung, including farmyard manure, is applied by 49% of agricultural holdings, which is 11% of the total utilized agricultural area (Statistical Office of the Republic of Serbia 2013). In terms of the use of hazardous and harmful substances, food production can be performed without risk in approximately 80% of Serbia. For the remaining 20%, food production should be organized in a way so as to reduce potential risk, for example by crop selection and/or by occasional to continuous control of soil and plant quality (Protić et al. 2005).

Forested areas have gradually been increasing since 1938 (Jović et al. 1992; Banković et al. 2009). When compared to 1979, forested areas have increased by 464,611 ha due to the afforestation of both state-owned and privately owned lands. In terms of the total forested area of the Republic of Serbia,

Fig. 1.11 Jerma river canyon, Stara planina Mt., eastern Serbia (*photo* Pavlović, P.)

coppice forests predominate (64.7%), while natural stands of seedling origin and artificially established forest stands occupy 27.5 and 7.8% of forests, respectively. Deciduous forests dominate in Serbia (59% of the total forested area), with the percentage of mixed deciduous–conifer and pure conifer forests significantly smaller (29.3 and 8.7% of the forested area) (Serbian Environmental Protection Agency 2007) (Fig. 1.19).

Fig. 1.12 Uvac river canyon, western Serbia (*photo* Milošević, M.)

The 'abandonment of managed land' category encompasses abandoned agricultural land, pastures, scrub and shrubberies, forest plantations and other lands under some form of management (Jović et al. 1992). According to the Spatial Plan of Serbia (1996), in 1990, the area for potential afforestation, covered by meadows and pastures, amounted to 162,224 ha. The 'forests and grasslands conversion' category includes the conversion of existing forests and natural grasslands to other forms of use, for example agriculture. After the Second World War in particular, forests in Serbia were cut down not so as to increase the land available for agricultural production, but for other purposes connected to infrastructure projects (Kadović 2013).

A detailed description of the current state and trends of land use change in Serbia is given in Chap. 10 of this book.

1.2 A Short History of Soil Science and Soil Survey

The pedological research in Serbia began in the middle of the nineteenth century in Principality of Serbia and continued latter in the Kingdom of Yugoslavia, when geological, physical and chemical investigations and classifications of soil were carried out. It was in Vojvodina, at that time part of the Austro-Hungarian Empire, that the first soil studies in

Fig. 1.13 Dojkinačka river, Stara planina Mt., eastern Serbia (*photo* Pavlović, P.)

Table 1.4 Most important lakes in Serbia

Lake	Area (km^2)	Altitude (m)	Water volume (M m^3)
Đerdap	253	69.5	5000
Vlasina	16	1213	165
Perućac	12.4	290	340
Gazivode	11.9	692.7	370
Zvornik	8.1	140	42
Zlatar	7.2	880	250
Potpeć	7.0	437	43
Palić	5.6	101	11
Belo	4.8	75	7

Fig. 1.14 Artificial reservoir Đerdap, eastern Serbia (*photo* Mitrović, M.)

our country were carried out. In Serbia, the founder of soil science, i.e. agriculture, is considered to be the great Serbian botanist, Josif Pančić (1814–1888), who was first appointed in 1853 as an adjunct professor at Belgrade Lyceum's Department of Natural History and Agronomy by decree of Prince Alexander Karadjordjević, before becoming a full-time professor of Natural History and Agriculture in 1854, as decreed by the Ministry of Education of the Principality of Serbia. During his fruitful university career, Pančić wrote the first textbook on Mineralogy and Geology in Serbia. In the period before the First World War and then in the 1920s, the 'Russian genetic school of soil science' was accepted with its new concept of soil science, established by Vasilii Vasilevich Dokuchaev (1846–1903), Nikolai Mikhailovich Sibirtsev (1860–1900) and Konstantin Dmitrievich Glinka (1867–1927). This new concept was based on soils as independent natural bodies, each with unique properties resulting from a unique combination of climate, living matter, parent material, relief and time. They hypothesized that the properties of each soil reflected the combined effects of the particular set of genetic factors responsible for the soil's formation (Glinka 1927).

Work on studying, classifying and mapping the soils of Serbia began with the founding of the Agricultural School in

Fig. 1.15 Perućac lake created by damming the Drina river, western Serbia (*photo* Pavlović, P.)

Belgrade in 1849 by Prince Aleksandar Karađorđević. It was later named 'Zavedenie ekonomije' and was allocated agricultural land and forests in Topčider suburb of Belgrade. This institution carried out the first soil experiments with manure application in Serbia, as well as the crossbreeding of domestic breeds of animals with those imported from abroad, and also the first meteorological measurements. It also had a nursery with apples, pears and grapes (Kanitz 1985).

With the adoption of the Law on Amendments and Supplements to the University Act, dated 27 February 1905 ('Official Gazette of the Kingdom of the Serbs, Croats and Slovenes', no. 85, year 1, 23 August 1919), and thanks to the determined efforts of Prof. Sima Lozanić (1847–1935), the first professor of Agricultural Chemistry and the Rector of Belgrade University at the time, with considerable support from the geologist and petrographer Prof. Jovan Žujović (1856–1939), then the Minister of Education, in 1920 the Faculty of Agriculture was founded in Zemun. In 1930, the Faculty was renamed the Faculty of Agriculture and Forestry, before being separated into the Faculty of Forestry and the Faculty of Agriculture of Belgrade University in 1949 (Fig. 1.20). After the founding of the Faculty of Agriculture and Forestry, soil research gained new impetus, thanks to the

Fig. 1.16 Artificial Borsko lake is a result of the construction of a dam of the rivers Valja Dzoni, Marec and part of Zlotska reka; eastern Serbia (*photo* Pavlović, P.)

work of Dr. Nikolaj Vasiljev, a professor of Agricultural Chemistry, and Dr. Aleksandar Stebut, a professor of Pedology and Plant Production and Selection. These academics contributed greatly to the formation of a group of research staff in the field of soil science, which was well-established at the Faculty from the first decades of the twentieth century.

Parallel to the growth of pedology, there were developments in agrogeology, too. In 1985, the Laboratory for Soil Mineralogy was set up at the Faculty of Agriculture, marking the beginning of systematic pedomineralogical research of soil samples and the application of biogeochemistry to tackle problems of environmental planning and soil protection when dealing with mining and industry. In 2008,

Fig. 1.17 Vlasina Lake is part of Vlasina river, southeast Serbia (*photo* https://www.google.rs/search?q=vlasinsko+jezero)

geology became a field of study in its own right at the Faculty of Agriculture, aimed at studying the alteration of minerals and development of clay minerals during rock weathering and soil formation. Equally, detailed research into forest soils continues in new laboratories at Belgrade University's Faculty of Forestry.

The period from 1930 to 1941 was significant for the development of soil science in agriculture and the Ministry of Agriculture of the Kingdom of Yugoslavia approved the founding of the Agricultural Research and Monitoring Station in 1938 (Fig. 1.21), which went through a series of transformations before eventually becoming the Institute for Field and Vegetable Crops in Novi Sad in 1995.

By 1920, a great amount of material on different soils, in particular agricultural, had already been collected, and soil science research had developed thanks to the work of Prof. Aleksandar Stebut, Prof. D. Todorović, Prof. Đ. Zloković, and Prof. V. Neugebauer, and also that of Prof. M. Gračanin and Prof. A. Sajfert. In fact, in 1945, Prof. Neugebauer

Fig. 1.18 Large areas of soil under extensive agriculture in northern parts of Serbia (Vojvodina plane), smaller fields surrounded by meadows and pastures and fields not utilized by agriculture in central Serbia (*photo* Pavlović, P., https://www.google.co.uk/search?q=vojvodina,+oranice)

Fig. 1.19 Oak, beech and pine forests, western Serbia (*photo* Pavlović, P.)

Fig. 1.20 University of Belgrade's Faculty of Agriculture and Forestry, founded in 1930 (*photo* http://www.agrif.bg.ac.rs/pages/fakultet/istorijat-fakulteta)

Fig. 1.21 Agricultural research and monitoring station in Novi Sad, founded in 1938 (*photo* http://www.nsseme.com/about/?opt=istorijat&cat)

(1897–1989) arrived at what was then known as the Institute for Agricultural Research in Novi Sad, where he led research connected to soil classification and mapping. He is thought of as the founder of the Yugoslav school of mapping and his soil classification was accepted by the Yugoslav Society of Soil Science. He oversaw the publication of soil maps of Yugoslavia (1:1,000,000) and Vojvodina (1:1,000,000 and 1:50,000).

The most intensive period of development in soil science was witnessed in the post-World War II, from 1945 in what was then the Federal People's Republic of Yugoslavia (FNRJ). A large number of researchers emerged and many faculties and institutes concerned with soil studies were founded. During 1951, all institutions in the People's Republic of Serbia engaged in soil science and agrochemistry, supported by similar institutions of FNRJ, initiated the founding of the 'Soil and Plant' magazine, a journal of soil science, plant nutrition, soil microbiology and soil fertilization. The inaugural issue of the magazine was printed in 1952 (Manojlović et al. 1997).

Researchers in the field of soil science first came together with the opening of the Yugoslav branch of the International Society of Soil Science and the first president of the branch was Prof. Mihovil Gračanin from 1931 to 1940. In 1953, the Yugoslav Society of Soil Science (YSSS) was set up and the first meeting held, attended by the academic Dr Siniša Stanković, chairman of the Academic Council; Dr George Barbier, Head of the Central Institute for Agricultural Research (Versailles, France); Prof. Jean Cuzin (University of Paris, France); Prof. Vukić Mićović, Rector of Belgrade University; Prof. Žarko Miletić, Dean of the Faculty of Forestry at the University of Belgrade and others. At this meeting, the lecture 'The development of an international society for soil science and its role in the world and at home' given by Prof. Todorović is viewed as being the most significant. It dealt with the forging of links between the different pedological institutions in our country and also with international associations, with a view to dealing with key issues in the field of soil science more efficiently. It was decided that the YSSS should join the International Society of Soil Science (ISSS) and that the Society's inaugural Congress would be held in 1955 (First YSSS Assembly 1954). The Society only changed its name in 1996 when it became the Serbian Society of Soil Science (SSSS), which was the legal successor to the Yugoslav Society of Soil Science (YSSS).

From its inception, the YSSS was a member of the ISSS. It later joined the International Scientific Centre of Fertilizers (CIEC) and, in cooperation with this organization, hosted the International Symposium on Microelements held in Belgrade in 1956, and the fourth CIEC-CICRA World Fertilizer Congress in Opatia (Croatia) in 1956. Members of these associations were chosen from the ranks of the YSSS; thus, Ć. Jelenić was CIEC President from 1972 to 1991, S. Manojlović was CIEC Secretary General from 1984 to 1996 and an honorary CIEC member, R. Kastori was Executive Secretary from 1980 to 1996 and S. Blagojević was Executive Secretary from 1982 to 1984 (Manojlović et al. 1997).

References

Banković S, Medarević M, Pantić D, Petrović N (2009) National Forest Inventory of the Republic of Serbia, Ministry of Agriculture, Forestry and Water Management, Forestry Commission (in Serbian)

Bogdanov N (2007) Need assessment of small rural households in Serbia. UNDP, Belgrade

First YSSS Assembly (1954) Conclusions from the First YSSS Assembly, 8–11 November 1953. YSSS Bulletin, No. 2, Belgrade

Glinka KD (1927) Dokuchaiev's ideas in the development of pedology and the cognate sciences. Russian pedological investigations no. 1, USSR Academy of Sciences, Leningrad

Jović D, Tomanić L, Banković S (1992) Forest growing stock of Serbia. In: Nikolić S (ed) Forestry and wood processing through the ages, Serbian Association of Forestry and Wood Processing Engineers and Technicians, Belgrade, pp 10–23 (in Serbian)

Kadović R (2013) The potential of Serbian forests for climate change mitigation and environmental protection. In: Andjelković M (ed) Energy and environment, Scientific meetings, Book CXLIII, Serbian Academy of Science and Arts (SANU), pp 493–514

Kanitz F (1985) Serbia. Country and population. Book 1, The Serbian literary cooperative—Rad, Belgrade [Kanitz F, Das Königreich Serbien und das Serbenvolk von der Römerzeit bis zur Gegenwart, 1904, Leipzig: Verlag von Bernh. Meyer]

Law on the Spatial Plan of the Republic of Serbia (1996) Official Gazette of the RS, no. 13/96

Manojlović S, Hadžić V, Antonović G (1997) Review of the Yugoslav society of soil science congress activities and some current problems. In: Dragović S (ed) Organising, using and preserving soils, Mala knjiga, Novi Sad (in Serbian)

Protić N, Martinović Lj, Miličić B, Stevanović D, Mojasević M (2005) The status of soil surveys in Serbia and Montenegro. European Soil Bureau—Research Report No. 9:297–315

Serbian Environmental Protection Agency (2007) In: Karadžić B, Mijović A (eds) Environment in Serbia: an indicator-based review, First edition. Serbian Environmental Protection Agency, Belgrade, p 168

Statistical Office of the Republic of Serbia (2013) Census of agriculture 2012—agriculture in the Republic of Serbia—labour force and activities of agricultural holdings in the Republic of Serbia, Belgrade (in Serbian)

Statistical Office of the Republic of Serbia (2014) Statistical pocketbook of the Republic of Serbia, Belgrade (in Serbian)

2 Soils as Natural Resources

The soil cover of Serbia is diverse due to heterogeneous conditions for the formation and pedogenesis of certain soil types. The crucial impact on the genesis and evolution of soils in Serbia was had by the climate, geomorphological structure of the terrain, geological substrates, vegetation, soil flora and fauna. Physical and chemical characteristics of different soil types vary significantly; therefore, particular types of soil have severe restrictions for use. For example, due to extremely low water-retention capacity, the aeolian sands (WRB, Arenosols, see Table 6.1, Chap. 6, which gives soil names according to the classification system used in Serbia and their equivalents in the WRB) have low productive values. Due to excessive leaching, the base exchange capacity as well as fertility of luvisoil is low. Other unfavourable processes (excessive acidification, translocation of iron and aluminium oxides with organic complexes to illuvial horizons, anaerobic conditions due to long-term waterlogged periods that produce gleyed soils, toxic effects of increased salt concentration in halomorphic soils, etc.) may significantly reduce soil fertility. Different measures such as soil drainage, irrigation, recultivation and technical regulations that reduce erosion risks may improve physical, chemical and biological properties of soil.

The most fertile soil in Serbia is Chernozem, within the Pannonian Plain and its surroundings (Fig. 2.1). The Pannonian Plain is also covered by the following: (1) alluvial deposits, on the fluvial terraces, on which fluvisols, humogley (WRB, Gleysol, humic), solontchak and solonetz develop; (2) loess terraces, on which gleyed chernozem soil and solontchak and solonetz develop; (3) aeolic sand with the types: arenosols, rendzina (WRB, Leptosol, calcaric) and chernozems. Soils on Fruška gora and the Vršac Mountains are heterogeneous, comprising rendzina (WRB, Leptosol, calcaric), ranker (WRB, Leptosol, eutric), eutric and distric cambisols, luvisols and colluvial soils on a heterogeneous substrate. The Pannonian region is characterized by a continental climate, and the dominant crops are wheat, maize, sunflowers, soya and sugar beet. All types of vegetables are farmed, but orchards and vineyards are somewhat less common. The degradation and pollution of the soils in this area is a result of numerous factors. Aeolian erosion occurs due to the powerful gusts of wind, which are particularly marked in the area of Deliblato and Subotica Sands. The strong winds often remove nutritious materials from the surface of agricultural land and when they blow during periods of severe drought, they even remove parts of the surface soil horizons.

The surface mining of coal is also a contributory factor in the degradation and permanent loss of agricultural land. Soil pollution occurs in areas where there are industrial plants and numerous livestock farms and is a result of localized acid rain and the deposition and release of pollutants. Besides industry, traffic contributes to the pollution of soils, above all through lead accumulation along motorways and other major roads. The salinization of soils is also becoming more widespread despite efforts to halt the process and to ameliorate salinized areas. Compaction arises from the use of heavy agricultural machinery when cultivating land, particularly at times when the soil is poorly drained after heavy rain or melting snow. There is periodic flooding, especially in the valleys of those rivers into which water from the Carpathian Mountains flows.

Tertiary lake sediments, limestone or sediments rich in alkali bases, predominate in the hilly area of central Serbia (Fig. 2.2). On structurally lighter—sandy—sediments, eutric cambisol (typic and lessive) soil has developed, which in Serbia is called gajnjača (WRB, Cambisol, eutric), in the sequence regosol–rendzina (WRB, Leptosol, calcaric)–eutric cambisol–luvisol. On the structurally heavy—clayey—sediments, vertisol is predominantly found. Chernozem can be found on small areas in central Serbia, mainly on loess and pseudogley in specific climatic and lithogenic conditions. In the southern part of this region, due to marked erosion, there are many areas of colluvial soils. On lower fluvial terraces,

P. Pavlović et al., *The Soils of Serbia*, World Soils Book Series,
DOI 10.1007/978-94-017-8660-7_2

Fig. 2.1 Agricultural fields in Pannonian Plain (*photo* http://www.rgz.gov.rs)

fluvisols, humic gleysol and eugley (WRB, Gleysol) are present. The climate is moderate continental and the xerothermic forests dominate in this region. Wheat, maize and sugar beet are grown on agricultural land, while vegetable farming is developed in the river valleys. Vineyards and orchards occupy a significant area, particularly in the highlands. The kinds of soil degradation and damage in this region are different from the first. Water erosion dominates in almost all parts of the region in spring and autumn, particularly on those slopes where there is no vegetation. Flooding is also very common as the region abounds in rivers whose drainage basins are far away in the mountain region, which results in the deposition of sand and gravel on the fertile areas of the river valleys. The meandering of rivers has a similar effect. A further problem is the large number of sand and gravel borrow pits in the river valleys. Surface mining is very common in this region and occupies large areas, which often results in the permanent loss of fertile cultivable soil. The exploitation of ore is accompanied by the formation of spoil tips, which reduces the amount of agricultural land yet further. Soil pollution is particularly marked in the vicinity of the numerous coal-burning industrial plants (thermal power stations), ore-processing plants (smelters, etc.) and industrial waste deposits, such as fly ash and slag from thermal power stations and smelters, and is the result of the toxic effects of acid rain and the dispersion of aerosol and particles of ash and slag over surrounding agricultural land.

The mountainous regions of western and eastern Serbia involve the Dinaric Alps, the Rhodope Mountains, the Šar-Pindus range and the Carpathian and Balkan Mountains. Kalkomelanosols (WRB, Leptosols) are present on the limestone and dolomite rocks at the highest altitudes, with the other areas dominated by kalkocambisols (WRB, no adequate WRB name) and luvisols. Rankers (WRB, Leptosols) occur on the acid silicate rocks in the higher zones and on the steepest slopes, while dystric cambisols dominate the other areas. There are small areas of podzols and brunipodzols (WRB, Podzol, humic) on quartzite rocks, with

Fig. 2.2 Agricultural fields, meadows and pastures in hilly area of central Serbia (*photo* Mitrović, M.)

rankers (WRB, Leptosols), eutric cambisol, luvisols and smonitza (WRB, Vertisol) found on the basic rocks. The climate is continental, characterized by warm summers and very cold winters with high precipitation, approximately 600 mm annually. The mountains are dominated by forests, pasture and natural meadows. In terms of agricultural cultivation, wheat, maize, barley, rye, oats and potatoes are grown. Fruit farming is very well developed and there are also vineyards. In this region, tobacco cultivation is very common.

The processes of soil degradation and pollution are also present in the mountainous region with water erosion at its most pronounced. All forms of soil removal are present next to the rivers and in the valleys beneath the mountains, particularly on those slopes where there is a lack of vegetation due to felling. Rocky wastelands occur in places on the limestone and also on the silicate rocks. There are frequent flash floods, which results in the removal of large amounts of soil. Pollution and the permanent loss of soil due to industrial activity occur only in and around mines, where ore and coal are excavated at great depths and flotation tailings and tailings disposed of on the surface, which results in the degradation or complete destruction of areas of soil for agricultural cultivation. Quarries also occupy a significant area and hence contribute to the significant loss of soil.

2.1 Soil Types and Their Production Value

The soils of Serbia are extremely heterogeneous as a result of the varied geological base, climate, vegetation and pedofauna. Some soils are naturally fertile, providing optimal conditions for high-, stable- and good-quality yields, such as carbonate chernozems on loess, while others have unfavourable characteristics to such an extent that production on them is not economically viable. There are also soil types with exceptionally unfavourable characteristics even for the formation of natural vegetation, such as solontchak and solonetz (Table 2.1).

Soils are divided into eight fertility classes, reflecting the relative suitability for agricultural production with classes 1–4 representing higher quality soils. However, about 45% of the total territory belongs to soil classes 4–8, which are not suitable for tillage and profitable crop production, and this land is used for semi-subsistence vegetable or fruit production or as meadows (Protić et al. 2005).

The most frequent soil types in the Republic of Serbia are as follows: (1) lithosol (WRB, Leptosol), an undeveloped soil composed of weathered rocks belonging to soil productivity class 8; (2) arenosol, an undeveloped sandy soil, which is not suitable for crop production and also belongs to soil productivity class 8; (3) kalkomelanosol (WRB, Lep-

Table 2.1 Main soil types and restrictions for land use in Serbia

Soil type Serbian class/WRB	Area (ha)	Restrictions
Lithosol/Leptosol	107,000	Unproductive soil
Arenosol	86,000	Severe restrictions due to excessive filtration; Poor to medium productive soil
Rendzinas/Leptosol, calcaric	~527,000	Severe to medium restrictions
Kalkomelanosol/Leptosol	~155,000	Severe restrictions
Ranker/Leptosol, eutric, dystric	572,000	Severe restrictions
Chernozem	1 2200,000	Fertile soil; No restrictions
Smonitza/Vertisol	780,000	Moderate restrictions
Kalkocambisol/No adequate WRB name	~350,000	Severe to medium restrictions
Eutric cambisol/Cambisol, eutric	560.000	Moderate restrictions
Dystric cambisol/Cambisol, dystric	~2 280.000	Severe to very severe restrictions
Luvisol	~510.000	Moderate to medium restrictions
Pseudogley/Stagnosol	538,000	Moderate to severe restrictions—conditionally productive soil
Podzol	~17,000	Severe to very severe restrictions
Fluvisol, Humofluvisol/Fluvisol, humic Humogley/Gleysol, humic; Eugley/Gleysol	~760,000	No restrictions to serious restrictions—conditionally highly productive soil
Solontchak and Solonetz	233,000	Severe restrictions
Histosol	~3.000	Moderate to severe restrictions
Deposol/Anthrosol	~50.000	Unproductive soil; Moderate to severe restrictions

Source Protić et al. (2005)

tosol) and kalkocambisol (WRB, no adequate WRB name) are categorized in productivity class 7 and are used mainly as natural pastures or are forested with only small areas used for cultivation; (4) ranker (WRB, Leptosol, eutric) is mainly used for pastures and meadows with smaller areas used for the cultivation of potatoes, barley and oats. Rankers belong to soil productivity class 7, meaning they are not suitable for intensive plant cultivation; (5) chernozem is mainly found in Vojvodina where it accounts for approximately 1,000,000 ha, with only 200,000 ha in other parts of Serbia. It is categorized in soil productivity class 1, which in conditions of optimal irrigation produces high, stable yields of cultivated crops; (6) vertisol is a soil type which belongs to productivity class 3. It has an unfavourable water, air and temperature regime, but has favourable chemical characteristics; (7) eutric cambisol belongs to soil productivity class 2. It is a medium heavy soil which has a very favourable water and air regime. It is suitable for crop, vegetable, fruit and viticulture production; (8) dystric cambisol belongs to productivity class 5 or 6, depending on the gradient of the terrain on which it is formed and solum thickness. In nature, this soil is covered with forest vegetation and small areas are used for fruit, crop and vegetable cultivation. Satisfactory yields can be obtained when growing potatoes, rye, oats and barley; (9) ranker (WRB, Leptosol, eutric) and eutric cambisol are characterized by unfavourable physical and chemical properties and are mainly used as pastures and meadows. Due to their significant limitations for plant cultivation, they belong to productivity class 7; (10) pseudogley (WRB, Stagnosol) is soil which is characterized by an unfavourable water, air and temperature regime and belongs to productivity class 4. More intensive use of pseudogley for plant cultivation is possible if accompanied by complex ameliorative measures; (11) fluvisol and humogley (WRB, Gleysol, humic) belong to productivity class 3, which comprises soils that are used for plant production, but with certain limitations. Due to the additional water content of the solum from flood and/or underground water and the

Table 2.2 Productivity structure of soils in Serbia

Classes	Area (km^2)				Structure (%)			
	Total Serbia	Vojvodina	Central Serbia	Kosovo and Metohija	Total Serbia	Vojvodina	Central Serbia	Kosovo and Metohija
1	11,650	9688	1675	287	14.40	51.40	3.20	2.80
2	9357	3284	5481	592	11.60	17.40	10.60	5.80
3	10,522	3823	5383	1316	13.00	20.30	10.50	13.00
4	8682	355	7133	1194	10.80	1.90	13.80	11.80
Total for cultivation	40,211	17,150	19,672	3389	49.80	91.00	38.10	33.40
5	11,073	531	9002	1540	13.70	2.80	17.40	15.20
6	20,144	889	17,185	2070	25.00	4.70	33.20	20.40
7	8069	193	5232	2644	10.00	1.00	10.10	26.10
8	1178	72	604	502	1.50	0.50	1.20	b4.90
Total unfavourable for cultivation	40,464	1685	32,023	6756	50.20	9.00	61.90	66.60
Total cultivable soil	80,675	18,835	51,695	10,145	100.0	100.0	100.0	100.0

Source Spatial Plan of the Republic of Serbia 2010–2020 (2010)

mechanical composition, fluvisols and humogleys (WRB, Gleysol, humic) are characterized by an unfavourable water, air and temperature regime; (12) solontchak and solonetz are used as natural, quite barren pastures due to their unfavourable chemical and physical properties. In terms of their suitability for plant production, solontchak and solonetz belong to productivity class 7. In order to be included in the reserves of cultivable soils, solontchak and solonetz need to be ameliorated.

In Serbia, soils that are favourable for cultivation are present in almost equal measure to those which are not. Limitations in terms of intensive agricultural production are least marked in Vojvodina and most pronounced in Kosovo and Metohija, which, like the majority of central Serbia, is characterized by wide variations in the natural fertility of soils along narrow geomorphological entities.

Since there are no significant climatic limitations when it comes to plant cultivation in the Republic of Serbia, if many of the soil types with low productivity were subject to certain ameliorative measures, which were then monitored, and if available fertilizers were used, significantly more favourable conditions could be achieved for a wide range of agricultural cultivation and the development of livestock breeding.

The scientific basis of the Soil Fertility and Fertilizer Use Control System was established in 1980 and was legislated in 1985. The system encompasses the control of all factors that determine soil fertility and fertilizer action, i.e., how soil affects the growth, development and yield of crops and which measures should be undertaken to ensure high, stable and economic yields and effective biosphere protection. Unfortunately, due to economic crises, the system was not fully exploited (Vidojević and Manojlović 2007).

2.2 Soil Under Agriculture

Data from the Spatial Plan of the Republic of Serbia 2010–2020 (2010) show that soil under agriculture makes up approximately 81% of the Republic's total territory (central Serbia 52%, Vojvodina 19% and Kosovo and Metohija 10%) (Table 2.2).

Intensive agricultural production is least restricted in Vojvodina and most restricted in Kosovo and Metohija. The latter territory, similar to that of central Serbia, has a wide range of natural fertility in narrow geomorphological units (Vidojević and Manojlović 2007).

References

Protić N, Martinović Lj, Miličić B, Stevanović D, Mojasevic M (2005) The status of soil surveys in Serbia and Montenegro. In: Jones RJA, Houšková B, Bullock P, Montanarella L (eds) Soil resources of Europe, 2nd edn. European Soil Bureau, Luxembourg, Research Report No. 9, pp 297–315

Spatial Plan of the Republic of Serbia 2010–2020 (2010) http://www.rapp.gov.rs/sr-Latn-CS/content/cid310/prostorni-plan-republike-srbije

Vidojević D, Manojlović M (2007) Overview of soil information and policies in Serbia. In: Hengl T, Panagos P, Jones A, Tóth G (eds) Status and prospect of soil information in southeastern Europe: soil databases, projects and applications, European Commission, Directorate-General Joint Research Centre, Institute for Environment and Sustainability, pp 87–189

3 Climate

Climate is a key factor in the development of different vegetation and soil types, although local flora and parent material are also of significance for soil and vegetation diversification (Tuhkanen 1980). The rates of different pedogenetic processes (the 'physical weathering' of parent material, leaching, podzolization, gleying, salinization, clay formation, the migration of sesquioxides, etc.) depend on climatic conditions. Consequently, similar soil types are distributed in regions with similar climates.

On the other hand, both vegetation and soils control climatic conditions. Part of the energy received from the sun is returned as long-wave (infrared) radiation, heating the atmosphere, while part of the absorbed solar energy serves to evaporate water, either in the soil or in plant leaves, sending water back into the atmosphere. Because evapotranspiration requires energy, both soil moisture and vegetation cover have strong influence on atmospheric temperature.

Due to the mutual dependence and functional interrelations of soil, vegetation and climate, global classifications of climate, soil and vegetation types are similar (1936; Holdridge 1947; Prentice et al. 1992). The correspondence between geographic patterns of climate, vegetation and soil is a consequence of the interaction and strong interdependence of particular components of ecosystems (Pianka 1983; Prentice et al. 1992; Peel et al. 2007).

The main characteristics of both the spatial and temporal variability of climate in Serbia have been described in numerous studies (Rakićević 1980, 1989; Dukić 1981; Ducić and Radovanović 2005; Milovanović 2005; Ducić et al. 2010; Unkašević and Tošić 2011).

3.1 Temperature

In terms of climate, geographic position of Serbia falls within a mild temperate (mesothermal) climatic zone (i.e. Köppen's 'C climate zone', where the average temperature is above 10 °C during the warmest months, and the coldest month average is above −3 °C).

Although Serbia spans a relatively narrow latitude range, the temperature conditions in the country are diverse due to the heterogeneous topography. Air density decreases with increasing elevation and, due to the air density gradient, both air pressure and air temperature (the kinetic energy of air molecules) decrease with altitude. Thermal lapse (the decrease in temperature with altitude) is similar to temperature lapse with increasing latitude. Air moisture significantly affects the temperature lapse rate. For dry air, temperature decreases at a rate of 1 °C per 100 m. However, the lapse rate for wet air is much lower (approximately 0.5 °C/100 m). In Serbia, the average altitudinal temperature lapse rate is approximately 0.53 °C per 100 m (Milovanović 2010). Mean annual temperatures in Serbia are presented in Fig. 3.1.

Compared to lowland regions, the mountainous regions of Serbia have lower temperature averages. On high mountains (i.e. at altitudes above 1500 m), the climate is 'boreal' or microthermal (i.e. Köppen's 'D climate'). These regions have an average temperature above 10 °C in their warmest months and a coldest month average below −3 °C.

The Tundra climate (i.e. Köppen's 'ET climate') is found only sporadically on the Balkan Peninsula, restricted to high mountain peaks above 2000 m (Peel et al. 2007). The average temperature of the warmest month in these regions

P. Pavlović et al., *The Soils of Serbia*, World Soils Book Series,
DOI 10.1007/978-94-017-8660-7_3

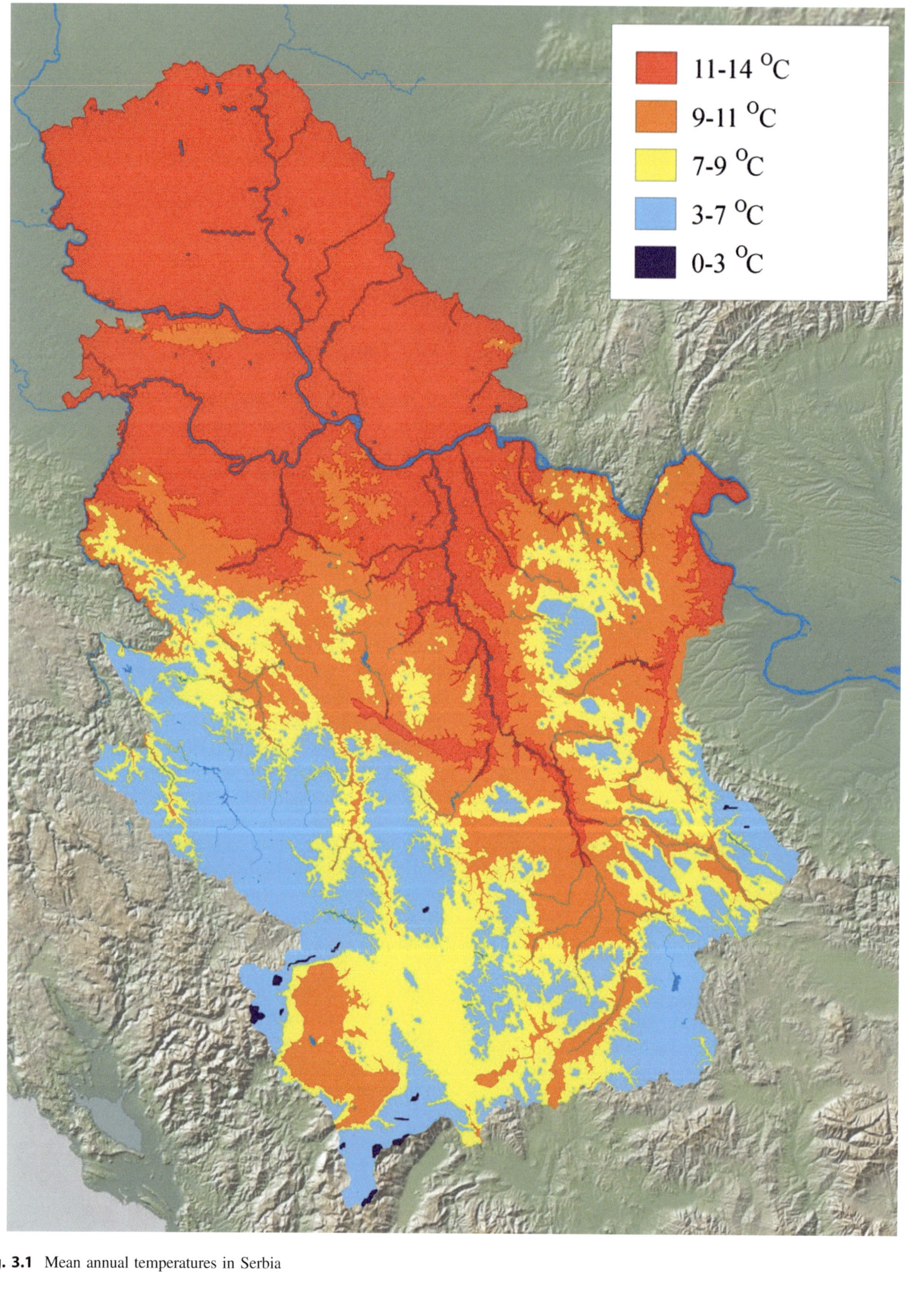

Fig. 3.1 Mean annual temperatures in Serbia

is between 0 and 10 °C, while this climate is also characterized by long cold winters and short summers.

The limits of the Tundra climate in Serbia correspond to the tree line. Most boreal tree species are adapted to survive the extremely low temperatures which occur ephemerally from time to time. Therefore, temperature extremes are not a key factor determining the tree line. The Arctic-alpine treeless zone does not correlate with the extremely low temperatures, but to the accumulated growing-season warmth or 'heat sum' (Tuhkanen 1980).

Mesothermal (temperate) conditions prevail in Serbia. The dominant 'C climate zone' can be divided into several variants:

(a) The 'Cfb' climate, where the letters 'fb' denote a climate without a dry season (f) and with a warm summer (b), is characteristic of mountain regions (between 500 and 1500 m in altitude). The letter 'b' denotes a warmest month average temperature of below 22 °C, with at least 4 months averaging above 10 °C.
(b) The lowlands below 300 m (the Pannonian plain, the Danube, Morava and Sava valleys) have a more arid climate (either a 'Csa' or 'Cfa' climate). The letter 's' denotes dry summers (precipitation during the driest summer month is less than one-third of the wettest winter month), while the letter 'a' denotes a warmest month average temperature of above 22 °C with at least 4 months averaging above 10 °C.
(c) The high mountains (those over 1500 m) cover just 206 km^2 or 0.23% of Serbian territory. According to Köppen's classification system, the microthermal climate of Serbia may be divided into the variants 'Dfb', 'Dfc' and 'Dsb', where the letter 'c' indicates a climate with 3 or fewer months with mean temperatures above 10 °C.
(d) Finally, the Tundra climate (Köppen's 'ET climate') is sporadically represented, only in those regions above 2000 m. In southern, central and eastern Serbia there are several mountain ranges with peaks above 2000 m (Table 3.1).

The amplitude of mean monthly temperatures on the Pannonian plain and in eastern lowland regions is wide (for example, the annual temperature range in Negotin is 23.1 °C). In other regions, temperature amplitudes are much less marked.

On the other hand, the amplitude of absolute temperatures varies much more widely. For example, at the Pešter Plateau the temperature range exceeds 70 °C and the absolute temperature extremes observed at the Sjenica weather station range from −38.4 to 34.7 °C (Šabić and Pavlović 2004). According to the Republic Hydrometeorological Service (RHMZ 2014), the highest temperature ever recorded was in Kraljevo (44.3 °C), while in other regions, the absolute temperature maximum ranges from 37.1 to 42.3 °C. The lowest ever temperature (−39.0 °C) was measured at Karajukica Bunari (the Pešter Plateau).

3.2 Precipitation

Climate depends not only on average temperature conditions, but also on the precipitation pattern. Precipitation, in combination with temperature, affects relative moisture or dryness. Besides the fact that precipitation increases with increasing altitude, mountain ranges may have more complex and profound effects on precipitation patterns.

The Dinaric Alps extends from north-west to south-east direction. Such a position is almost orthogonal with respect to the prevailing westerly winds. Therefore, the Dinaric Mountains form an orographic barrier that affects the precipitation pattern of the Balkan Peninsula. On the windward side, air currents are forced to ascend. As air rises windward, it cools adiabatically and water vapours condense, forming precipitation. The west-facing side of the Dinarides receives large amounts of rainfall. According to Ducić et al. (2012), the village of Crkvice on Mount Orjen in Montenegro is the wettest place in Europe, with the highest annual precipitation total (over 5000 mm). Due to orographic rain, relatively dry air reaches the eastern, continental regions of the Balkan Peninsula. Therefore, the amount of precipitation gradually decreases from the Adriatic coast eastwards.

Analysis of annual precipitation totals at 27 weather stations in Serbia for the period 1951–2010 showed the lowest values of mean annual precipitation at the stations in northern Serbia, while the highest values were at the higher altitude stations and those in western and south-western Serbia (Stanojević 2012). Generally, the total annual amount of precipitation decreases from west to east. Map of mean annual precipitation in Serbia is presented in Fig. 3.2.

Average annual precipitation totals ranged from 554 mm at the Kikinda weather station to 972.5 mm at the Zlatibor station and 977.1 mm at Kopaonik. The highest values were found at those stations located at higher altitudes, hence the values at Kopaonik and Zlatibor.

Kikinda and generally other stations located in northern Serbia (Palić 560.3 mm, Sombor 607.2 mm, Zrenjanin 583.7 mm, Rimski Šančevi 625.6 mm, Vršac 659.6 mm and Sremska Mitrovica 629.2 mm) had the lowest levels of precipitation, followed by stations in the east and south-east of Serbia (Veliko Gradište 637 mm, Negotin 652.6 mm, Zaječar 611.1 mm, Niš 592.9 mm, Dimitrovgrad 645.7 mm, Leskovac 627.7 mm, and Vranje 611.8 mm), while relatively higher values were recorded at weather stations in

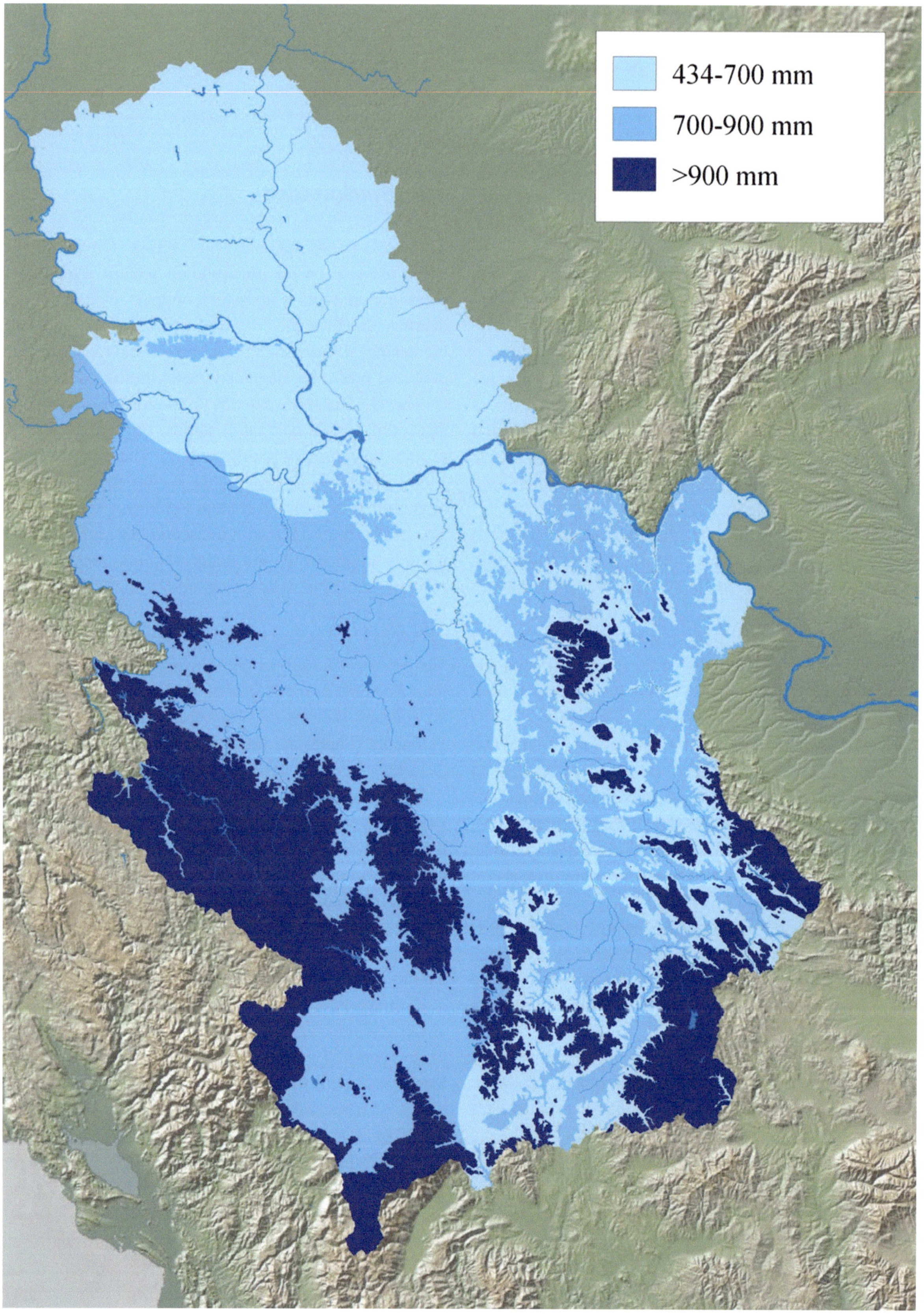

Fig. 3.2 Mean annual precipitation in Serbia

central Serbia (Belgrade 698.2 mm, Kragujevac 635.8 mm, Smederevska Palanka 646 mm, Kraljevo 763.7 mm and Kruševac 648.4 mm). The highest values were in the west and south-west of Serbia (Loznica 844.8 mm, Valjevo 806.6 mm, Požega 737.1 mm and Sjenica 734.5 mm).

Mean annual temperature and precipitation averages alone are insufficient to describe the climatic conditions of a region. More elaborate climate parameters describe the seasonality of temperature and precipitation. For example, Walter (1973) and Walter and Lieth (1964) outlined a simple method for describing the temperature, moisture and continentality conditions of a region. This method uses monthly temperature and precipitation averages to present the seasonality of climate conditions. Temperature and precipitation values are scaled so that 1 °C ~ 2 mm, and in certain cases even 1 °C ~ 3 mm. A water deficit occurs when temperature exceeds precipitation; a water surplus exists when the situation is reversed. He referred to any period during which the temperature curve lies above the precipitation curve on the 1:2 scale as a period of drought, and any period in which this happens on the 1:3 scale as a semi-dry period. During a period of water surplus, some water may be stored by plants and some may accumulate in the soil as soil moisture. During the dry season, stored water may be used by plants and released back into the atmosphere. Thus, Walter's diagrams contain a large amount of information on the climate characteristics of a region and we used them to describe the climate variability of Serbia.

Climate data was taken from the annual reports of the Republic Hydrometeorological Service of the Republic of Serbia. In addition, the WorldClim database (Hijmans et al. 2005) was used for some mountainous regions to access average values of climatic elements over a 50-year period (1950–2000, Table 3.1) using the DIVA-GIS package (Hijmans et al. 2012).

Walter's climate diagrams for lowland regions in Serbia (regions below 150 m a.s.l.) are presented in Fig. 3.3. These regions cover the Pannonian plain and valleys along large rivers (the Danube, Morava and Sava) and their tributaries. The average annual temperatures of Sombor, Palić, Kikinda (the Pannonian plain), Loznica (the Drina valley) and Zaječar (the Timok valley) are similar and range from 10.4 to 11 °C. Temperature peaks occur in the summer months, while the average January temperature is below 0 °C. The mean annual precipitation total on the Pannonian plain is lower than in other regions, while total annual precipitation decreases from west (820.3 l/m^2 in Loznica) to east (610.5 l/m^2 in Zaječar). Rainfall is unevenly distributed, but a precipitation peak occurs in June.

This precipitation pattern indicates that the climate in lowland regions is continental. The Pannonian region is characterized by a semi-arid season lasting from July to October and these semi-arid conditions have a profound effect on both vegetation and soil on the Pannonian plain. In eastern lowland regions (the Danube and Timok valleys), seasonal drought is more severe, with the dry period lasting from July to September. On the other hand, the climate of the western lowlands (the Drina valley) is more humid. A huge amount of precipitation ensures water surplus in this region.

Climatic conditions change with altitude. In comparison to lowland regions, the climate of hilly and low mountainous regions (from 300 to 500 m a.s.l.) is colder. In Serbia, the average annual temperature of these regions is usually below 10 °C (Fig. 3.4). However, due to the Mediterranean influence that spreads through the Drim valley, mean annual temperatures in Kosovo and Metohija are much higher. For example, Peć is located at 498 m a.s.l. but despite this relatively high altitude, its mean annual temperature (11 °C) is greater than that of the northern lowland regions, which is a result of the Mediterranean influence.

The precipitation patterns of mountainous regions in Serbia vary significantly. Požega is located in western Serbia at 310 m a.s.l. and its mean annual precipitation is 737 l/m^2. Peak precipitation occurs in May, while the amount of precipitation throughout the year is sufficient to prevent the occurrence of a (semi)dry season. On the other hand, Peć has much greater levels of precipitation (850.7 l/m^2). Despite this, the climate of Peć is characterized by a semi-arid period. A precipitation minimum occurs during the summer months (July and August) and a precipitation maximum in November. This sub-Mediterranean precipitation pattern results in a semi-arid period during the summer months.

The climate of Dimitrovgrad (south-eastern Serbia) is also characterized by a semi-dry period, which, unlike the climate of Peć, is a consequence of the relatively low amount of precipitation (635.5 l/m^2). Although Peć and Dimitrovgrad are at similar altitudes, the amount of precipitation decreases significantly from west to east. The distribution of precipitation in Dimitrovgrad corresponds to a continental precipitation pattern, with maximum precipitation in the summer months.

High mountainous regions (i.e. regions above 1000 m a.s.l.) have cold climates. The average temperature on Zlatibor (1028 m) and Crni Vrh (1037 m) is 7.1 and 6.7 °C, respectively (Fig. 3.5). At higher altitudes, average temperatures are much lower. For example, the climate diagram for Kopaonik (1710 m a.s.l.) indicates that temperatures are above 0 °C for only 7 months of the year. Moreover, only two months (July and August) have temperatures above 10 °C. Such a short vegetation period is string selective pressure for most trees.

The amount of precipitation on high mountains is much greater than in the surrounding lower regions. The summer months are perhumid since the amount of precipitation on Kopaonik and Zlatibor exceeds 100 l/m^2. As with lower

Table 3.1 Climate in different regions of Serbia

Locality		Months												
		I	II	III	IV	V	VI	VII	VIII	IX	X	XI	XII	Mean
Lowland region														
Sombor	Temperature	−1.3	1.3	5.7	11.1	16.3	19.3	20.7	20.0	16.2	10.9	5.3	0.9	10.5
	Precipitation	37.1	31.7	35.8	51.0	56.9	79.0	60.7	51.8	37.4	36.8	51.8	43.5	573.5
Palić	Temperature	−1.6	1.1	5.6	11.1	16.3	19.4	21.0	20.2	16.5	11.0	5.2	0.7	10.5
	Precipitation	35.6	31.3	33.7	42.7	55.0	73.9	56.7	54.0	36.4	27.7	46.3	46.0	539.3
Kikinda	Temperature	−1.5	1.2	5.9	11.4	16.6	19.6	21.1	20.4	16.7	11.3	5.5	0.9	10.8
	Precipitation	34.0	31.4	35.1	46.3	52.7	75.7	51.3	50.8	37.4	31.3	43.6	45.6	535.2
Zaječar	Temperature	−1.4	0.8	5.1	11.1	16.0	19.2	20.9	20.4	16.4	10.4	5.2	0.8	10.4
	Precipitation	39.8	44.0	48.1	55.3	71.7	71.2	55.1	37.2	37.8	40.8	58.0	51.5	610.5
Loznica	Temperature	−0.2	2.3	6.5	11.4	16.1	19.2	20.7	20.1	16.5	11.3	6.2	1.8	11.0
	Precipitation	51.0	49.2	57.7	70.0	82.7	99.5	84.4	76.3	61.9	53.7	70.3	63.6	820.3
Mountain region														
Požega	Temperature	−1	0.8	4.2	9.9	14.8	18.1	18.2	19.1	15.2	9.1	4.3	0.5	9.1
	Precipitation	44.8	45.6	52.2	70.2	79.5	84.5	80.2	70.8	59	57.3	62.0	64.1	737
Peć	Temperature	−0.5	2.1	6.4	11.2	15.9	19.0	21.1	20.8	17.2	11.8	5.9	1.2	11.0
	Precipitation	85.9	71.5	65.2	67.2	68.2	53.0	54.7	48.0	52.1	75.3	118.2	91.4	850.7
Dimitrovgrad	Temperature	−1.3	0.9	4.9	10.0	14.6	17.5	19.3	19.0	15.4	10.4	5.2	0.7	9.7
	Precipitation	42.2	40.5	46.5	51.1	74.9	87.1	60.7	44.1	38.9	39.1	61.4	49.0	635.5
Altimontane region														
Zlatibor	Temperature	−3.3	−1.5	2.0	6.6	11.5	14.4	16.3	16.3	13.1	8.4	3.2	−1.5	7.1
	Precipitation	68.0	60.8	64.0	76.8	100.0	110.0	96.0	78.3	83.4	66.6	85.4	75.0	964.3
Crni Vrh	Temperature	−4.2	−3	1	6.4	11.5	14.1	16	15.8	12.6	7.3	1.8	−2.3	6.4
	Precipitation	42.6	46.3	52.2	63.5	104.6	121.9	90.2	64.3	63	53.3	58.4	49.8	810.1
Kopaonik	Temperature	−5.6	−5.5	−3.2	1.2	6.4	9.0	11.3	11.4	8.2	4.1	−0.7	−3.9	2.7
	Precipitation	62.0	58.7	64.8	74.7	108.8	124.7	92.1	94.3	66.7	46.2	68.9	58.9	920.8
Stara Planina	Temperature	−5.8	−4.9	−1.8	2.4	7	10.35	12.3	12.5	9.45	5	0	−3.6	3.6
	Precipitation	59	53	58	68	89	95	72	63	46	45	63	65	776.0

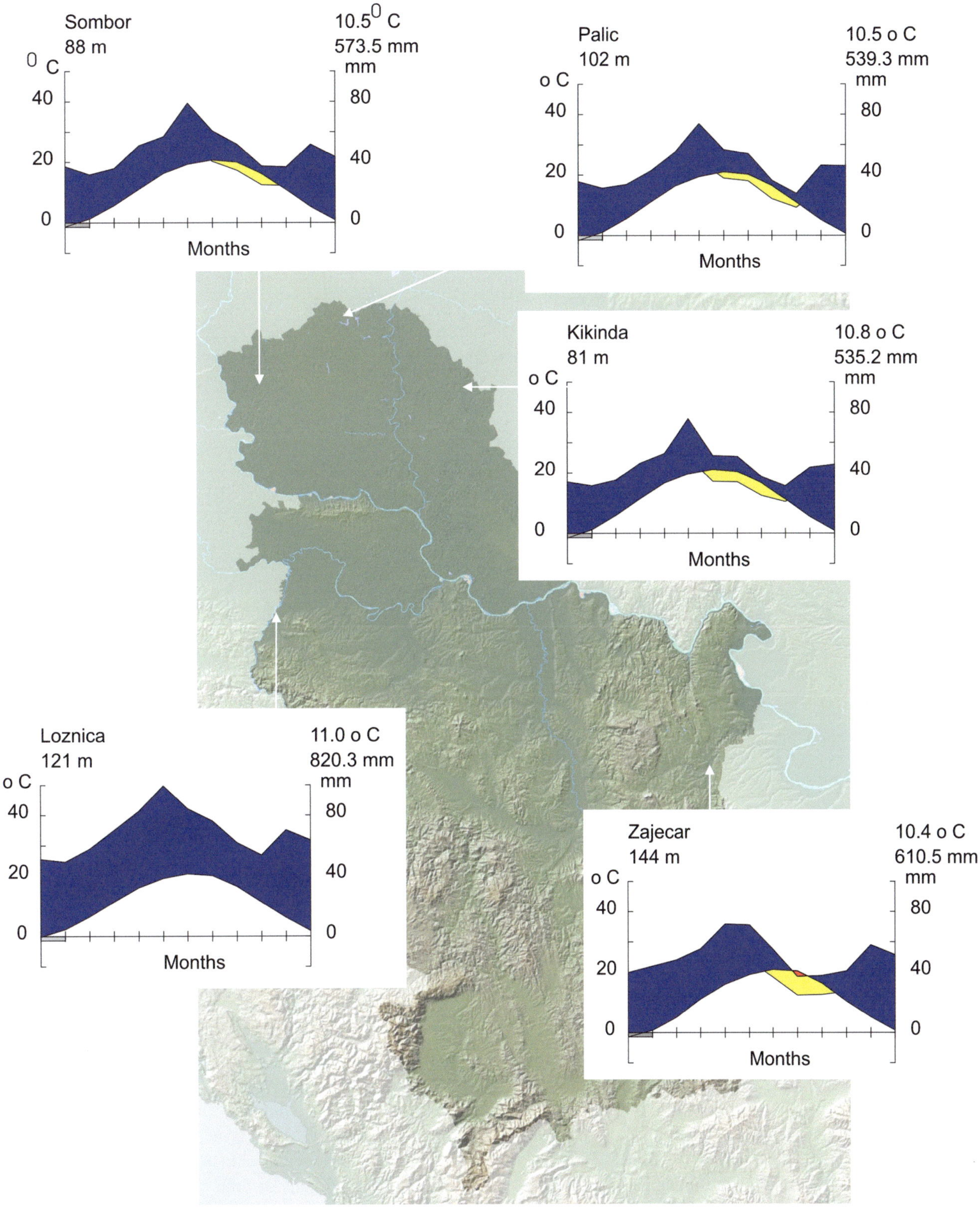

Fig. 3.3 Walter's climate diagrams for lowland regions in Serbia

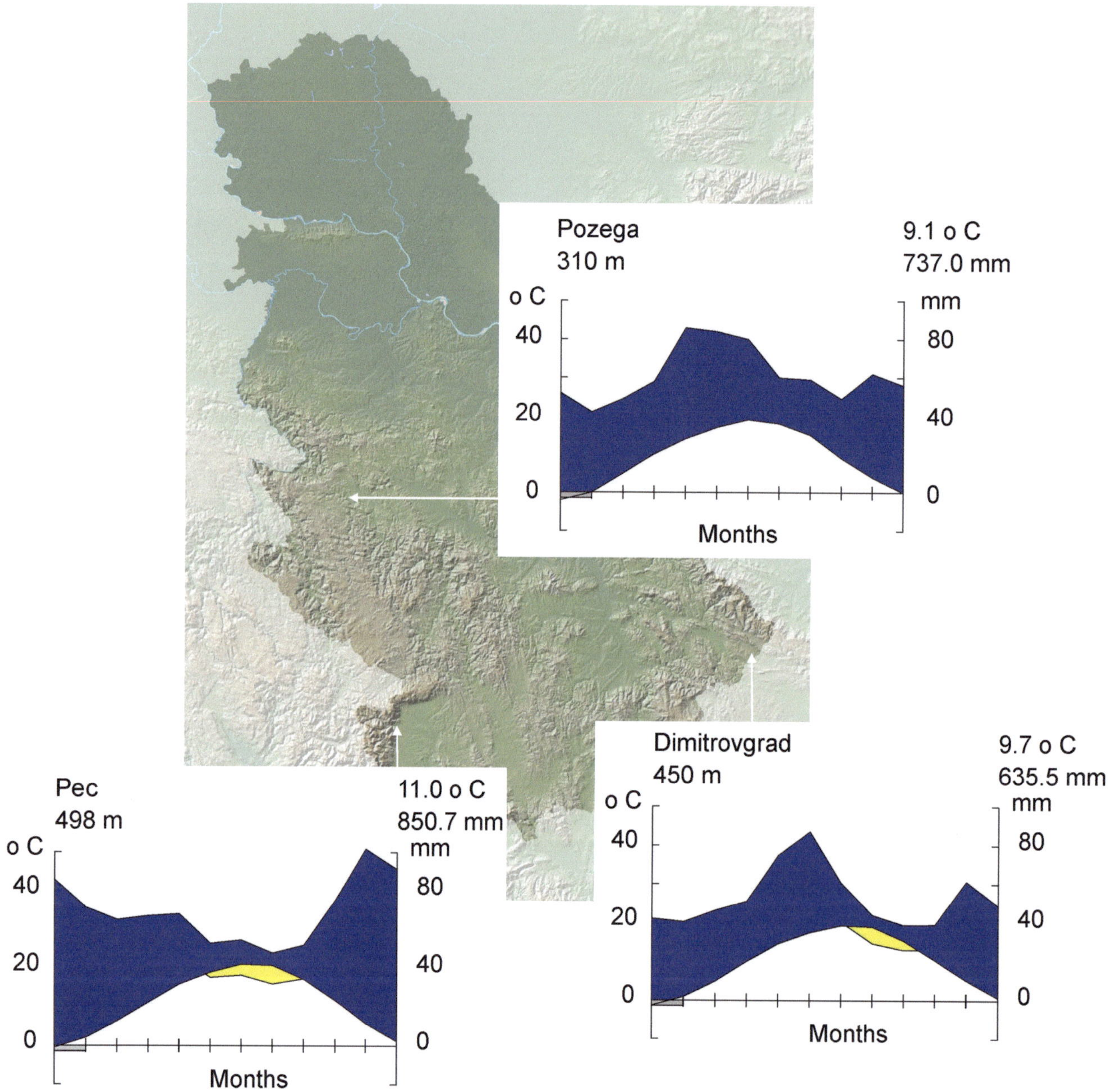

Fig. 3.4 Walter's climate diagrams for low mountainous regions in Serbia (altitude of 300–500 m a.s.l.)

regions, the amount of precipitation reduces from west to east.

The climate of the high mountains in Kosovo is modified by a sub-Mediterranean influence. While the distribution of precipitation on the high mountains in central and eastern Serbia corresponds to a continental precipitation pattern, the climate of the Prokletije mountains, Hajla, Koritnik and Šar Planina has the characteristics of an oro-sub-Mediterranean climate with precipitation peaks during the autumn and winter months and precipitation minima during the summer months. An oro-sub-Mediterranean climate is cold and, unlike a boreal continental climate, dry during the summer months. Such conditions of both physiological drought (i.e. drought caused by extremely low temperatures that prevent

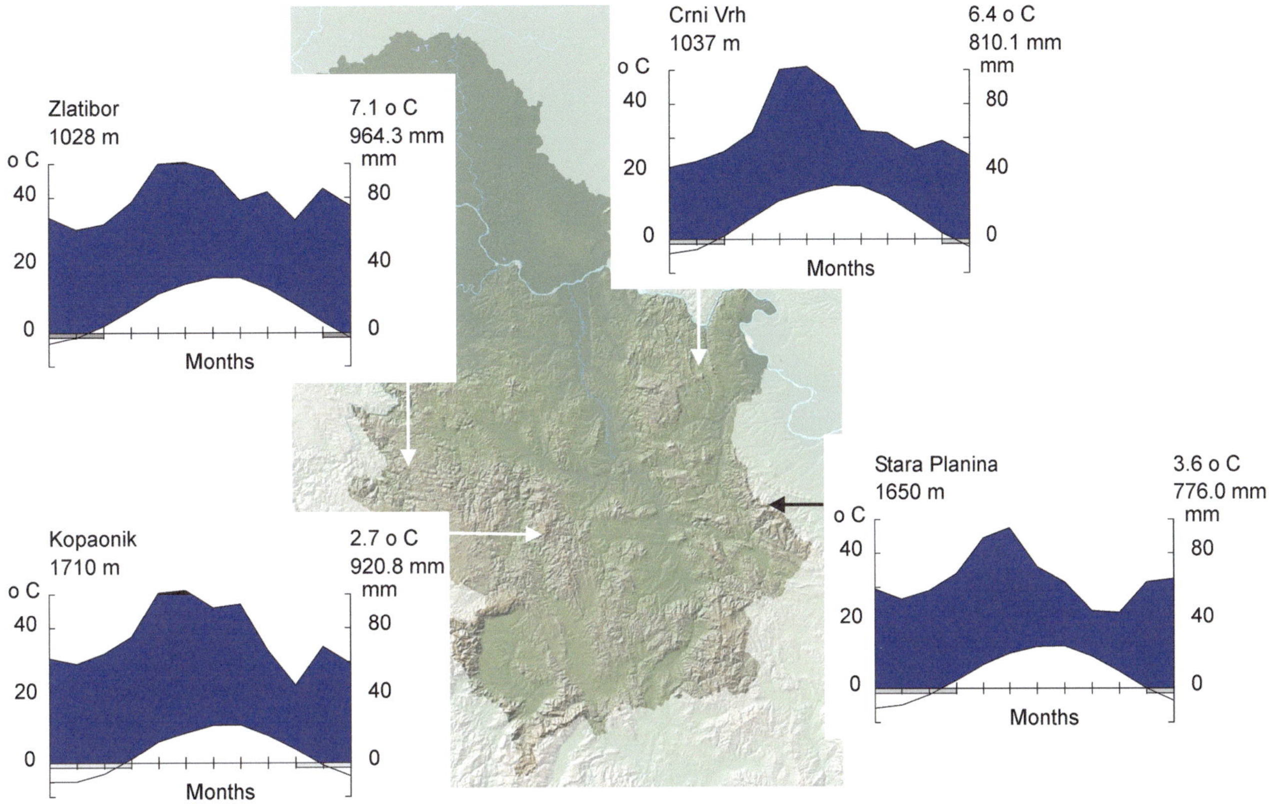

Fig. 3.5 Walter's climate diagrams for high mountainous regions (above 1000 m a.s.l.) in Serbia

plant activity) during the winter months and physical drought during the summer have led to the development of coniferous forests with paleo-endemic Balkan pines *Pinus heldreichii* H.Christ and *Pinus peuce* Griseb.

References

Ducić V, Radovanović M (2005) Climate of Serbia. Institute for textbook publishing and teaching aids, Belgrade (in Serbian)

Ducić V, Luković J, Stanojević G (2010) Atmospheric circulation and the precipitation variability in Serbia for the period 1949–2004. Bull Ser Geogr Soc 90(2):85–107

Ducić V, Luković J, Burić D, Stanojević G, Mustafić S (2012) Precipitation extremes in the wettest Mediterranean region (Krivošije) and associated atmospheric circulation types. Nat Hazard Earth Sys Sci 12(3):687–697

Dukić D (1981) Climatology. Scientific Book, Belgrade (in Serbian)

Hijmans RJ, Cameron SE, Parra JL, Jones PG, Jarvis A (2005) Very high resolution interpolated climate surfaces for global land areas. Int J Climatol 25:1965–1978

Hijmans RJ, Guarino L, Mathur P (2012) DIVA-GIS version 7.5 manual http://www.diva-gis.org

Holdridge LR (1947) Determination of world formations from simple climatic data. Science 105:367–368

Milovanović B (2005) Statistical precedures application and results of research of precipitation on mountain Stara planina. J Geograph Ins 'Jovan Cvijić', Serbian Acad Sci Art, Belgrade 54:33–44

Milovanović B (2010) Climate of the mountain Stara planina. J Geogr Inst 'Jovan Cvijić' SANU, 137 (in Serbian)

Pianka ER (1983) Evolutionary ecology. Harper and Row, New York

Peel MC, Finlayson BL, McMahon TA (2007) Updated world map of the Köppen-Geiger climate classification. Hydrol Earth Syst Sci 11:1633–1644

Prentice C, Cramer W, Harrison SP, Leemans R, Monserud RA, Solomon AM (1992) A Global Biome Model Based on Plant Physiology and Dominance, Soil Properties and Climate. J Biogeogr 19:117–134

Rakićević T (1980) Climatic zoning in the socialist Republic of Serbia. Collection of papers from the Geographical Institute of the Faculty of Sciences (Belgrade) 27:29–42 (in Serbian)

Rakićević T (1989) Fundamental laws in the geographical distribution of rainfall in the Socialist Republic of Serbia. Collection of papers from the Geographical Institute of the Faculty of Sciences (Belgrade) 26(5–18):42 (in Serbian)

Republic Hydrometeorological Service, Republic of Serbia (RHMZ) (2014) Daily hydrometeorological bulletin. Republic hydrometeorological service, Republic of Serbia, www.hidmet.gov.rs, www.meteoalarm.rs (in Serbian)

Stanojević G (2012) Analysis of annual sums of precipitation in Serbia. J Geogra Inst 'Jovan Cvijić' SANU 62(2):1–13

Šabić D, Pavlović M (2004) Climate characteristics of the Sjenica region. Bull Ser Geogr Soc 84:38–45

Tuhkanen S (1980) Climatic parameters and indices in plant geography. Acta Phytogeog Suecica 67:1–105

Unkašević M, Tošić I (2011) A statistical analysis of the daily precipitation over Serbia: trends and indices. Theor App Climatolog 106(1–2):69–78

Walter H (1973) Vegetation of the earth in relation to climate and the eco-physiological conditions. Springer-Verlag, USA

Walter H, Lieth H (1964) Klimadiagramm-Weltatlas. Gustav Fischer Verlag, Jena

4 Vegetation

Vegetation is one of the key factors in soil development. In Serbia, the wide range of climate, bedrock and soil types have resulted in heterogeneous vegetation, and almost all European biomes (steppes, deciduous forests in lowland, hilly and mountain zones, sub-Mediterranean forests and scrubs, boreal forests in subalpine mountain belts and fragments of heath formations) are present.

Zonal vegetation reflects regional climatic conditions. Climate zones are the main determinants of vegetation diversification and soil-forming processes. The lowlands (the Pannonian Plain and the Danube, Morava and Sava river valleys) have a (semi)-arid climate (either a 'Csa' or 'Cfa' climate). This zone is covered either by steppe fragments, steppe forests, or riparianforests along large rivers. The colline (hilly) and low-mountain regions (between 300 and 500 m) are more humid, but also very dry, especially in eastern Serbia. This climate zone is covered by thermophilous oak forests. The mountain regions (between 500 and 1000 m) are characterized by humid periods during all the seasons, with at least 4 months averaging above 10 °C (a moderate 'Cfb' climate). Beech (*Fagus sylvatica* L.) forests on moist soil dominate within this climate zone. Coniferous forests dominate in the altimontane habitats (between 1000 and 1500 m) with a microthermal boreal climate (Köppen's 'Dfb', 'Dfc' and 'Dsb' climates). Finally, herbaceous vegetation dominates within the (sub)alpine zone with a tundra climate (Köppen's 'ET climate').

Within each climate zone, local topography may significantly modify climate, resulting in the appearance of '*extrazonal*' plant communities, adapted to the conditions of other climate zones. For example, thermophilous oak forests in the central and northern parts of Europe represent extrazonal vegetation, thriving mainly on a southern aspect and carbonated bedrock. Increased drought and warmth in such habitats resemble the ecological conditions of southern European colline regions, where communities of thermophilous oaks represent zonal vegetation.

Plant communities with a similar floristic composition may be distributed within different climate zones. Such '*azonal*' vegetation is determined primarily by specific soil and hydrological conditions (waterlogged habitats, halomorphic soils, sand dunes, etc.).

Distribution maps of the main vegetation types in Serbia were obtained using the Serbian vegetation databank, syntaxonomy checklists in Serbia (Horvat et al. 1974; Ranđelović and Rexhepi 1980; Jovanović et al. 1986; Parabućski et al. 1986; Kojić et al. 1998; Lakušić 2005), and the CORINE Land Cover database (Moss et al. 1996). The vegetation databank was created using the FLORA package (Karadžić 2013), the latest version of which is a culmination of a programming project that has been running continuously since 1998 (Karadžić et al. 1998, 1999, 2003). The syntaxonomic databank of Serbian vegetation is compatible with other European databanks created using the TURBOVEG package (Hennekens 1995).

Vegetation in Serbia comprises forests, scrub and (semi)-natural herbaceous communities. The main types of ***forest vegetation*** in Serbia include riparian forests (*Salicetea purpureae* Moor 1958), swampy forests (*Alnetea glutinosae* Br.-Bl. et Tx. 1943), broad-leaved forests on xeric (dry) habitats (*Quercetea pubescentis* Doing-Kraft ex Scamoni and Passarge 1959), broad-leaved forests on mesic (moist) habitats (*Carpino betuli-Fagetea sylvaticae* Passarge in Passarge and Hofmann 1968), acidophilous beech forests, oak, oak-birch, oak-pine and oak-chestnut forests on nutrient-poor soils, coniferous forests on eutric soils (*Erico-Pinetea* Horvat 1959) and coniferous forests on dystric soils (*Vaccinio-Piceetea* Br.-Bl. in Br.-Bl. et al. 1939).

Scrub vegetation in Serbia is represented either by successional stages of forest communities (communities of the orders *Fraxino orni-Cotinetalia* Jakucs 1961 and *Prunetalia spinosae* Tx. 1952, which represent successional stages of thermophilous and mesophilous forest vegetation, respectively) or by relatively stable scrub communities that are

P. Pavlović et al., *The Soils of Serbia*, World Soils Book Series,
DOI 10.1007/978-94-017-8660-7_4

determined by extreme climatic conditions (either extremely dry or extremely cold habitats).

Natural ***herbaceous communities*** include *psammophytic vegetation* (*Festucetea vaginatae* Soó 1968 emend. Vicherek 1972), *chasmophytic vegetation* (communities of the classes *Asplenietea trichomanis* Br.-Bl. 1934 corr. Oberd. 1977, *Thlaspietea rotundifolii* Br.-Bl. et al. 1947, and *Drypetea spinosae* Quezel 1967), *halophytic* vegetation (*Thero-Salicornietea* Pignatti 1953 *emend.* R. Tx. 1955), *steppe vegetation* (*Festuco-Puccinellietea* Soó 1968), *subalpine herbaceous vegetation* of the classes *Juncetea trifidi* Hadač 1944 (siliceous alpine and boreal grasslands), *Elyno-Seslerietea* BR.-BL. 1948 = *Festuco-Seslerietea* Barbero *et* Bonim 1969 (alpine and subalpine calcareous grasslands), *Betulo-Adenostyletea* Br.-Bl. et R. Tx. 1943 (subarctic-subalpine to alpine tall-grass and tall-herb grasslands and related krummholz) and heath vegetation of the class *Nardo-Callunetea* Preising 1949 = *Calluno-Ulicetea* BR.-BL. *et* R. Tx. *ex* Westhoff et al. 1946 (temperate and boreal heathlands and related grasslands on nutrient-poor soils).

4.1 Forest Vegetation

Hydrological conditions (the intensity and duration of flooding and the groundwater level) and characteristics of alluvial deposits are the main factors that affect the floristic differentiation of hygrophilous forest vegetation (Horvat et al. 1974; Karadžić et al. 2015).

Pioneer riparian forests (*Salicetea purpureae* Moor 1958) colonizing the altimontane and montane rivers with a 'torrential' regime belong to the alliances *Alnion incanae* Pawl. 1978 (mountain alder scrub and forest vegetation on coarse gravel alluvia), *Salicion eleagno-daphnoidis* (Moor 1958) Grass 1993 (willow scrub of the banks of streams of the montane to subalpine belts of the Alps and Carpathians) and *Salicion triandrae* (Malcuit 1929) Müller *et* Görs 1958 (willow scrub of montane river banks with sandy alluvia). At higher altitudes, riparian forests are replaced by altimontane herbaceous/scrub vegetation on pebbly alluvia and riverine gravel (communities of the alliances *Epilobion fleischeri* G. Br.-Bl. and J. Br.-Bl. 1931 and *Adenostylion alliariae* Br.-Bl. 1926).

Riparian forests of poplars and willows colonizing the *muddy shores* of large rivers (communities of the alliances *Salicion albae* Soó 1930 and *Populion albae* Br.-Bl. 1931) are regularly and periodically flooded. These communities grow at the edge of rivers on different variants of fluvisol.

Compared to poplar and willow forests, the *temporarily flooded wetland forests* of ash, black alder and common oak that occupy topographic depressions are less frequently exposed to fluvial floods. These forests include pedunculate oak, black alder and narrow-leaved ash (the alliance *Anlo-Quercion roboris* Horvat 1938), willow communities in muddy swamps (communities of the alliance *Salicion cinereae* Th.Müll. et Görs 1958) and swamp forests of black alder (the alliance *Alnion glutinosae* Malcut 1929), which develop on temporarily inundated habitats and in topographic depressions with long-lasting floods, on hydromorphic, gleyed, nutrient-rich alluvial soils.

All pan-European deciduous forests are, traditionally, grouped into a single class (*Querco-Fagetea* Br.-Bl. *et* Vlieg. 1937), within which the floristic-coenological differences between oak woods, mixed forests and beech woods are expressed at the level of order or alliance (Horvat et al. 1974; Scoppola et al. 1995; Rivas-Martínez et al. 2001). More recent classification systems (Quézel et al. 1980; Chytry 1997; Sanda et al. 2008) separate the xerothermic woods of south-eastern Europe in the class *Quercetea pubescentis* Doing-Kraft *ex* Scamoni and Passarge 1959 from shady forests included in the class *Carpino betuli-Fagetea sylvaticae* Passarge in Passarge and Hofmann 1968.

Besides different thermal and moisture conditions, the communities included in these latter two classes also differ with respect to light climate. Shade gradually increases from xerothermic oak communities to hornbeam-dominated communities and beech-dominated forests. Due to the dense foliage and specific shape of the tree crowns, these species significantly reduce the amount of light reaching the interior of mesic forests. Such dysphotic (light shortage) conditions represent a strong selective pressure for many plants. The shade-adapted plants may be divided into two categories: *shade-evaders* and *shade-tolerant* species.

Lowland and colline (hilly) regions, outside the flooding area, are covered by ***thermophilous oak forests***. These forests are climazonal in warmer parts of Europe (the Balkans, Iberia and the Apennines) and extrazonal in the northern, cooler regions (Blasi et al. 2004; Knollová and Chytrý 2004; Roleček 2005; Rousakova and Tzonev 2003; Bergmeier and Dimopoulos 2008).

The oak forests in Serbia occupy colline regions and parts of the Pannonian Plain (Fig. 4.1). Thermophilous oak communities are ordered along a moisture gradient from extremely dry and semi-dry habitats (forests of *Quercus trojana* Wbb, *Quercus pubescens* Willd., *Quercus frainetto* Ten., and *Quercus cerris* L.) to relatively moist habitats (forests of *Quercus cerris* L. and *Quercus petraea* (Mattuschka) Liebl.). The ecological differentiation between the dominating oaks in thermophilous forests may be seen not only with respect to moisture gradient, but also with respect to thermic and continentality gradients (Kojić et al. 1997; Ellenberg and Leuschner 2012).

According to Čarni et al. (2009), thermophilous oak forests in Serbia belong to the alliances *Quercion petraeae-cerris* (Jovanović et Lakušić 1980) Čarni et al. 2009,

Quercion frainetto Horvat 1954, *Quercion pubescenti-sessiliflorae* Br.-Bl 1932 and *Aceri tatarici-Quercion* Zólyomi *et* Jakucs 1957. Within the same group of forests (the order *Quercetalia pubescenti-petraeae*), we can also include thermophilous forests dominated by *Ostrya carpinifolia* Scop. and *Carpinus orientalis* Miller (the alliances *Carpinion orientalis* Horvat 1954, *Syringo-Carpinion orientalis* Jakucs 1959 and *Fraxino orni-Ostryion carpinifoliae* Tomažič 1940). Due to the major human impact (cutting, fragmentation of habitats and the transformation of forests into pastures and soil under agriculture), thermophilous oak forests in Serbia and other south European countries are highly endangered.

Quercion pubescenti-sessiliflorae Br.-Bl 1932 is an alliance with a Central European distribution pattern, extending to the edge of the Pannonian Basin at its most southerly point. These forests occur over a wide area, which includes southern extra-Mediterranean and south-eastern France, northern Spain, stretching across to southern Switzerland, the Illyric parts of Italy, arid parts of the Vienna Basin, peri-Pannonian regions of Hungary, southern Slovakia, northern Serbia and Romania. The forests of this group inhabit calcareous (compact limestone, marly limestone or calcarenites), or less frequently, siliceous substrates, while the soils are often shallow rendzinas (WRB, Leptosol, calcaric). These relatively mesic communities appear in moist habitats on deeper soils.

Aceri tatarici-Quercion Zólyomi *et* Jakucs 1957 is the most continental alliance of thermophilous oak forests. These communities of Pontic xeric taxa are frequently described as *forest steppes* (meadow steppes or dry grasslands alternating with deciduous broad-leaved forests or xerophytic scrub) and are distributed across the Pannonian Plain (Jovanović 1997).

The alliance *Quercion frainetto* Horvat 1954 includes climate zonal communities in the eastern part of the Balkan Peninsula and extends to the south towards Greece and Turkey (Kavgaci et al. 2010). The communities of this alliance are found on deep soils, where we can find a certain degree of acidity. They thrive in the central part of the Balkan Peninsula, where the majority of precipitation is at the beginning of the year followed by a long dry and hot period. The combination of dry and hot summers and cold winters is very unfavourable for plants (hot summers for mesophilous and cold winters for thermophilous), so these forests are rather poor in species (Čarni et al. 2009). These communities are restricted to low altitudes and develop on vertisols and eutric cambisols.

The alliance *Quercion petraeae-cerris* (Jovanović et Lakušić 1980; Čarni et al. 2009) occupies habitats at altitudes above *Quercion frainetto* communities, in contact with beech forests (Kojić et al. 1998). Compared to the alliance *Quercion frainetto*, the alliance *Quercion petraeae-cerris* is much richer in species. Numerous species from mesophilous and acidophilous oak, oak-birch and pine forests on nutrient-poor soils of Europe (class *Quercetea roboris* Br.-Bl. *ex* Oberd. 1957) occur in these forests, which develop on gajnjača (WRB, Cambisol, eutric).

Besides oak-dominated forests, thermophilous forests of *Ostrya carpinifolia* and *Carpinus orientalis* also occur in Serbia. *Ostrya carpinifolia* Scop. is a characteristic species of the alliance *Fraxino orni-Ostryion*, while *Carpinus orientalis* Miller is characteristic for the coastal (*Carpinion orientalis*) as well as for extrazonal continental communities (*Syringo-Carpinion*).

The alliance *Carpinion orientalis* Horvat 1954 includes amphi-Adriatic xeric forests that form a belt in coastal regions above the evergreen vegetation of *Quercetea ilicis*. In comparison with oak-dominated forests, these forests thrive in areas with a mild climate during winter (Čarni et al. 2009). Relevés from this group are found along the coast of the Adriatic Sea and in regions where the maritime influence extends deep inland, along the Una, Neretva, Drim and Vardar rivers. In Serbia, these communities are fragmentarily distributed in Metohija.

The alliance *Fraxino orni-Ostryion* Tomažič 1940 is sometimes treated as a synonym for the alliance *Carpinion orientalis* Horvat 1954. However, Čarni et al. (2009) demonstrated that these two alliances can be separated. The *Carpinion* alliance is distributed in Adriatic coastal areas and also, extrazonally, in inland areas (on steep southern slopes, around lakes where the climate is mitigated by the water body, etc.). Communities of the alliance *Fraxino orni-Ostryion* Tomažič 1940 occupy more humid habitats, within the inner part of the mountain chains along the Adriatic coast at higher altitudes, and extrazonally within areas with a maritime sub-Mediterranean climate. These forests are distributed in Kosovo and Metohija and in numerous canyons and gorges in western Serbia.

The alliance *Syringo-Carpinion orientalis* Jakucs 1959 is extrazonal Mediterranean vegetation that thrives in the continental part of the central Balkans (Mišić 1979, 1981). The communities of this alliance inhabit steep southern slopes in eastern Serbia.

Mesophilous shady forests in Serbia are represented by oak-hornbeam (*Carpinus betulus* L.) communities, (sub)-montane beech forests, altimontane mixed beech-fir (*Fagus sylvatica-Abies alba*) forests and subalpine beech forests with *Acer heldreichii*. Due to the specific floristic composition and ecological conditions (shady habitats with moist, relatively deep soil), these forests may be grouped into the vegetation class *Carpino-Fagetea* Passarge in Passarge and Hofmann 1968.

Although the Central European species *Carpinus betulus* L. has a narrow ecological amplitude, especially in regard to moisture conditions (a moisture-demanding species), it inhabits different forest types. Its distribution is limited

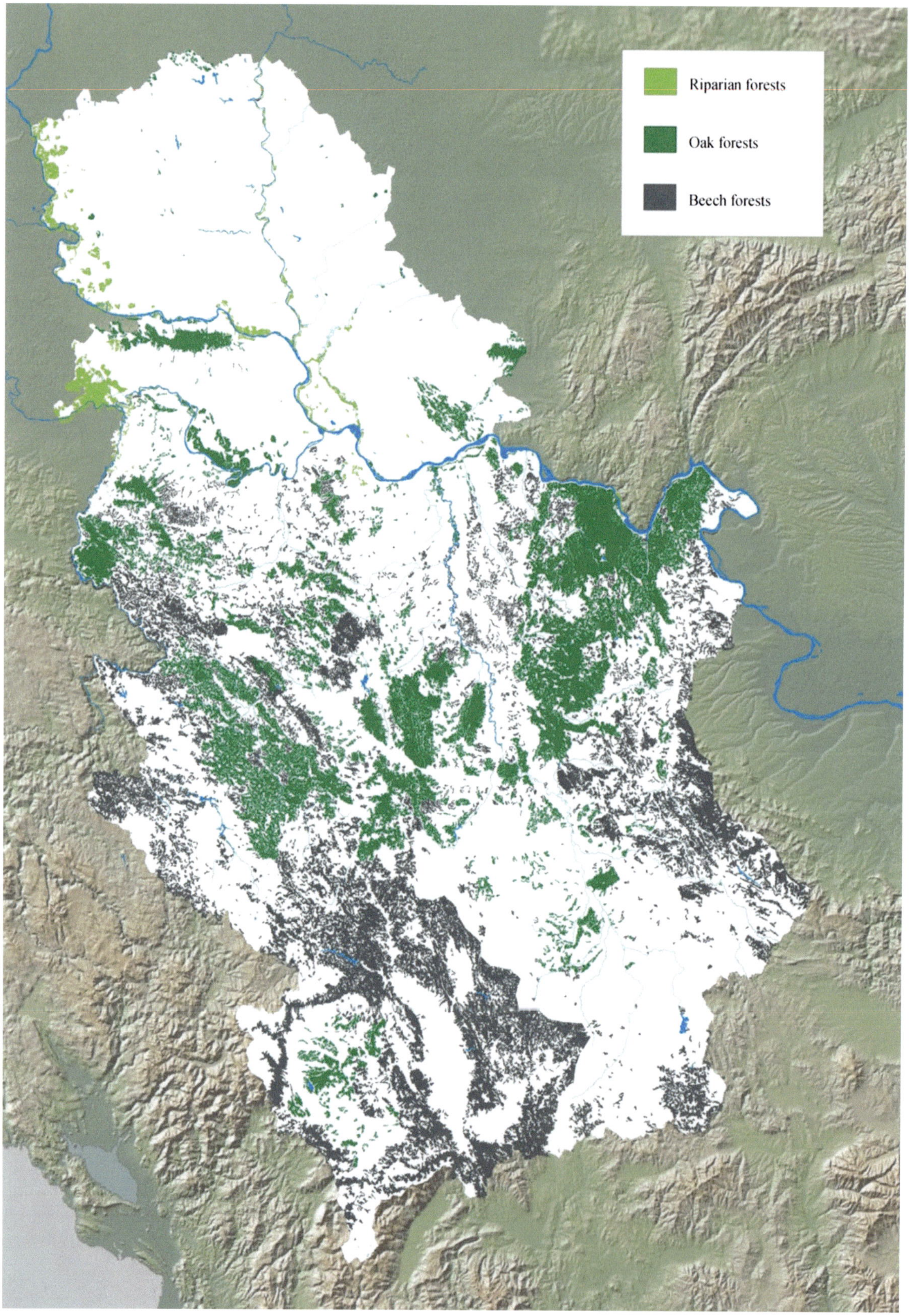

Fig. 4.1 Deciduous forests

towards the east and south of Europe by the semi-arid climate with seasonal droughts. In the hilly regions of the western part of the Balkans, the oak-hornbeam forms a particular climazonal vegetation belt, represented by the alliance *Erythronio-Carpinion betuli* Marinček in Wallnöfer et al. 1993. According to numerous authors (Janković and Mišić 1960; Jovanović 1967; Vukićević 1976; Dinić 1977, 1998), *Carpinus betulus* L. can be found in the mesophilous zone of beech forests, in thermophilous oak-dominated forests (the alliance *Quercion petraeae-cerris* Jovanović and Lakušić 1980; Čarni et al. 2009) and in temporarily flooded forests (the alliance *Anlo-Quercion roboris* Horvat 1938). It rarely forms mono-dominated forests with the exception of the colline regions of western Serbia. Relatively large complexes of oak-hornbeam-dominated forests are preserved in peri-Pannonian low mountains.

Fagus sylvatica L. (beech) is the most abundant broadleaved tree in Central Europe and in mountain regions of southern Europe. It is highly competitive in many types of zonal forest as it is shade-tolerant when juvenile and casts deep shade as a canopy tree. Dominating plants in beech forests are also adapted to dysphotic (light shortage) stress (Popović et al. 2005, 2006, 2015). Beech is a species adapted to a moist climate. It thrives well in regions with mild winters and humid summers without a pronounced dry season. With its shallow roots, it cannot outcompete deeply rooted oak species. Therefore, in southern Europe, beech can find advantageous conditions only in mountain regions.

Research into beech forests in Nordic countries (Diekmann et al. 1999), central and western parts of Europe (Dierschke 1990, 1997; Ellenberg and Leuschner 2012), the Apennines (Pignatti 1998) and the Balkans (Soó 1964; Török et al. 1989; Horvat et al. 1974; Dzwonko et al. 1999; Dzwonko and Loster 2000) has accumulated knowledge for a precise syntaxonomy of European beech forests. In terms of the Balkan Peninsula, beech-dominated forests were traditionally grouped into the regional alliances *Fagion illyricum*, *Fagion moesiacum*, *Fagion dacicum* and *Fagion orientalis*. The names of these groups were generated according to the former names of the ancient Roman provinces of Illyria, Dacia and Moesia. *Fagion illyricum* (*nomen invalidum*) covers the beech forests in Slovenia, Croatia, Bosnia and Herzegovina, Montenegro, Albania and Greece (Horvat et al. 1974; Soó 1964; Bergmeier and Dimopoulos 2001), while *Fagion dacicum* (*nomen invalidum*) those in the Carpathian Mts. of Romania, south-eastern Serbia, Bulgaria and northern Greece (Borhidi 1964; Soó 1964). Moesian beech forests (*Fagion moesiacum*) are distributed in the area with a more continental climate in the eastern Balkans (eastern Serbia and Bulgaria). Finally, the alliance *Fagion orientalis*, which includes communities of *Fagus orientalis* Lipsky, is found in south-eastern Bulgaria, eastern Greece, and the European part of Turkey, the Anatolian plateau and the Caucasus (Quézel et al. 1980, 1992).

Beech forests in Serbia are traditionally grouped into the geographic alliance *Fagion moesiacae* Blečić *et* Lakušić 1970 (Kojić et al. 1998; Jovanović et al. 1986; Lakušić 2005; Tomić 2006). A more accurate syntaxonomy of Serbian beech forests should be based on ecological groups of species along climate gradient and edaphic patterns. These forests are distributed in colline regions (with a warm and relatively dry climate), montane regions (with a moist and moderately warm climate), altimontane regions (with a humid and moderately cold climate) and subalpine regions (with a humid and microthermal climate). Moreover, they are distributed in habitats with different edaphic conditions (nutrient-rich, eutric soils and dystric, nutrient-poor soils). Finally, beech forests in Serbia occupy ravine habitats, where they form polydominant, species-rich forests. Diverse ecological conditions affect the floristic differentiation of Serbian beech forests, which come under one of the following alliances: *Heleboro odori-Fagenion* Soó et Borhidi 1960 (submontane beech forests), *Galio odorati (Asperulo)-Fagion* Tx. 1955 (montane beech forests), *Abieti-Fagenion* sensu lato (altimontane mixed fir-beech forests), *Aceri heldreichii-Fagion* B. Jovanović 1957 (subalpine beech forests), *Ostryo-Fagion* B. Jovanović 1976 (ravine beech forests in western Serbia), *Fago-Coluion colurnae* Borhidi 1964 (ravine beech forests in eastern Serbia) and *Luzulo luzuloides-Fagion* Lohmeyer et Tx. in Tx. 1954 (pan-European acidophilous beech forests).

The floristically poor acidophilous beech forests occur on silicate acid soils, from the montane to the altimontane belts. Equivalent acidophilous communities within the belt of thermophilous forests (the class *Quercetea pubescentis* Doing-Kraft *ex* Scamoni and Passarge 1959) belong to the alliances *Castaneo-Quercion* Soó 1962 (acidophilous chestnut-oak forests) and *Betulo pendulae-Quercion petraeae* Gillet 1986 ex Julve 1993 (acidophilous oak-birch forests) and are mainly distributed in perhumid peri-Atlantic regions. These floristically poor acidophilous oak, oak-birch and oak-chestnut forests are floristically similar to acidophilous beech forests (*Luzulo luzuloides-Fagion* Lohmeyer et Tx. In Tx. 1954). However, the former are more diverse due to more favourable light conditions.

Acidophilous oak and mixed oak forests on nutrient-poor soils are sporadically and extrazonally distributed in southern Europe. In Serbia, these forests are limited to edaphic enclaves, mainly in the colline regions of western Serbia and Metohija, where the climate is humid and relatively maritime. They develop on luvisols and dystric cambisol. Differentiation of Serbian deciduous forests with respect to altitude and soil nutrients is presented in Fig. 4.2.

Holarctic **coniferous forests** are distributed in boreal regions. In southern Europe, these forests form an altimontane vegetation belt above deciduous and mixed coniferous-broad-leaved forests. Braun-Blanquet et al. (1939) grouped

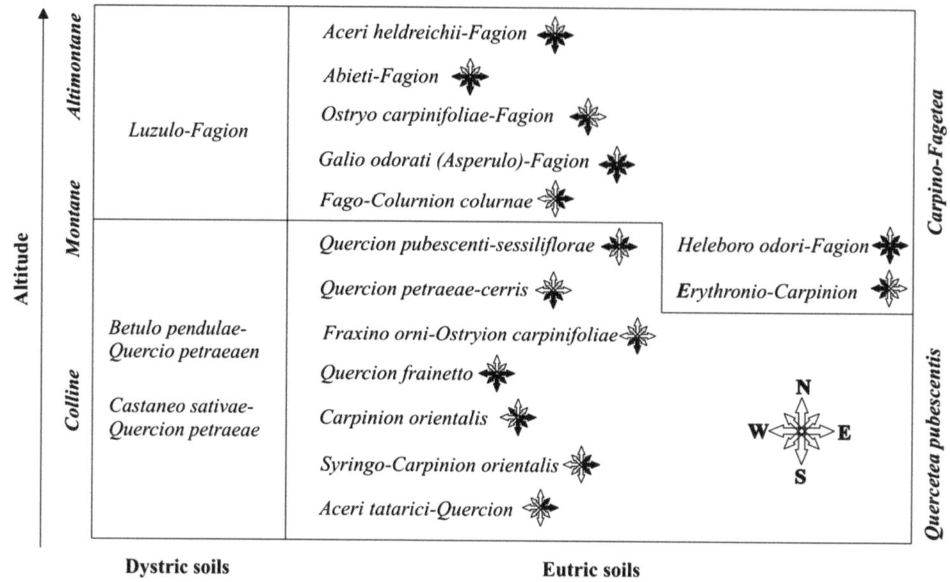

Fig. 4.2 Syntaxonomic differentiation of Serbian deciduous forests

the boreo-montane spruce and pine forests of nutrient-poor habitats into the vegetation class ***Vaccinio-Picetea*** Braun-Blanquet et al. 1939. This class in Serbia comprises three ecologically and floristically different groups of coniferous forests (the alliances *Vaccinio-Piceion* Br.-Bl. (1938, 1939), *Pinion peucis* Ht. 1950 and *Piceion omorikae* Tregubov 1941).

The alliance *Vaccinio-Piceion* Br.-Bl. (1938) 1939 includes acidophilous spruce (*Picea abies* (L.) Karsten) forests distributed in altimontanae and subalpine regions with a 'boreal', microthermal climate (i.e. Köppen's 'D climate'). These climazonal forests occupy soils on serpentine, calcareous and silicate rocks. Irrespective of the type of geological substrate, soils in these forests are base-deficient and strongly acid because of the intense leaching processes caused by the perhumid climate.

Shady acidophilous forests of the Balkan endemic pine *Pinus peuce* Grisebach (forests of the alliance *Pinion peucis* Ht. 1950) are distributed in altimontane regions of Kosovo and Metohija. The climate of altimontane regions in central and eastern Serbia is cold and continental, with a precipitation peak during the summer months. However, altimontane regions in Kosovo and Metohija have precipitation peaks during the autumn and winter months. The paleo-endemic Balkan pines, *Pinus heldreichii* Christ. and *Pinus peuce* Grisebach, are adapted to both *physiological* drought (i.e. drought caused by extremely low temperatures during the winter months) and *physical* drought (i.e. drought caused by low precipitation during the summer). Therefore, these two species replace the more continental European spruce in sub-Mediterranean altimontane regions.

Shady acidophilous forests of the alliance *Pinion peucis* Ht. 1950 are distributed in the Prokletije Mts. and Mt. Šar, from 1500 to 2100 m a.s.l., occupying mainly silicate soils, but they also occur on calcareous and serpentine soils.

The alliance *Piceion omorikae* Tregubov 1941 comprises floristically rich forests of Serbian spruce. These forests occupy ravine habitats in west and south-west Serbia (Tara Mt., the Drina River gorge, and Mileševka gorge).

Besides relatively shady acidophilous forests, the Serbian spruce forms more open communities on skeletal calcareous and serpentine soil. The basophylous, open forests of *Picea omorika* (Pančić) Purkyne occupy slopes in gorges or at the foot of slopes with colluvial, unstable soil. These forests are extremely diverse since their floristic composition is a mixture of species from relic deciduous forests of the alliance *Ostryo carpinifoliae-Fagion* B. Jovanović 1976, from shady beech forests (*Fagetalia* species) and from basophylous relic pine forests (*Orno-Ericion* Ht. 1958). Compared to central and north European regions, coniferous forests in Serbia, and in southern Europe more generally, are more diverse. Besides the boreo-montane spruce and Scots pine forests of nutrient-poor habitats (the class *Vaccinio-Picetea* Braun-Blanquet et al. 1939), coniferous forests in Serbia are represented by relict pine forests on calcareous and ultramafic soils (the vegetation class *Erico-Pinetea* Horvat 1959). They are distributed mainly in western parts of Serbia (Fig. 4.3) and are developed on Ranker (WRB, Leptosol, eutric), regosol and lithosol (WRB, Leptosol) on basic serpentine rocks.

The vegetation class ***Erico-Pinetea* Ht. 1959** includes sparse forests of Scots Pine (*Pinus sylvestris* L.), Austrian Pine (*Pinus nigra* Arnold) and Balkan Pine (*Pinus heldreichii* Christ.) on basic and ultrabasic soils. These species-rich, endemic communities are floristically and ecologically clearly separated from the acidophilous Scots

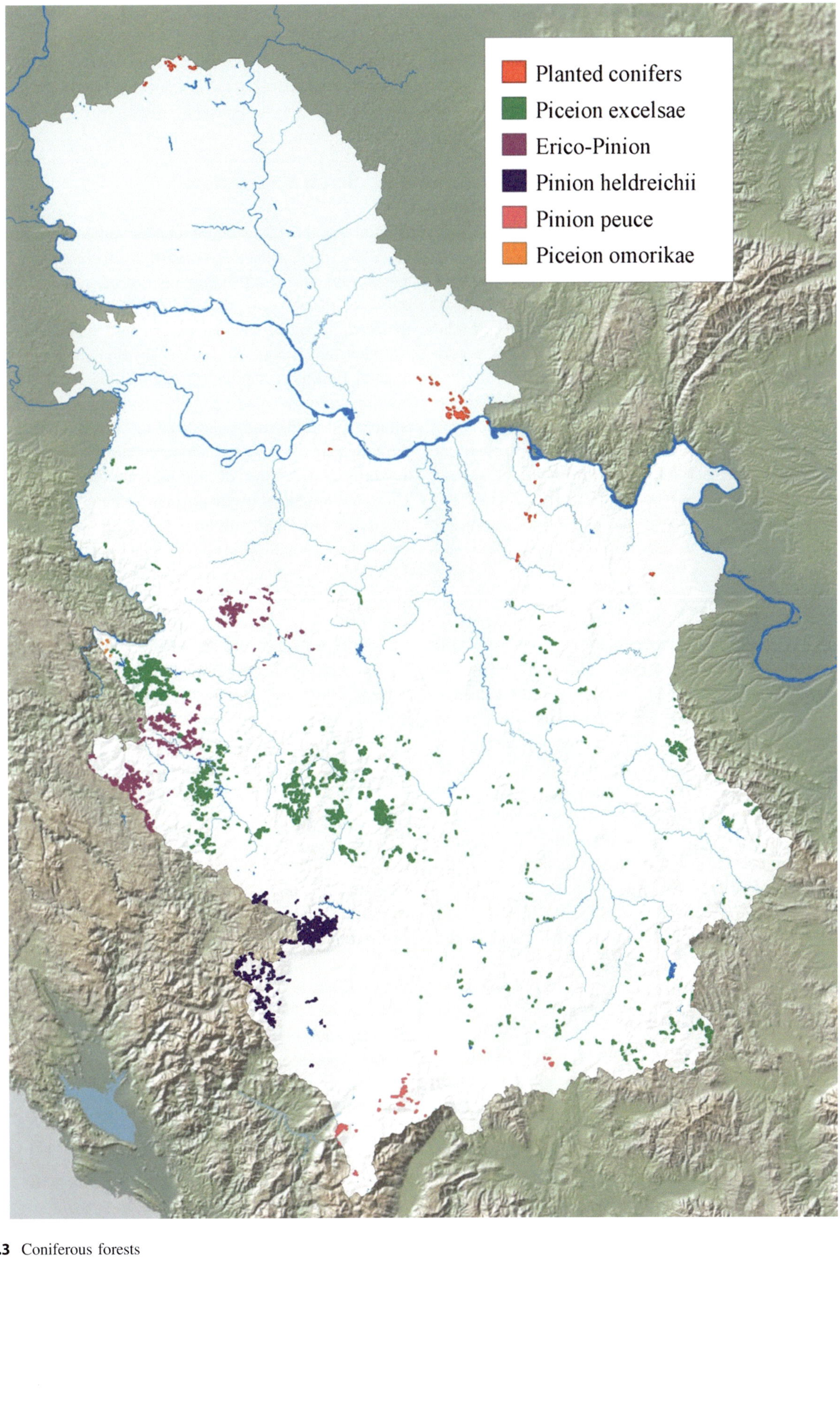

Fig. 4.3 Coniferous forests

Pine forests of Central and northern Europe (alliance *Dicrano-Pinion* (Libbert 1932) Matuszkiewicz). The relatively shallow soils of these forests represent different developmental stages of eutric soils: from regosol (WRB, Leptosol) and ranker (WRB, Leptosol eutric) to eutric cambisol.

In Serbia, forests of the class *Erico-Pinetea* Ht. 1959 are differentiated into the alliances *Fraxino orni-Pinion nigrae-sylvestris* (Horvat 1953) Zupančič 2007 and *Pinion heldreichii* Ht. 1946.

The alliance *Fraxino orni-Pinion nigrae-sylvestris* (Ht. 1959) Zupančič 2007 comprises sparse forests of Scots Pine (*Pinus sylvestris* L.) and Austrian Pine (*Pinus nigra* Arnold) on basophylous, ultra-basophylous and neutral soils. A relatively large complex of these forests is distributed in eastern Bosnia and western Serbia. Zupančič (2000, 2007) divided them into two distinct groups (communities on calcareous and dolomite soils and communities on serpentine soils).

Serpentine soils are characterized by low levels of essential plant elements (nitrogen, phosphorus, potassium and calcium), as well as high levels of iron, magnesium and manganese, and toxic levels of chromium, cobalt and nickel (Kruckeberg 1986; Proctor 1999; Pepper and Norwood 2001). Due to 'serpentine stress' (the toxic effects of heavy metals, nutrient shortages and droughts), most plant species avoid serpentine soils. A small percentage of serpentine-tolerant taxa has evolved morpho-anatomical and physiological adaptations that allow them to survive in extremely unfavourable conditions. The strong selective pressures of serpentine soils and the spatial isolation of serpentine regions have resulted in a high percentage of endemic serpentinophyte taxa in the Balkans, on the Scandinavian Peninsula, in Britain, the Urals, etc.

The largest serpentine areas of Europe are on the Balkan Peninsula (Epirus, Thessaly and Sterea Ellas in Greece, north, north-central and south-eastern Albania, central and eastern Bosnia, western, south-western and central Serbia, the eastern and central Rhodope mountains in Bulgaria, and the gorge of the River Pčinja and the south eastern slopes of Mt. Šar in the north-eastern part of the Republic of Macedonia). In Serbia, there are several serpentine zones located primarily in central, western and south-western Serbia and in Kosovo. These areas include the following: Brdjanska gorge, Mt. Suvobor, Mt. Maljen, Mt. Tara, Mt. Zlatibor, Mt. Mokra Gora, Mt. Goč, Mt. Stolovi, Mt. Kopaonik, the Ibar valley, Peć and the surrounding area, Orahovac and Koznik, and Mt. Ostrovica (Jakovljević et al. 2011). The serpentine flora on the Balkan Peninsula was investigated by Novak (1926), Krause and Klement (1958), Ritter-Studnička (1970), Babalonas (1988, 1989), Tatić and Veljović (1992), Pavlova et al. (1998), Vasić and Diklić (2001) and Kabaš et al. (2013).

The endemo-relic pine forests of the alliance *Pinion heldreichii* Ht. 1946 are distributed in southern parts of the Balkan Peninsula and in southern Italy. In Serbia, these forests develop on calcareous and serpentinous rendzinas (WRB, Leptosol, calcaric), rankers (WRB, Leptosol eutric) and eutric cambisols in the altimontane region (from 1500 to 1800 m a.s.l.) of the Prokletije Mts.

4.2 Scrub Vegetation

Natural shrubby vegetation occurs either in cryothermic climates (*ecotones between boreal forests and grassy tundra vegetation*) or in arid climates (*ecotones between xerophilous forests and steppe vegetation*), where trees are replaced by more adapted shrubs and herbaceous plants. Xerothermic deciduous scrub is distributed across subcontinental European regions (the Pannonian Basin and neighbouring regions, including the Transylvanian Plateau and the adjacent foothills and valleys of the Eastern and Southern Carpathians, the lower Danubian Plain, the Dobrogea and the hills and valleys of the northern Balkan Peninsula). These communities occur on both calcareous and siliceous substrates, forming mosaic-like vegetation with steppe grassland. Cryophilic shrubs are zonally distributed north of the boreal region and in the altimontane belts of Europe's high mountain ranges (the Alps, Dinarides, Carpathians and Rhodopes).

Both types of natural shrubby vegetation are distributed in Serbia. Xero-thermophilic scrub forms a wide zone in regions with a (semi)-arid and subcontinental climate (the peri-Pannonian region and Danube lowlands). These xeric ecotonal communities, thriving on sunny, steep slopes, belong to the alliance *Pruno tenellae-Syringion* (Jovanović 1979; Čarni et al. 2009) of the vegetation order *Fraxino orni-Cotinetalia* Jakucs 1961.

Natural cryophilic shrubby communities occupy a zone above the treeline in altimontane regions. Their specific form (i.e. cushion-like shrubs with short internodia and intense lateral branching) is an excellent adaptation of (sub)alpine shrubs to severe climate (extremely low temperatures, strong winds and long-lasting snow cover). Most plants of the genera *Erica* L., *Rhododendron* L., *Vaccinium* L., *Loiseleuria* Desv., *Juniperus* L., *Genista* L. and *Empetrum* L. have xeromorphic ericoid leaves.

Arctic-boreal tundra scrub and (sub)alpine dwarf heath in Serbia are represented by communities that belong to the alliances *Pinion mugo* Pawl. 1928, *Juniperion sibiricae* Br.-Bl. 1939, *Loiseleurio-Vaccinion* Br.-Bl. in Br.-Bl. *et* Jenny 1926, *Rhododendro ferruginei-Vaccinion myrtilli* A. Schnyd. 1930 and *Bruckenthalion spiculifoliae* Ht. 1949.

Bushes with *Pinus mugo* Turra occupy the upper treeline of the altimontane regions in Serbia. These communities are grouped into the alliance *Pinion mugo* Pawl. 1928 of the class *Roso pendulinae-Pinetea mugo* Theurillat 1995 in

Theurillat et al. 1995 and involve both acidophilous and basophilic, shrubby vegetation, dominated by *Pinus mugo* Turra. Due to strong mountain winds and deep and long-lasting snow, this species has flattened physiognomy. Due to its high competitive ability, *Pinus mugo* Turra frequently forms almost monodominant dense scrub with a species-poor understory. More open communities are rich with both forest species and (sub)alpine grasslands and heaths.

Subalpine shrubby heaths in wind-exposed habitats are represented by communities of the alliances *Juniperion sibiricae* Br.-Bl. 1939, *Bruckenthalion spiculifoliae* Ht. 1949, *Loiseleurio procumbenti-Vaccinion* Br.-Bl. in Br.-Bl. *et* Jenny 1926 and *Rhododendro ferruginei-Vaccinion myrtilli* A. Schnyd 1930.

Alpine-Arctic hygrophilous scrub communities of willows on long-lasting snowbeds and slopes irrigated by melt waters are represented by communities of the alliance *Alnion viridis* Aichinger 1933 and *Salicion herbaceae* Br.-Bl. (1924, 1926). In specific topographic conditions (sinkholes and karst valleys, leeward from strong mountain winds), snow remains for a long time. Vegetation in the proximity of long-lasting snowbeds develops on shallow humic soils. Besides plants that are connected exclusively with glaciers, there are some species that belong to neighbouring vegetation types, such as alpine and subalpine pastures. These communities have rich moss flora. In Serbia, these communities are distributed in altimontane regions between 1280 and 1900 m a.s.l. on Mt. Stara planina, Kopaonik, Šar and the Prokletije Mts. (Mišić 1960; Mišić et al. 1978; Jovanović 1980; Ranđelović and Rexhepi 1984; Rexhepi 1994; Amidžić et al. 2003).

Besides natural, relatively persistent shrubby vegetation, ephemeral shrub communities that represent particular succession stages from disturbed habitats to climax forest vegetation are also present in Serbia. Different successional stages within the zone of thermophilous forests are represented by communities of the alliances *Prunion spinosae* Soó 1940 and *Rhamno-Paliurion* Trinajstić (1978) 1996, while successional communities within the zone of mesophilous forests belong to the alliances *Crataego-Coryllion avelanae* Fukarek 1969 and *Sambuco racemosae-Salicion capreae* Tuxen et Neumann ex Oberd 1957.

A large area of thermophilous, climazonal oak forest in Serbia is degraded. Successional stages of oak forests are represented by shrub communities of the alliance *Prunion spinosae* Soó 1940. Shrubby vegetation within the zone of mesophilous forests is represented by communities of the alliance *Genistion (Cytisanthion) radiate* Ranđ. *et* Rexh. 1979, which develop in canyon habitats, often on scree. In such conditions, shrubs are more competitive than most trees (with the exception of *Ostrya carpinifolia* Scop.), and therefore, shrubby communities are relatively stable and persistent type of vegetation. In Serbia, they are distributed in ravine habitats in the west and south-west, within montane and altimontane regions. More ephemeral succession stages within mesophilous forests are represented by communities of the alliances *Crataego-Coryllion avelanae* Fukarek 1969 and *Sambuco racemosae-Salicion capreae* Tuxen *et* Neumann ex Oberd. 1957.

A separate group of shrubby communities is represented by **forest fringe vegetation** (class *Trifolio-Geranietea sanguinei* T. Muller 1962). Forests used to cover large areas in Serbia, and more generally in Europe. However, the long-lasting and severe human impact (forest cutting, transformation of forest areas into arable fields and pasture grasslands and mowing) have transformed forests into degraded ecosystems. The recovery of degraded ecosystems by vegetation succession is a long-lasting process. During succession, ecosystems turn into very dynamic types of vegetation that contain species of both meadows and forests habitats (Čarni 1997).

In places, forest edges support herb-rich vegetation forming a belt-like saum, bordering either directly onto forests or a so-called mantle (a shrubby fringe of forests) on the one hand, and usually mown grasslands of the *Molinio-Arrhenatheretea*, dry pastures of the *Festuco-Brometea*, or heathlands and nutrient-poor pastures of the *Calluno-Ulicetea* on the other. The saum, dominated by tall forbs and grasses, is represented in Serbia by communities of the alliances *Trifolion medii* T. Muller 1962 (mesophilous and subxerophilous fringe vegetation on nutrient-poor soils) and *Geranion sanguinei* R. Tx. in T. Muller 1962 (xerophilous fringe and tall-herb vegetation). This vegetation comprises tall herbs and, sporadically, shrubs.

Trifolion medii communities occupy mesic habitats, usually fringing the woods of shady deciduous forests (*Carpino betuli-Fagetea* Passarge in Passarge and Hofmann 1968). The neighbouring meadows are often mown and belong to *Molinio-Arrhenatheretea*. In further successional stages, these communities develop into ephemeral shrubby communities of the alliances *Crataego-Coryllion avelanae* Fukarek 1969 and *Sambuco racemosae-Salicion capreae* Tuxen et Neumann ex Oberd 1957. The *Geranion sanguinei* communities fringe xericoak forests in dry and warm habitats (*Quercetea pubescentis* Doing-Kraft ex Scamoni and Passarge 1959).

4.3 Herbaceous Vegetation

Herbaceous vegetation in Serbia is represented by natural (stable and persistent) steppe, chasmophytic and alpine cryophilic communities that replace forest and shrub vegetation due to specific climatic (either extreme drought or extreme cold climates) and specific edaphic conditions (halomorphic soil, undeveloped soil on bare rock and scree, unconsolidated sand soil and ultrabasic, shallow soil in serpentine areas).

Steppe vegetation occurs only in (semi)-arid regions, where more competitive xeric graminoid herbs replace trees and shrubs (the Pannonian Plain and the Danube lowland). Due to the severe human impact, most steppe habitats have been transformed into soil under agriculture. Native habitats with steppe vegetation are only sporadically preserved.

At higher altitudes, extrazonal steppe vegetation is distributed fragmentarily in specific edaphic conditions (dry habitats with shallow soil). Communities of mountain steppes belong to the alliance *Festucion valesiacae* Klika 1931.

Natural subalpine herbaceous vegetation develops in altimontane regions, where herbs replace the trees and shrubs which cannot survive in a microthermic climate. Subalpine herbaceous communities belong to the vegetation classes *Juncetea trifidi* Hadač 1944 (siliceous alpine and boreal grasslands), *Elyno-Seslerietea* BR.-BL. 1948 = *Festuco-Seslerietea* Barbero *et* Bonim 1969 (alpine and subalpine calcareous grasslands), *Betulo-Adenostyletea* Br.-Bl. *et* R. Tx. 1943 (subarctic-subalpine to alpine tall-grass and tall-herb grasslands and related krummholz) and heath vegetation of the class *Nardo-Callunetea* Preising 1949 = *Calluno-Ulicetea* BR.-BL. et R. Tx. ex Westhoff et al. 1946 (temperate and boreal heathlands and related grasslands on nutrient-poor soils).

Psammophytic vegetation is represented by tuft-forming grasses or suffrutescents, together with their associated therophyte communities developed on mobile or fixed sands (alluvial sands and subfossil dune systems). This vegetation is distributed in the province of Vojvodina (Deliblato Sands, Subotica-Horgoš Sands, and Selevenj Heath,) and along the Danube lowland (Golubac Sands). Communities on unconsolidated, mobile sands belong to the alliance *Festucion vaginatae* Soó 1929. Successional stages on mobile sands are highly correlated with pedogenetic processes on sand. Early successional stages initiate humification processes. Primary production of older successional stages with abundant vegetation increases. Older successional stages are represented by graminoid, rhizomatous hemicryptophytes with a strong root system. These plants play an important role in the stabilization and consolidation of mobile sandy dunes and sand enrichment with humus. Therefore, they may be used for the biotic recultivation of ash deposits produced by coal combustion at thermoelectric power plants (Pavlović et al. 2004; Mitrović et al. 2008, 2012).

Pannonic steppic grasslands on chernozem occupy soil on the loess terraces of Srem, Bačka, Banat and north-eastern Serbia (from Kladovo to Radujevac and Negotin). These communities, rich in perennial grasses and herbs, develop on typical chernozem with a deep, humus-accumulative A horizon and a calcareous loess C horizon. Species-rich steppic grasslands covered a large area of the Pannonian Plain in the past. Recently, their distribution is restricted to specific land forms like loess ridges formed by fluviatile erosion and accumulation. These communities belong to the alliance *Festucion rupicolae* Soó 1940.

Salt steppes are distributed on the Pannonian Plain within extremely arid spots. These communities are highly influenced by a Pannonic climate with extreme temperatures and aridity in summer. During the hot season, ground water with soluble salts rises by capillarity and evaporates. The soluble salts from the evaporated water are accumulated through the soil profile. Capillary strength is indirectly proportional to soil porosity, and heavy, clayey soils have the strongest capillary forces. Soils with such a texture in arid areas are sensitive to salinization processes. In sandy soil, surface layers dry out rapidly. A break in the capillary connection between the surface and the deeper water of sandy soils prevents water rising from the deeper horizons. Therefore, sandy soils of (temporarily) arid areas are less sensitive to salinization processes.

In order to absorb water from a saline solution, some plants have a high concentration of osmotically active organic compounds. Other species absorb salts along with water until the osmotic forces of the soil solution and cell sap are in equilibrium. However, an excessive concentration of salts in cells may be toxic. Some halophytes eliminate excessive salts through salt glands (*Atriplex, Chenopodium, Salsola*).

Halophytic vegetation in Serbia is represented by graminoid salt steppes on solonetz (the class *Festuco-Puccinellietea* Soó 1968) and also by euhaline herbaceous communities with annual succulent halophytes on solonchak (the alliances *Thero-Salicornion* Br.-Bl. 1933 emend. R. Tx. 1955 and *Cypero-Spergularion salinae* Slavnić 1948 of the class *Thero-Salicornietea* Pignatti 1953 emend. R. Tx. and Oberd. 1958). Most species that inhabit halomorphic soil belong to the IUCN red list of threatened species. Therefore, these species and their habitats are protected.

Montane meadow steppe vegetation is represented by extrazonal communities of the alliance *Festucion valesiacae* Klika 1931. These communities occupy shallow, dry calcareous and serpentine soils on southern slopes.

Chasmophytic vegetation (*Asplenietea trichomanes*) is distributed on the rocky habitats of the paleo-temperate region (Pignatti et al. 1995). Despite its wide distribution, chasmophytic vegetation is fragmented and isolated. The gene flow between populations in fragmented and geographically isolated habitats is obstructed. Even within the same saxatile region, the effective gene flow between chasmophytic populations may be considerably inhibited through the frequent failure of seeds to establish new individuals (Davis 1951).

Reduced gene flow, in combination with both strong natural selection in adverse saxatile habitats and genetic drift due to reduced effective population size, has resulted in amazing diversification of chasmophytic flora. Most chasmophytic species are *endemics*. Strong selective pressures and reduced gene flow due to both the geographical isolation of saxatile areas and ecesis problems resulted in the fast speciation of chasmophytic taxa with a relatively narrow distribution. Most chasmophytic taxa belong to the IUCN red list of threatened species (Stevanović 1996, 1999; Lakušić et al. 2006; Tomović et al. 2007; Jakovljević et al. 2011; Rakić et al. 2012). Therefore, these species and their habitats are protected.

All chasmophytic communities in Serbia belong to the vegetation classes *Asplenietea trichomanis* Br.-Bl. 1934 corr. Oberd 1977 (cliff vegetation occupying narrow holes, fissures and crevices in rocks), *Thlaspietea rotundifolii* Br.-Bl. et al. 1947 (vegetation colonising scree, moraines and colluvial deposits) and *Adiantetea capilli-veneris* Br.-Bl. 1948 (fern-rich chasmophytic communities of water-splashed habitats).

Chasmophytic vegetation is very complex since it involves communities on calcareous, serpentine and silicate rocks, at different altitudes, from colline (usually canyon) habitats to nival regions (Lakušić and Karadžić 2010). These communities are intra-zonal since they appear in all vegetation zones.

Chasmophytes in colline canyon habitats usually belong to the group of (paleo)endemic rare species. Canyons and gorges offer certain plants a refuge from unfavourable climatic conditions, competition with hillside communities, grazing and human-induced disturbances. High biodiversity and the presence of many endemic and rare species are the main characteristics of these communities, which may be explained by the stability of environmental conditions within canyon habitats, in both recent and historical terms. Both human-induced and climate-induced disturbances of habitats are minimized in canyons (Karadžić et al. 2001).

Nival chasmophytic communities are distributed in high mountains (Prokletije, Šar planina, Kopaonik and Stara planina).

Scree vegetation is distributed across Europe, from the submontane to the alpine belts. The majority of calcareous and serpentine scree plant communities occur in southern Europe, in particular in the peri-Mediterranean mountain ranges, while in northern Europe (Scandinavia) acidophilous communities dominate. Scree vegetation in Serbia is mainly represented by the alliances *Achnatherion calamagrostis* Jenny-Lips 1930 and *Corydalo ochraleucae-Geranion macrorhyzi* Lakušic 1975, *Teucrion montani* Csürös *et* Pop 1965.

Cryophilic subalpine tussocky grassland communities are distributed in the Arctic zone and in the alpine belt of high mountain ranges (the Alps, Pyrenees, Carpathians, Dinarides, Rhodopes and the Scrado-Pindic Mts.) above the treeline. These communities occupy heterogeneous soils at different pedogenetic stages. Some alpine communities occupy bare rocks, alpine cliffs and rocky cracks (communities of the class *Asplenietea trichomanis* Br.-Bl. 1934 corr. Oberd. 1977). Other nival communities colonize stony screes, gravel deposits and boulder screes. Finally, some communities develop on skeletal and shallow humic soils, rendzinas (WRB, Leptosol, calcaric) and rankers (WRB, Leptosol, eutric).

The climate of the alpine belt is characterized by a short growing season, severe winters, long-lasting snow cover and strong, cold winds. Such conditions are unfavourable for trees and shrubs. The dominant plants are cryophilic herbs and tussocky grasses.

The alpine grasslands in Serbia belong to the vegetation classes *Juncetea trifidi* Hadač in Klika and Hadač 1944 (alpine acidophilous grasslands) and *Elyno-Seslerietea* Br.-Bl. 1948 (calciphile alpine grasslands). These communities have species common to other alpine vegetation classes, such as *Mulgedio-Aconitetea* Hadač and Klika in Klika and Hadač 1944 (alpine scrub and tall-herb vegetation in habitats moistened and fertilized by percolating water), *Salicetea herbaceae* Br.-Bl. 1948 (Arctic and alpine-subnival vegetation of long-lasting snowbeds and slopes irrigated by melt waters) and *Asplenietea trichomanis* Br.-Bl. 1934 corr. Oberd 1977 (chasmophytic vegetation). The grasslands in lower alpine belts are often grazed or mown. Due to the strong human impact, the treeline of Serbian mountains has been lowered. Calcareous alpine grasslands (*Elyno-Seslerietea*Br.-Bl. 1948) occupy altimontane, wind-exposed habitats on calcareous soils, rendzina (WRB, Leptosol, calcaric), regosol, lithosol (WRB, Leptosol) on limestone and carbonate rocks. Acidophilous alpine grasslands (*Juncetea trifidi* Hadač in Klika & Hadač 1944) occupy humic, siliceous acid soils ranker (WRB, Leptosol dystric), regosol, lithosol (WRB, Leptosol) on siliceous rocks in altimontane regions.

Wet anthropogenic meadows and pastures are extremely heterogeneous. These pan-European communities belong to the class *Molinio-Arrhenatheretea* Tüxen 1937 and include secondary mesic and wet grasslands on nutrient-rich soils. They have developed due to regular mowing or grazing on sites of deciduous, mixed or coniferous forests.

There is little consensus as to the classification of lowland wet meadows. They occur across a broader geographic gradient from the suboceanic to subcontinental areas within Central and Western Europe. These communities represent unstable successional stages. Moreover, communities that belong to different syntaxa have a high proportion of common species. Overlapping species distribution prevents the unambiguous delimitation of different syntaxonomic units. Such a situation requires a revision of the syntaxonomic relations of wet meadows and pastures at the pan-European level (Karadžić et al. 2015).

These communities represent different successional stages that are primarily affected by the frequency and intensity of mowing and/or grazing. Besides biotic factors, wet meadows and pastures are affected by climate, the water regime of habitats (frequency and intensity of flooding and groundwater level) and soil conditions (exchangeable basic cation content, fertilization level, salinization level, sediment deposition, soil texture, etc.). Soil texture refers to the proportion of minerals of varying sizes that comprise the solid fraction of the soil. Gravel and coarse sand corresponds to parent material with a particle size greater that 2 mm. Smaller particles are denoted as fine sand, silt and clay. The physical properties of soil such as water retention capacity, aeration and water permeability depend on the soil texture. The physical properties of soils in wet meadows and pastures are usually degraded by permanent trampling and treading.

Depending on the duration of the flooding period, groundwater level and soil properties, these communities may be divided into alliances *Calthion palustris* Tüxen 1937, *Cnidion dubii* Bal.-Tul. 1966, *Molinion caeruleae* W. Koch 1926, *Alopecurion pratensis* H. Passarge 1964, *Potentillion anserinae* Tüxen 1947 and *Deshampsion caespitosae* Horvatić 1958, which differ with respect to the duration of periodic flooding, the type of alluvial deposits, soil salinity and climate continentality, haying and fertilization frequencies.

Due to seasonal flooding and the periodic disturbance of habitats, wetland plant communities have an unstable species composition, and such a situation complicates the classification of wetland vegetation yet further. Communities of periodically flooded habitats are ephemeral. On the other hand, numerous plant communities represent different successional stages with variable structure.

Semi-natural calcareous dry grasslands on limestone, chalk and calcareous loess belong to the class *Festuco-Brometea* Br. Bl. *et* Tx. 1943. These communities have an almost pan-European distribution (the peri-Atlantic region, central European and sub-Mediterranean region). In the north, calcareous grasslands extend to the southern part of Scandinavia and in the south to northern Spain. In the Mediterranean region, calcareous grasslands are replaced by the more open, thermophilous communities of *Ononido-Rosmarinetea*, while they change into *Seslierietea* in subalpine and alpine zones (Royer 1991). Due to changes in agricultural practices (urbanization, the abandonment of grazing practices, and fertilization), the extent of dry grasslands has decreased dramatically in Europe.

References

Amidžić L, Janković MM, Jakšić P (2003) High altitude vegetation: Prokletije of Metohia—The natural and cultural heritage. Institute for Nature Conservation, Belgrade, pp 474 (in Serbian)

Babalonas D (1988) Zur Kenntnis der Flora und Vegetation auf Serpentin stadorten Nordgrichelands II Serpentinvegetation auf submontanen Stufen. Bot Jahrb Syst 110:145–156

Babalonas D (1989) Beitrag zur Flora des serpentinischen Vourinos-Gebirges (Notdgriecheland). Willdenowia 18:387–399

Bergmeier E, Dimopoulos P (2008) Identifying plant communities of thermophilous deciduous forest in Greece: species composition, distribution, ecology and syntaxonomy. Plant Biosyst 142:228–254

Borhidi A (1964) Die Zonologie des Verbandes *Fagion illyricum*. II. Systematische Teil. Pflanzensystem und geobotanisches Institut Universitet, Budapest

Braun-Blanquet J, Sissingh G, Vlieger J (1939) Klasse der *Vaccinio-Piceetea*. Prodromus der Pflanzengesellschaften. Comité International du Prodrome Phytosociologique 6:1–123

Diekmann M, Eilertsen O, Fremstad E, Lawesson JE, Aude E (1999) Beech forest communities in the Nordic countries—a multivariate analysis. Plant Ecol 140:203–220

Dierschke H (1990) Species-rich beech woods in mesic habitats in central and western Europe: a regional classification into suballiances. Vegetatio 87:1–10

Dierschke H (1997) Syntaxomomical survey of European beech forests: some general conclusions. Annali di Botanica 55:17–26

Dinić A (1977) Hornbeam associations (Carpinetum betuli serbicum ass. nova) on alluvial and diluvial river and stream terraces in Serbia. Arch Biol Sci Belgr 29:9P–10P

Dinić A (1998) The hornbeam (*Carpinus betulus* L.) in relict forest communities on karst terrains in Đerdap. Ecol, Belgr 33:85–90

Dzwonko Z, Loster S (2000) Syntaxonomy and phytogeographical differentiation of the *Fagus* woods in the Southwest Balkan Peninsula. J Veg Sci 11(5):667–678

Dzwonko Z, Loster S, Dubiel E, Drenkovski R (1999) Syntaxonomic analysis of beechwoods in Macedonia (former Republic of Yugoslavia). Phytocoenologia 29:153–175

Ellenberg H, Leuschner C (2012) Vegetation Mitteleuropas mit den Alpen. UTB GmbH, Stuttgart

Hennekens SM (1995) TURBO(VEG). Software package for input, processing, and presentation of phytosociological data. User's guide. Instituut voor Bos en Natuur, Wageningen and Unit of Vegetation Science, University of Lancaster, Lancaster

Horvat I (1959) Systematic relations of the thermophilous oak and beech forests of Southeastern Europe. Biol Gaz 12:1–40

Horvat I, Glavač V, Ellenberg H (1974) Vegetation Südosteuropas. Gustav Fischer Verlag, Jena

Jakovljević K, Lakušić D, Vukojičić S, Tomović G, Šinžar-Sekulić J, Stevanović V (2011) Richness and diversity of Pontic flora on serpentine of Serbia. Cent Eur J Biol 6:260–274

Janković M, Mišić V (1960) Forest vegetation of Fruška Gora Mt. Matica Srpska. J Nat Sci 19:26–97

Jovanović B, Jovanović R, Zupančič M (1986) Natural potential vegetation of Yugoslavia. Map and commentary to the Map 1: 1,000,000. 18th IUFRO Congress Yu, Ljubljana, p 86

Kabaš EN, Alegro AA, Kuzmanović NV, Jakovljević KM, Vukojičić SS, Lakušić DV (2013) *Stipetum novakii ass. nova*—a new association of serpentine rocky grassland vegetation (*Halacsyetalia sendtneri*) in Serbia. Acta Bot Croat 72:169–184

Karadžić B (2013) FLORA: a software package for statistical analysis of ecological data. Water Res Manag, Belgr 3:45–54

Karadžić B, Šašo-Jovanović V, Jovanović Z, Popović R (1998) 'FLORA' a database and software for floristic and vegetation analyzes. In: Tsekos I, Moustakas M (eds) Progress in botanical research. Kluwer Academic Press, Dordrecht, pp 69–72

Karadžić B, Šašo-Jovanović V, Jovanović Z, Karadžić D (1999) On detrending in correspondence analysis and principal component analysis. Ecoscience 6:110–116

Karadžić B, Mijović A, Popović R, Perišić S, Marinković S (2001) Forest vegetation in West Serbian canyons: biodiversity hotspots. In: Ragodlou K (ed) Forest research: a challenge for an integrated European approach. NAGREF-Forest Research Institute, Thessaloniki, Greece, pp 513–518

Karadžić B, Marinković S, Kataranovski D (2003) Use of the β-function to estimate the skewness of species responses. J Veg Sci 14:799–805

Karadžić B, Jarić S, Pavlović P, Mitrović M (2015) Aquatic and wetland vegetation along the Sava River. In: Sčančar J, Paunović M, Milačič R (eds) The Sava River. Springer, Heidelberg, pp 249–316

Knollová I, Chytrý M (2004) Oak–hornbeam forests of the Czech Republic: geographical and ecological approaches to vegetation classification. Preslia 76:291–311

Kojić M, Popović R, Karadžić B (1997) Vascular plants of Serbia as bioindicators. Institute for Biological Research "Siniša Stanković", University of Belgrade, Belgrade (in Serbian)

Kojić M, Popović R, Karadžić B (1998) Syntaxonomic review of Serbian vegetation. Institute for Biological Research 'Siniša Stanković', University of Belgrade, Belgrade (in Serbian)

Krause W, Klement O (1958) Zur Kenntnis der Flora und Vegetation auf Serpentinstandorten des Balkans. 3. Felsflechten-Gesellschaften im Gostovic-Gebiet (Bosnien) und Zlatibor-Gebirge (Serbien). Vegetatio 8:1–19

Kruckeberg AR (1986) An essay: the stimulus of unusual geologies for plant speciation. Syst Bot 11:455–463

Lakušić D, Karadžić B (2010) New associations of serpentine chasmophytic vegetation (*Asplenietea trichomanis* Br.-Bl. 1934 corr. Oberd. 1977) on Kopaonik Mt in Serbia. Bot Serb Belgr 34:67–79

Lakušić D, Novčić R, Kučera J, Marhold K (2006) *Cardamine pancicii (Brassicaceae),* a neglected species of the Balkan Peninsula. Willdenowia 36:177–191

Lakušić D (ed) (2005) Habitats of Serbia, results of the project "Harmonization of National Habitats Classification Nomenclature with European Community Standards", Institute of Botany and the "Jevremovac" Botanical Garden, Faculty of Biology, Belgrade University, Ministry of Science and Environmental Protection of the Republic of Serbia, pp 684, http://habitat.bio.bg.ac.yu/ (in Serbian)

Mitrović M, Kostić O, Jarić S, Karadžić B, Djurdjević L, Gajić G, Lj Oberan, Pavlović D, Pavlović M, Pavlović P (2012) Photosynthetic efficiency of four woody species growing on fly ash deposits of a Serbian 'Nikola Tesla-A' thermoelectric plant. Pol J Environ Stud 21(5):1339–1347

Mitrović M, Pavlović P, Lakušić D, Stevanović B, Djurdjević L, Gajić G, Kostić O (2008) The potential of *Festuca rubra* and *Calamagrostis epigejos* for the revegetation on fly ash deposits. Sci Total Environ 407(1):338–347

Mišić V (1979) Relict polydominant forest communities in Serbia. Matica Srpska, Novi Sad (in Serbian)

Mišić V (1960) Ecological study of subalpine bush vegetation on Mt. Kopaonik. Institute of Biology, National Republic of Serbia, Belgrade, Special edition 6:1–47 (in Serbian)

Mišić V (1981) The forest vegetation in gorges and canyons in Eastern Serbia. Institute for Biological Research "Siniša Stanković", Belgrade (in Serbian)

Mišić V, Jovanović-Dunjić R, Popović M, Borisavljević Lj, Antić M, Dinić A, Danon J, Blaženčić Ž (1978) Plant communities and habitats of Stara Planina. Serbian Academy of Sciences and Arts (SANU), Special edition 511, Department of Natural and Mathematical Sciences 49: 1–389, Belgrade (in Serbian)

Moss D, Davies C, Roy D (1996) CORINE biotopes sites, database status and perspectives 1995, topic report 27/96, European topic centre on nature conservation, supervision of Ulla Pinborg, Project Manager, European Environment Agency, 1995

Novak FA (1926) Ad florae Serbiae cognitionem additamentum primum. Preslia 4:37–56

Parabućski S, Stojanović S, Butorac B, Pekanović V (1986) Prodromus vegetation of Vojvodina. Matica Srpska: J Nat Sci 71:5–40 (in Serbian)

Pavlova D, Kožuharova E, Dimitrov D (1998) The serpentine flora of the Eastern Rhodopes Mountains (Bulgaria). In: Tsekos I, Moustakas M (eds) Progress in botanical research. Kluwer Academic Publishers, Dordrecht, pp 133–136

Pavlović P, Mitrović M, Djurdjević L (2004) An ecophysiological study of plants growing on fly ash deposits from the 'Nikola Tesla-A' thermal power station in Serbia. Environ Manage 33:654–663

Pepper AE, Norwood LE (2001) Evolution of *Caulanthus amplexicaulis* var. *barbarae* (*Brassicaceae*), a rare serpentine endemic plant: a molecular phylogenetic perspective. Am J Bot 88:1479–1489

Pignatti S (1998) I boschi d'Italia. Sinecologia e biodiversità.UTET, Torino

Pignatti S, Oberdorfer E, Schaminee JHJ, Westhoff V (1995) On the concept of vegetation class in phytosociology. J Veg Sci 6:143–152

Popović Z, Mijović A, Karadžić B (2005) Dry matter distribution patterns and photosynthetic traits of two vernal ephemeroids in a deciduous forest. Period Biol 107(3):287–292

Popović Z, Mijović A, Karadžić B, Mijatović M (2006) Response of growth dynamics of two spring geophytes to light regime in a lime-beech forest. J Integrat Plant Biol 48(5):527–535

Popović Z, Bojović S, Matić R, Stevanović B, Karadžić B (2015) Comparative ecophysiology of seven spring geophytes from an oak-hornbeam forest. Braz J Bot. doi:10.1007/s40415-015-0204-4

Proctor J (1999) Toxins, nutrient shortages and droughts: the serpentine challenge. Trends Ecol Evol 14:334–335

Quézel P, Barbéro M, Akman Y (1980) La végétation forestiére d'Anatolie septentrionale. Phytocoenologia 8:365–519

Quézel P, Barbéro M, Akman Y (1992) Typification de syntaxa décrits en région méditerranéenne orientale. Ecol Medit 18:81–87

Rakić T, Živković I, Šinžar-Sekulić J, Stevanović B, Stevanović V, Lakušić D (2012) Morphological variation within the *Edraianthus graminifolius* complex (*Campanulaceae*) from the central Balkan Peninsula—evidence from multivariate statistical analysis. Flora 207:354–364

Ranđelović N, Rexhepi RF (1980) Prodromusi i fitocenozave te Kosoves. Biotehnika, Priština 8:213–222 (in Albanian)

Ranđelović N, Rexhepi F (1984) Meadow and pasture vegetation of Koritnik. J Ecol Soc Bosnia Herzegovina B 2(1):271–274

Rexhepi F (1994) Vegetacioni i Kosovës 1. Universiteti i Prishtinës, Fakulteti i shkencave të natyrës, Prishtinë (in Albanian)

Ritter-Studnička H (1970) Die Vegetation der Serpentinvorkommen in Bosnien. Vegetatio 21:75–156

Rivas-Martínez S, Fernandez-Gonzàlez F, Loidi J, Lousã M, Penas A (2001) Syntaxonomical checklist of vascular plant communities of Spain and Portugal to association level. Itinera Geobot 14:1–341

Roleček J (2005) Vegetation types of dry-mesic oak forests in Slovakia. Preslia 77:241–261

Rousakova V, Tzonev R (2003) Syntaxonomy of the oak forests in the Pleven district (Danube plain in Bulgaria). Fitosociologia 40:23–31

Royer JM (1991) Synthèse eurosiberienne, phytosociologique et phytogeographique de la classe des *Festuco-Brometea.* J. Cramer, Stuttgart, Berlin, p 296

Sanda V, Öllerer K, Burescu P (2008) Fitocenozele din România. Sintaxonomie, structură, dinamică şi Evoluţie. Universitatea din Bucuresti (in Romanian)

Scoppola A, Blasi C, Abbate G, Cutini M, Di Marzio P, Fabozzi C, Fortini P (1995) Analisi critica e considerazioni fitogeografiche sugli ordini e le alleanze dei querceti e boschi misti a caducifoglie dell'Italia peninsulare. Ann Bot (Roma) 51:81–112

Soó R (1964) Die regionalen *Fagion*-Verbände und Gesellschaften Südosteuropas. Stud Biol Acad Sci Hung 1:1–104

Stevanović V (1996) Analysis of the Central European and Mediterranean orophytic element on the mountains of the W. and Central Balkan Peninsula, with special reference to endemics. Bocconea 5:77–97

Stevanović V (1999) The red data book of Flora of Serbia, 1. Extinct and critically endangered taxa. Ministry of Environmental Protection, Biological Faculty of Belgrade University, Institute for Nature Protection, Belgrade

Tatić B, Veljović V (1992) Distribution of serpentinized massives on the Balkan peninsula and their ecology. In: Robertis BS, Proctor J (eds) The ecology of areas with serpentinized rocks. A world view. Kluwer Academic Publishers, Dordrecht, pp 199–215

Tomić Z (2006) Revision and domination of Balkan beech plant communities in Serbia. Bull Fac For Belgr 94:29–82 (in Serbian)

Tomović G, Vukojičić S, Niketić M, Zlatković B, Stevanović V (2007) *Fritillaria (Liliaceae)* in Serbia: distribution, habitats and some taxonomic notes. Phytologia Balcanica 13:359–370

Török K, Podani J, Borhidi A (1989) Numerical revision of the *Fagion illyricum* alliance. Vegetatio 81:169–180

Vasić O, Diklić N (2001) The flora and vegetation on serpentinites in Serbia—a review. Bocconea 13:151–164

Vukićević E (1976) Forest phytocoenoses of Mount Gučevo. Bull Fac Forest, Belgrade A50:109–132 (in Serbian)

Zupančič M (2000) Some syntaxonomic problems of the class Vaccinio-Piceetea Br.-Bl. in Br.-Bl. et al. 1939. Acta Bot Croat 59:83–100

Čarni A, Košir P, Karadžić B, Matevski V, Redžić S, Škvorc Ž (2009) Thermophilous deciduous forests in Southeastern Europe. Plant Biosyst 143(1):1–13

Župančič M (2007) Syntaxonomic problems of the classes *Vaccinio-Piceetea* and *Erico-Pinetea* in Slovenia. Fitosociologia 44:3–13

Geology 5

5.1 Geomorphology

The territory of Serbia is situated in the Alpine realm in southern Europe. It comprises various geological units of composite structure and complex tectonic interrelations. The territory of Serbia, from the geological aspect, can be divided into several units, based on the geographic position of the units, their morphological expression in the relief, the age of the formation and orogenesis. The position of the geotectonic units and significant morphostructures is shown on the map in Fig. 5.1.

According to Ćirić (1996) and Dimitrijević (1997a, b), the subdivision of the units, from the south-west to the north-east, is as follows:

- The Dinarides with the East Bosnian-Durmitor Block, Ophiolite Belt, the Drina–Ivanjica Element and a small part of the Hellenides
- The Vardar Zone with its external, central and internal subzones
- The Serbian Massif with the Lower and Upper (Vlasina) complexes
- The Carpatho-Balkanides with the Supragetticum, Geticum, Lower and Upper Danubicum, and Krajina Zone
- The Pannonian and the Dacian depressions.

Central Serbia belongs to the Serbian Massif, eastern Serbia to the Carpatho-Balkanides or to the northern limb of the Alpine orogen, and western Serbia to the Dinarides with characteristic units of the inner zones of the southern limb of the Alpine orogen. The Vardar Zone consists of blocks of different ages, origin and composition represented by flysch and pelagic formations, basalts, and an ophiolitic mélange, which are squeezed between the Dinarides and the Serbian Massif. The northern part of Serbia is covered with sediments of the Pannonian Sea, a remnant of the Paratethys Ocean. Only the peripheral part of the Walachian-Pontian basin is situated to the east of the Carpatho-Balkan mountain arc.

The geological composition and structure of Serbia should be seen as the result of the complex tectonic and magmatic processes in the transition from the Palaeozoic to the Mesozoic, which led to the creation of the Tethys, and sedimentation, igneous and metamorphic processes, which took place through the Mesozoic and the Paleogene in the central parts of today's Balkan Peninsula. It is also important to consider the very powerful and complex tectonic movements, resulting from the subduction of the African and European continents, which led to the uplift of the Dinaric and Carpatho-Balkan mountain ranges and the development of the Pannonian Sea and numerous freshwater lakes. Younger erosion processes of denudation led to the destruction and modelling of pre-existing structural forms and the emergence of young Neogene and Quaternary geosculptured landforms.

5.1.1 Dinarides

The Dinarides on the Balkan Peninsula comprise several tectonic zones, but only three of them are found within the borders of Serbia today. From the north-east to the south-west, they are as follows: the Drina–Ivanjica Element, the Ophiolite Belt and the East Bosnian-Durmitor Block. The mountains are lined up in parallel chains and ridges with karstic valleys (polja) between them and are built up from complex nappe and anticline structures. Intense older Upper Jurassic mobility in this zone is represented by the Ophiolitic mélange. Younger underthrusting left numerous windows, inliers and klippen as far as 40 km towards the south-west. In the north-east, the zone was thrusted under the Ophiolite Belt.

Only a small section of the East Bosnian-Durmitor Block appears in the mapped area, and it is found in south-western Serbia. It is marked by a series of large and irregular brachy-forms with cores of Palaeozoic rocks extending almost through the middle of the Dinarides. The oldest rocks in this zone are Silurian limestone and Devonian dolomite,

P. Pavlović et al., *The Soils of Serbia*, World Soils Book Series,
DOI 10.1007/978-94-017-8660-7_5

Fig. 5.1 Morphostructural map

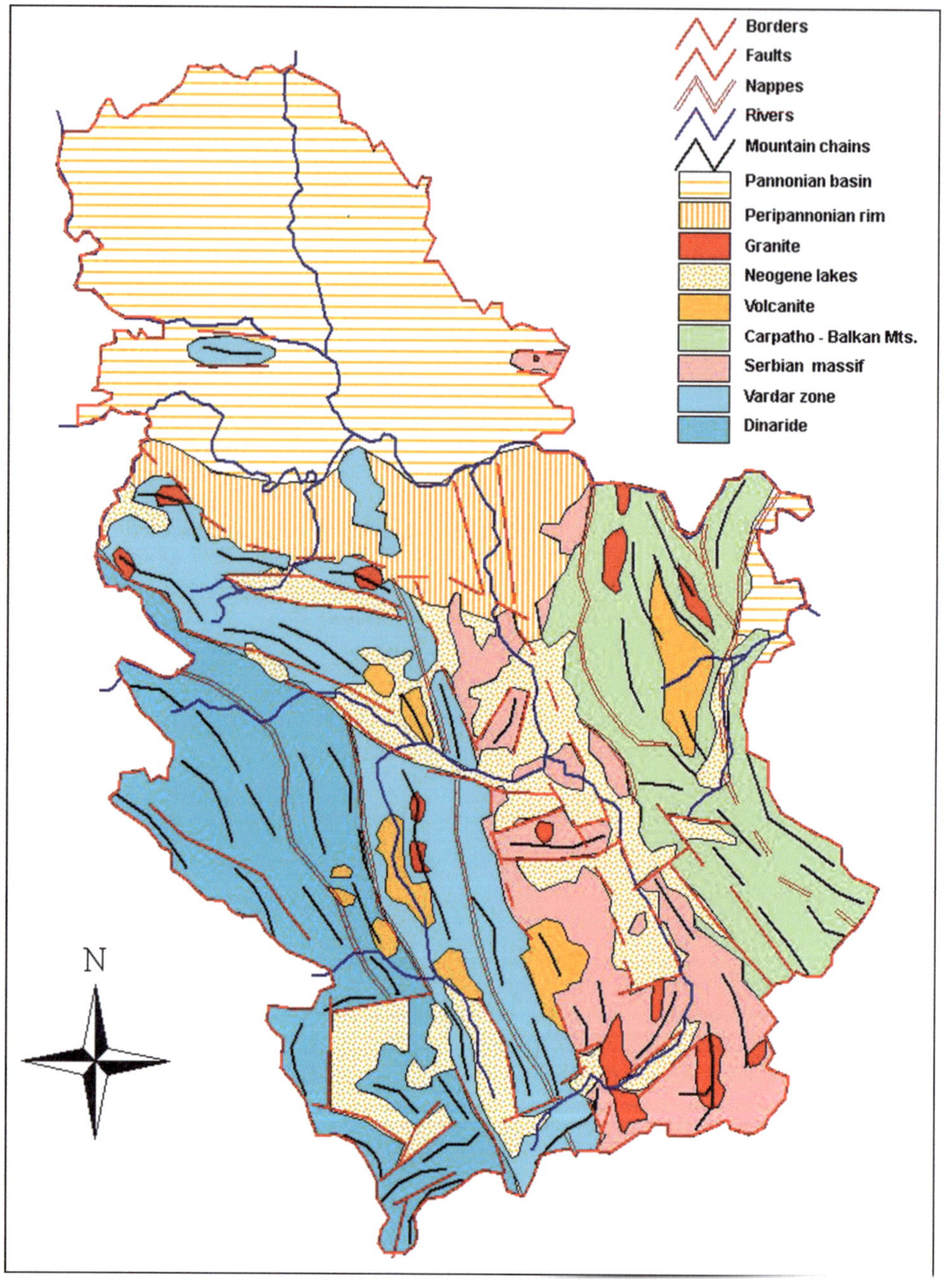

marbled limestone and argiloschist. The Permian and Lower Triassic sediments are composed of sandstone, dolomite and limestone, with red Bulog limestone and volcanics at the top. The volcanogenic-sedimentary formation begins with chert and porphyrite tuffs, associated with thin cherty limestone, dolomite and reef limestone, followed by a massive succession of limestone and dolomite in the Upper Triassic. Only thin Jurassic cherty limestone was deposited, while Cretaceous sediments are mostly absent. Neogene sediments fill only small lake basins.

The Ophiolite Belt is one of the most outstanding units in the Dinarides. It stretches from Bosnia into western Serbia and Old Raška to Metohija and further into Albania, where it is known as the Mirdita Zone. It has a base of Middle Triassic to Upper Jurassic limestones, overlaid with siliceous shale and chert, followed by the Ophiolitic mélange with inclusions of oceanic rocks and the large uplifted ultramafic masses of Zlatibor, Jadovnik and Ozren mountains. Over these lie Cretaceous clastics and Senonian flysch. The younger rocks are Tertiary andesite and Neogene lake sediments.

The internal structure of this zone is very complex, shown in particular by the fact that big ultramaphite bodies are thrusted over the diabase-chert formation and amphibolite.

This big and complex nappe is thrusted over the East Bosnian-Durmitor Block to the south-west, forming many tectonic windows and some klippen. The zone is thrusted under the Drina–Ivanjica Element from the north-eastern and eastern sides.

The Drina–Ivanjica Element is situated between the Ophiolite Belt in the south-west and the Vardar Zone in the north-east and east. It consists of Palaeozoic greenschists, argiloschist, meta-sandstone, Permian conglomerate and Lower Triassic clastics, which are overlaid with limestone. In the western and north-western parts of the belt, marine Upper Cretaceous clastics occur, and the eastern boundary of the element is fringed with flysch deposited in the basin in the boundary towards the Vardar Zone. Tertiary andesite, Golija granitoid and Neogene lake deposits are of local significance (Fig. 5.2).

The structural framework comprises large Drina and Ivanjica anticlinoria. The Palaeozoic base is intensely tectonized and overlaid by a succession of Variscan forms and Alpine plicative structures. Generally, the Drina–Ivanjica Element is thrusted towards the south-west over the Ophiolite Belt and lies under the Vardar Zone in the north and east.

Only a small part of the Hellenide unit appears in the mapped zone—in the southern area of the map. The oldest rocks are the Devonian/Carboniferous greenschists and phyllite with quartzite, glaucophane schist and the Triassic limestone and dolomite, which are intruded with syenite, diorite, keratophyre and rhyolite in a very intricate relationship. The area also exhibits well-developed glacial and fluvio-glacial geomorphology.

5.1.2 Vardar Zone

From the geological point of view, the Vardar Zone is the most complex domain of the Balkan Peninsula and comprises several blocks differing in composition, history and origin. In a west-east direction, it can be divided into the external, central and internal subzones. The external subzone comprises three complex blocks:

The Srem block crops out at Fruška Gora Mt. only, showing normal and metamorphosed Triassic rocks, Cretaceous flysch and shallow-marine clastics in two blocks separated by serpentinite. The blocks were later covered by Miocene sediments and Quaternary loess.

The Jadar block has a base of Palaeozoic strata with a complex structure. Triassic formations are followed by ultramafics (Maljen and Suvobor) and an Ophiolitic mélange, Cretaceous flysch and paraflysch, which are intruded by several Tertiary granitoids and volcanics in Brajkovac, Borač and the mountains of Rudnik and Kotlenik. From the structural aspect, the Jadar block is an anticlinorium with a widely exposed core of Palaeozoic rocks with Alpine folds and a Hercynian base.

Fig. 5.2 Mt. Golija (*photo* Koščal M., Menković Lj.)

The Kopaonik block is divided by the Kopaonik-Željin granitoid area into two belts: the western is regarded as a marginal basin and the eastern has a specific development of Cretaceous strata. Both belts are rich in peridotite (the Ibar valley, and Stolovi and Goč mountains), and in the central area, post-Triassic/Cretaceous metamorphism is characteristic (Kopaonik and Trepča). The central part of the subzone around the Ibar valley is rich in intermediate volcanics (Rogozna and Kopaonik mountains). In the southern part of the subzone, there is a flysch belt and the large Kosovo Neogene basin.

In terms of the other two subzones, the outstanding feature of the Central Vardar Subzone is Lower Cretaceous Gledići Paraflysch, over 150 km long, above metamorphics and an Ophiolitic mélange. The Internal Vardar Subzone is found in the border area of the Serbian Massif. It consists of crystalline schists of unknown age, which are overlaid with Senonian flysch, and on Mt. Radan, they are covered with massive deposits of Tertiary volcanics with abundant pyroclastics.

5.1.3 Serbian Massif

The Serbian Massif occupies the central part of what is now Serbia and is recognized as one of the primary geotectonic units. It extends from the Vršac Mountains in the north along the Velika and Južna Morava rivers down to Leskovac and Vranje in the south and further on to Macedonia, Bulgaria and Greece. It has a complex internal structure, formed over a long period of evolution. Basically, the Massif is composed of two crystalline complexes:

The lower complex of Precambrian age, in the west, is composed of rocks of the amphibolite facies: two-mica gneiss, micaceous schist, quartzite, some marble, particularly in the lower sections, and Lower Palaeozoic granitoids with pegmatites and abundant migmatization. In several places, it is covered by shallow-marine Cretaceous and Eocene clastics.

The Upper or Vlasina Complex lies unconformably over the lower complex. It mainly comprises Riphean-Cambrian rocks metamorphosed to greenschist facies: chlorite-epidote, chlorite-actinolite and amphibole schists, and is covered by weakly metamorphosed Ordovician to Carboniferous formations. It is intruded by the Palaeozoic granitoids of Vlajna and Tertiary granodiorite of Surdulica, which stitch it to the lower complex.

The Serbian Massif also includes magmatic rocks, such as Lower Palaeozoic-Caledonian Vlajna granite, Hercynian Bujanovac granite and the Alpine granitoids of Jastrebac and Surdulica, as well as Tertiary andesite and dacite of Lece. Serpentinite is located only in the form of small lenticular bodies within the dislocations. Polyphase magmatic activity, which varied in terms of chemistry and intensity, progressively affected adjacent rocks, leading to intensive migmatization, which advanced to anatexite and the formation of gneiss-granite. Younger Senonian and Paleogene sediments are preserved in small areas, while Neogene deposits have an extensive distribution, particularly in the Velika Morava valley and in the basins connected to it.

The central part of the Serbian Massif is characterized by polyphase folding, which produced several large anticlinoria and a big synclinorium with the axis of symmetry in the submeridian direction, which is found along the Velika Morava valley in this part of the Massif. During the Cenozoic Era, the Serbian Massif underwent heavy rupture deformation, which led to its present striking 'chess-board' tectonic pattern. In this way, many local structures, represented by horsts, basins and troughs, were formed. The horsts are surrounded by Neogene sediments and rise from them as 'island mountains' (the mountains of Jastrebac, Juhor and Crni Vrh), while graben structures can be even larger than the horsts, e.g. the Velika Morava valley. The Serbian Massif has tectonic relations with the Dinarides/Vardar Zone and the Carpatho-Balkanides, with the crystalline rocks of the Massif thrusted over adjacent tectonic units.

5.1.4 Carpatho-Balkanides

The Carpatho-Balkanides is the mountain arc which extends from Romania over the Danube and the Timok valley in the north to Niš and to Mt. Stara planina in the south-east and then further east to Bulgaria. The Laramian stage of Alpine orogenesis was responsible for the tectonic development of these mountains, while the final formation of the terrain and thrusting of the nappes occurred at the Styrian stage during the Miocene. The general geological structure of the Carpatho-Balkanides reveals a crystalline core as the basis over which numerous nappes of mainly Mesozoic sediments are thrusted, composed of the following tectonic units: Suprageticum, Geticum, Danubicum and Krajina Zone.

The Suprageticum comprises Silurian and Devonian meta-clastites and mica-schists with lenses of amphibolite and masses of marble, Permian red sandstone, Jurassic and Cretaceous limestone, dolomite and clastics with numerous breaks in sedimentation, which are thrusted over the Geticum autochthon.

The Geticum Zone comprises mainly Jurassic and Cretaceous limestone and dolomite, which lie over Cambrian (Rifej) and Palaeozoic schists, greenschists and the volcano-sedimentary series of southern Kučaj, which are invaded by gabbro and keratophyres in the lower parts and have granite intrusions in the upper parts. There are significant masses of Devonian to Lower Carboniferous and

Upper Devonian flysches, Permian red sandstone, Permian/Triassic variegated clastics, thick Jurassic and Cretaceous carbonates with numerous sedimentation breaks, Upper Jurassic Ruj Flysch, Jurassic/Lower Cretaceous carbonates and thin layers of Upper Cretaceous sediments. The zone is invaded by several large granitoids of various ages near Neresnica, Gorjane, Brnjica, Ziman and Plavna. Along the eastern flanks of the zone, there is the large Timok Eruptive Area (TEA) with a very intricate relationship between several volcanic formations, ranging from Senonian to Paleogene in terms of age.

The Upper Danubicum (Infrageticum) consists of different zones. In the north, it consists of Cambrian to Lower Cretaceous rocks, which have a complex tectonic relationship with Deli Jovan gabbro; then, in some areas, it is composed of the whole Palaeozoic and Jurassic and Cretaceous, which vary in thickness but have many breaks. Finally, in the south, on Mt. Stara planina, it comprises of semi-metamorphic Palaeozoic, Permian red sandstone, and Mesozoic up to the Cretaceous. These rocks are invaded by several smaller granitoid masses of various ages and Zaglavak gabbro. The Lower Danubicum or the Vrška Čuka Zone comprises Cambrian greenschists, very thin Carboniferous and Permian, and thin Jurassic and Cretaceous sediments, which extend to the north into the Krajina Zone, composed of Sinaia Beds and Mokranje Flysch covered with Neogene deposits and Quaternary loess.

From the tectonic point of view, the Carpatho-Balkan Mountains are characterized by thrusting to the east and north-east, with active pressure from the Serbian Massif, the Rhodope Mts. and the Mesian plate. The mountains have distinct anticlinal shapes with many examples of folding and faulting structures: horst—anticline, nappes and rare accretions. While there are many such structures, a few of the higher order nappes are worth mentioning: Morava, Rtanj-Kučaj, Tupižnica, Miroč and Đerdap. There are also many forms of volcanic activity: sequences of volcanic cones and craters with thick layers and plates of pyroclasts.

5.1.5 Pannonian Basin

The Pannonian basin is a new, large, intermountain depression between the Dinarides and the Serbian Massif in the south, and the Alpine-Carpatho-Balkanide mountain arc in the east. The basin was formed in the early Miocene by the tectonic processes of rupturing and the differential rising and sinking of the central Tisza mass. The resulting structure has highly diverse paleo-relief, which was flooded in the Middle Miocene by seawater. The basin is filled with a thick complex of marine and lacustrine Neogene deposits and covered with continental Quaternary.

South of the Pannonian basin, there are a few smaller tectonic basins (kotline), which are formed by vertical movements (faulting). All of them are filled by Neogene molasses and Quaternary deposits. They are separated by uplifted blocks (horsts), mainly made of crystalline schists. The most prominent horsts are the mountains of Avala, Kosmaj, Bukulja, Venčac, Crni Vrh, Juhor and Jastrebac.

The Walachian-Pontian (Dacian) basin is the Carpatho-Balkanide foreland of the hilly country in the east of Serbia, known as Timočka Krajina. It marks the extreme west periphery of the large Dacian basin. Neogene sediments are up to 2 km thick, starting with transgressive Miocene sediments, then delta deposits, and ending with Lower Pliocene and Eopleistocene deposits with well-developed river terraces.

5.2 Lithology

The lithological composition of Serbia is very complex with 28 mapping units of the soil-parent material (substrate) type, which are presented on the map. The data show a roughly equal proportion of hard sedimentary and soft, unconsolidated rocks. According to the lithological map (Fig. 5.3) and Table 5.1, fluvial (Holocene) sands, loams and clays are the most common mapped units and extend over about 19% of the Serbian land mass. This is followed by Pleistocene aeolian sands and loess (approximately 16%), and then metamorphic gneiss, schists and slates (approximately 14%). Lacustrine sedimentary unconsolidated clastics (sands, loams and gravels) extend over 13% of the territory, while consolidated carbonates (limestone, dolomite and marlstone) extend over 9%, consolidated clastics (sandstones, conglomerate and shale) cover approximately 5%, and flysch deposits approximately 6% of the territory. Unconsolidated marls and bog-lime extend over only 3%. All igneous rocks extend over 11% of the landmass, while mixed igneous, metamorphic and sedimentary rocks of complex lithology are found across about 4% of the land mass. All other types of parent material are found in <1% of Serbia.

The lithological map for the soil-parent material database was made using the 76 Basic Geological Maps (BGM) at a scale of 1:100,000 with corresponding explanation booklets. The number of mapped units in the booklets totalled over 3800. All of these were filtered through the European Soil Bureau classification of parent material (Finke et al. 2001) to obtain the lithological map presented in Fig. 5.3.

In terms of the lithological properties of the rocks, the customary distinction between crystalline (igneous, metamorphic and sedimentary) rocks is made. The first group of igneous rocks is represented by acid, intermediate, basic and ultrabasic classes, each in plutonic and volcanic forms. The

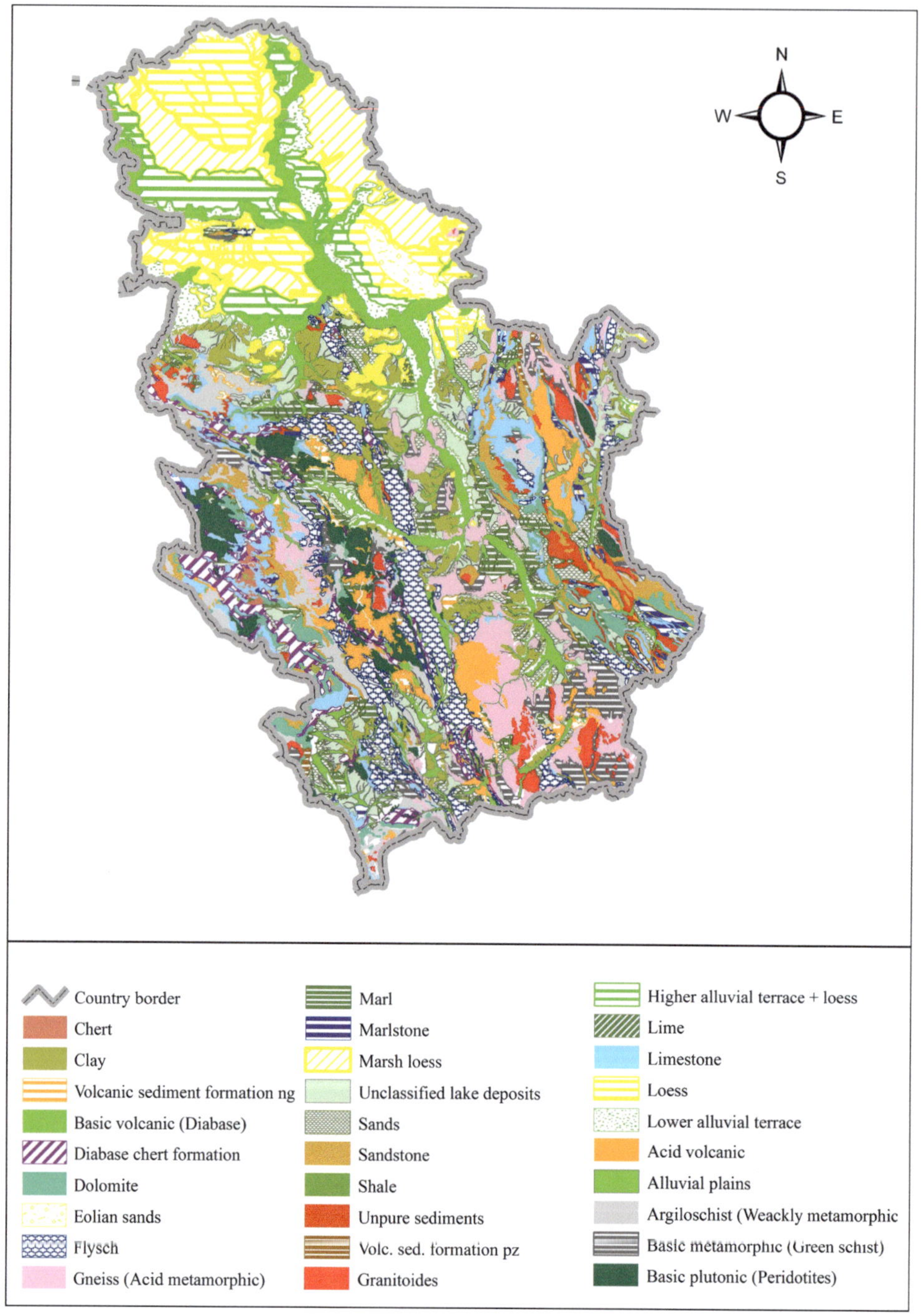

Fig. 5.3 Lithological map (Kostić 2012)

addition of a group of mixed rock formations of igneous, metamorphic and sedimentary rocks is made, represented by diabase-chert and volcano-sedimentary formations.

The second group comprises three classes: acid, basic and weakly metamorphic rocks; the third group contains various classes of sedimentary rocks, which are divided into hard consolidated and unconsolidated soft rocks. Hard (indurated) rocks are further divided into silico-clastic, calcareous, siliceous and evaporite, according to the dominant process of deposition. Unconsolidated rocks are not hard rocks and cannot be classified by their lithology in the strict sense. Their classification is a combination of lithologies, depositional environments and morphologies, e.g. marine, fluvial, lake, aeolian, glacial, residual, anthropogene deposits and organic materials.

In terms of **landforms**, they are described primarily by their morphology and not by their genetic origin nor by the processes responsible for their formation and shape. The regional slope class is the main differentiating criterion, followed by relief intensity. At the highest level of

Table 5.1 Classification of the main types of rocks

Rocks type	Subtype	Class	Area (ha)
Metamorphic rocks		Acid regional metamorphic rocks	547,713
		Basic regional metamorphic rocks	344,792
		Weakly (anchi- and epi-) metamorphic rocks	362,948
Igneous rocks		Acid and intermediate plutonic rocks	165,705
		Basic and ultrabasic plutonic rocks	354,515
		Acid and intermediate volcanic rocks	378,583
		Basic volcanic rocks	36,448
Complex lithology formations		Volcano-sedimentary formation	72,278
		Diabase-chert formation	234,045
Sedimentary rocks	*Consolidated rocks*		
	Clastic rocks	Sandstone, arkose, greywacke	251,359
		Undivided rocks	133,205
		Shale and mudstone	15,167
	Chemical precipitates or of biogenic origin	Limestone	396,987
		Dolomite	255,663
		Marlstone	115,270
		Chert	19,267
	Facies bound rocks	Flysch and molasse	527,668
	Unconsolidated rocks		
	Lake deposits	Sand and gravel	283,507
		Undivided deposits	503,364
		Marls and bog-lime	283,792
		Clays, loams and mud	405,087
	Aeolian deposits		1436,673
	Fluvial deposits		1,679,340
Water bodies	Rivers, lakes and swamps		32,724
Total area			8,836,100

separation, three groups of landforms are distinguished: level (plains and flats), sloping (hills) and steep (mountains) lands. Hence, from this point of view, the landscapes of Serbia could be divided into the following categories:

- the Pannonian basin and the Peripannonian rim with hills up to 500 m a.s.l.
- the mountain and valley region with hills and mountains over 500 m a.s.l.

Both regions are about the same size—the Pannonian basin and Peripannonian rim—cover 43,705 km^2, while mountains and valleys extend over 44,656 km^2, or 49.46 and 50.54%, respectively, of a total 88,361 km^2.

The **Pannonian part** consists of all of Vojvodina, as well as the regions around the Sava, Velika Morava and Danube rivers. The Pannonian area comprises lowlands and hilly parts of an expansive plain between 100 and 200 m a.s.l. It is represented by the wide plains of Vojvodina, with two low mountains [Fruška Gora (538 m) and the Vršac Mts. (639 m)] rising from it; however, the Pannonian depression is not entirely flat between these mountains. Across the lowest-lying land, several large rivers flow: the Danube, Sava, Tisa and Tamiš. Alluvial plains with river terraces stretch along both sides of all the rivers, and above them, there are various sizes of loess plateaus and sand plains with specific aeolian morphology.

Peripannonian Serbia comprises predominantly rolling hill terrains, with altitudes between 200 and 500 m. a.s.l., which extend along the valleys of the Sava, Danube, Velika Morava and Mlava rivers. In addition, in eastern Serbia, along the Danube River, a small part of the expansive Dacian or Walachian-Pontian basin enters Serbian territory. It is similar to the Pannonian basin, a depression with the lowest-lying land in Serbia, only 28 m a.s.l. Alluvial plains lie along the Danube and Timok rivers; further away from the rivers, there are terraces and hills, locally covered by loess and sand plains.

The **high mountain and valley** region (from 500 to 2656 m a.s.l.), or the southern part of Serbia, has more complex geomorphological and geostructural relationships. This part of the country belongs to the Dinarides and Carpatho-Balkanide mountain range, and the composite mountain terrain of the Serbian Massif lies in between. In southern Serbia, there are 30 mountain peaks over 2000 m a.s.l., with the highest peak in Serbia, Djeravica (2656 m), in the Prokletije Mts. on the border with Albania.

5.2.1 The Pannonian Basin and Peripannonian Rim Region

The Pannonian basin and Peripannonian rim region comprises the most fertile soils in Serbia, developed on loess plateaus and alluvial plains and terraces.

Alluvial plains are the lowest-lying lands in Vojvodina, and the most extensive plains, from 5 to 15 km wide, are located along the Danube, Sava, Tisa and Velika Morava rivers. They are composed predominantly of multi-layered sandy and silt-sandy deposits. In Vojvodina, alluvium is composed mainly of fine sand and silty-sand, silty clays and sandy-clayey-silt deposits, as a sorted or semi-sorted material supported by gravels and sands as in the Sava and Morava valleys.

Another important fluvial form is the **alluvial terrace**, as an indicator of climatic conditions, or may be even neotectonic activity. It is composed of unconsolidated deposits, including gravel, produced by downcutting of the flood plain or valley floor. Two levels of alluvial terraces are found in Vojvodina. The lower alluvial terrace, less than 10 km wide, flanks the left side of the Danube downstream to Deliblato Sands. Similarly, the lower terrace of the Tisa runs along its left side, where it appears in the form of discontinuous terrace flats of varying sizes. The relative height of the lower terrace is 4–6 m above the river. In the hilly/mountainous region, the grain-size composition of alluvial terraces is very variable. They comprise gravelly sandy deposits, particularly along the Velika, Zapadna and Južna Morava rivers and their tributaries, while more pebbly-gravels are found in the terraces of the rivers flowing from the higher mountains (Fig. 5.4).

The **higher alluvial terrace** is almost 10 km wide in places. It is formed on the left side of the Danube and the right side of the Tisa, sloping downstream to the south-east. A terrace scarp with a relative height of between 7 and 12 m borders the lower terrace and alluvial plains. The higher alluvial terrace is covered with loess and was therefore referred to by the researchers in the past as a loess terrace or 'Varoš terrace' (town terrace), since many towns and villages are built on it. The deposited material from alluvial plains and terraces is certainly very important for the development of fertile soils, as well as being potential aquifers essential for the water supply of the population.

During the Pleistocene, when a cold and dry climate prevailed, aeolian processes played an important part in the sculpturing of landforms in Vojvodina. Wind action formed **loess plateaus**, sand covers and plains, predominantly deposited on the initial fluvio-paludinal floor of the Pannonian basin (Fig. 5.5). Above the alluvial plains and terraces, there is loess cover, which is dissected by rivers into a number of loess plateaus, including Bačka loess plateau, the largest in Serbia, which covers about 2800 km^2; Titel loess mound, also found in Bačka; Banat and Tamiš loess plateaus in Banat; and Srem loess plateau surrounding Fruška Gora and extending south-east to Belgrade (Figs. 5.6, 5.7 and 5.8). Loess is also found to the south of the River Danube in small oases along the River Morava and in eastern Serbia. The thickness of the loess varies from a few metres to as much as 50 m in parts of Vojvodina.

The loess plains comprise sandy silts and sandy-clayey silts, and occasionally, as in the Bačka area, they can be more clayey-silt up to silty clays, which is known as 'gleyed loess'. Particle size analyses of loess have shown: a sand fraction of 4–46%, silt 50–94% and clay 0.5–21%, with well sorted and semi-rounded grains. Chemical analysis has shown variable amounts of calcium carbonate (2–38% $CaCO_3$), with calcite and dolomite minerals present (Kostić and Protić 2000).

The fluvio-paludinal floor of the Pannonian basin is a landform identified at the bottom of the Pannonian basin in Vojvodina, with elevations from 80 to 90 m. This was an area of water bodies formed by excessive flooding and with aeolian dust deposited into a fluvial–lacustrine environment forming so-called **marsh loess**. When water receded, loess, dust and sand continued being deposited on dry land.

In Vojvodina, there are two large **sand plains**, the Deliblato or Banat and Subotica sand plains, as well as smaller sand plains of kilometric dimensions which occur along the right side of the Danube in an area between Ram and Golubac and further on in eastern Serbia around Kladovo and Negotin.

Deliblato Sands occupy part of south-east Banat, stretching across an area of about 380 km^2 (Fig. 5.9). The plains comprise typical parallel, longitudinal dunes with an average dune altitude between 80 and 120 m, and the valleys between the dunes are about 20 m deep. It could be divided into three distinct area units: area of low dunes, area of high dunes and area of sand cover.

Subotica Sands occupy an area between Subotica and Horgoš and the state border with Hungary. The plain consists of two units: an area of low sand dunes on the Bačka loess plateau and an area of sand cover over the

Fig. 5.4 Fluvial deposits

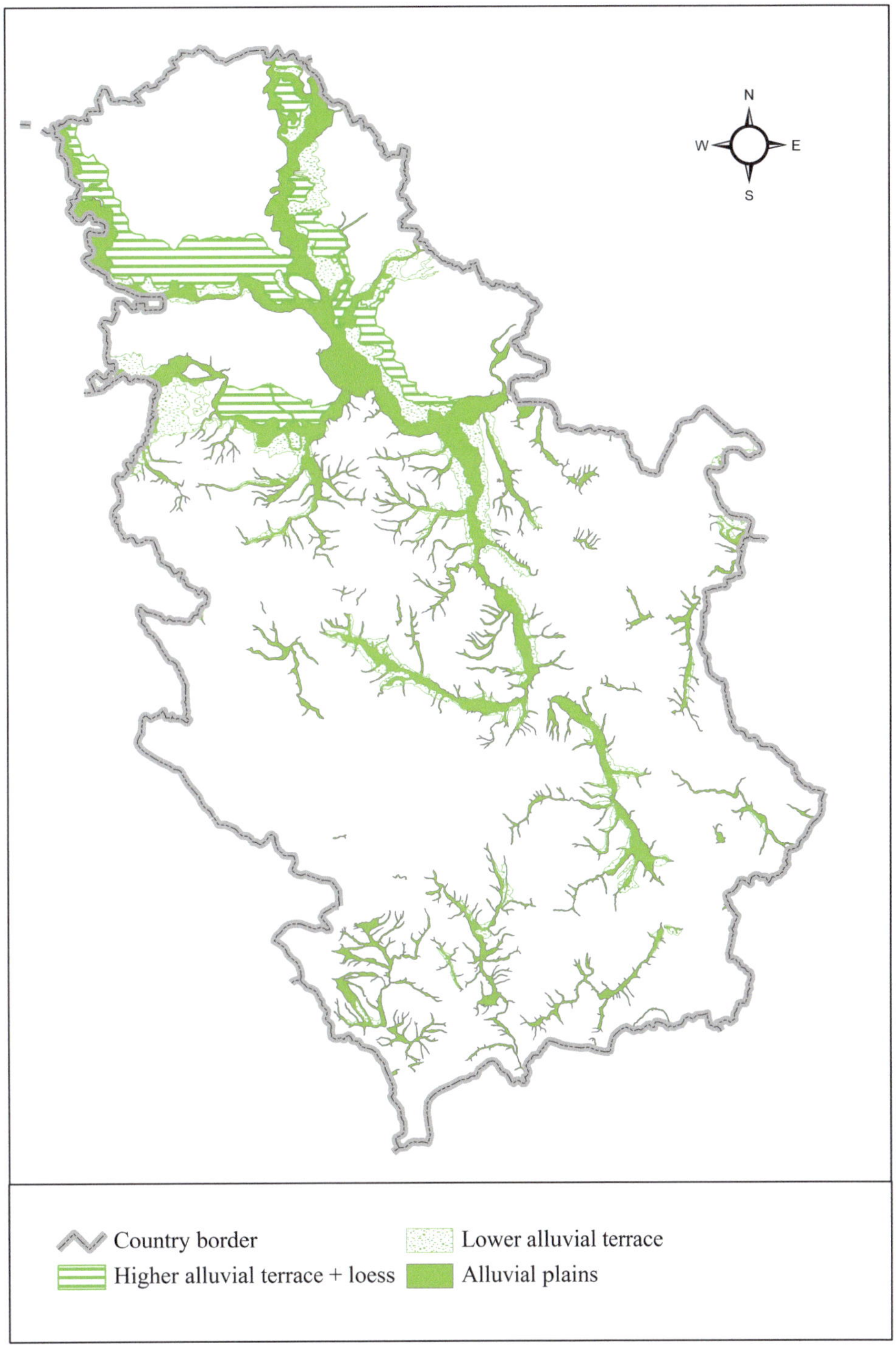

Tisa terrace. The sand plains are composed of fine- to medium-grained sands, fine-grained silty-sands and rarely clayey-silty-sands. Particle size analyses of sands have shown a sand fraction of 80–90%, silt 10–20% and traces of clay, with well to moderately sorted and rounded grains. Chemical analysis has shown variable amounts of calcium carbonate (5–31% $CaCO_3$), with two carbonate minerals present: calcite and dolomite.

Lake and marine deposits cover a large part of the territory south of the Sava and Danube rivers. They outcrop along the Peripannonian margins of the Paratethys, as well as in the mountainous area, where deposits formed in an interconnected system of mountain lakes from the Oligocene to Pliocene Ages.

The Pannonian basin is a depression formed in the Miocene and was invaded by the waters of the Pannonian

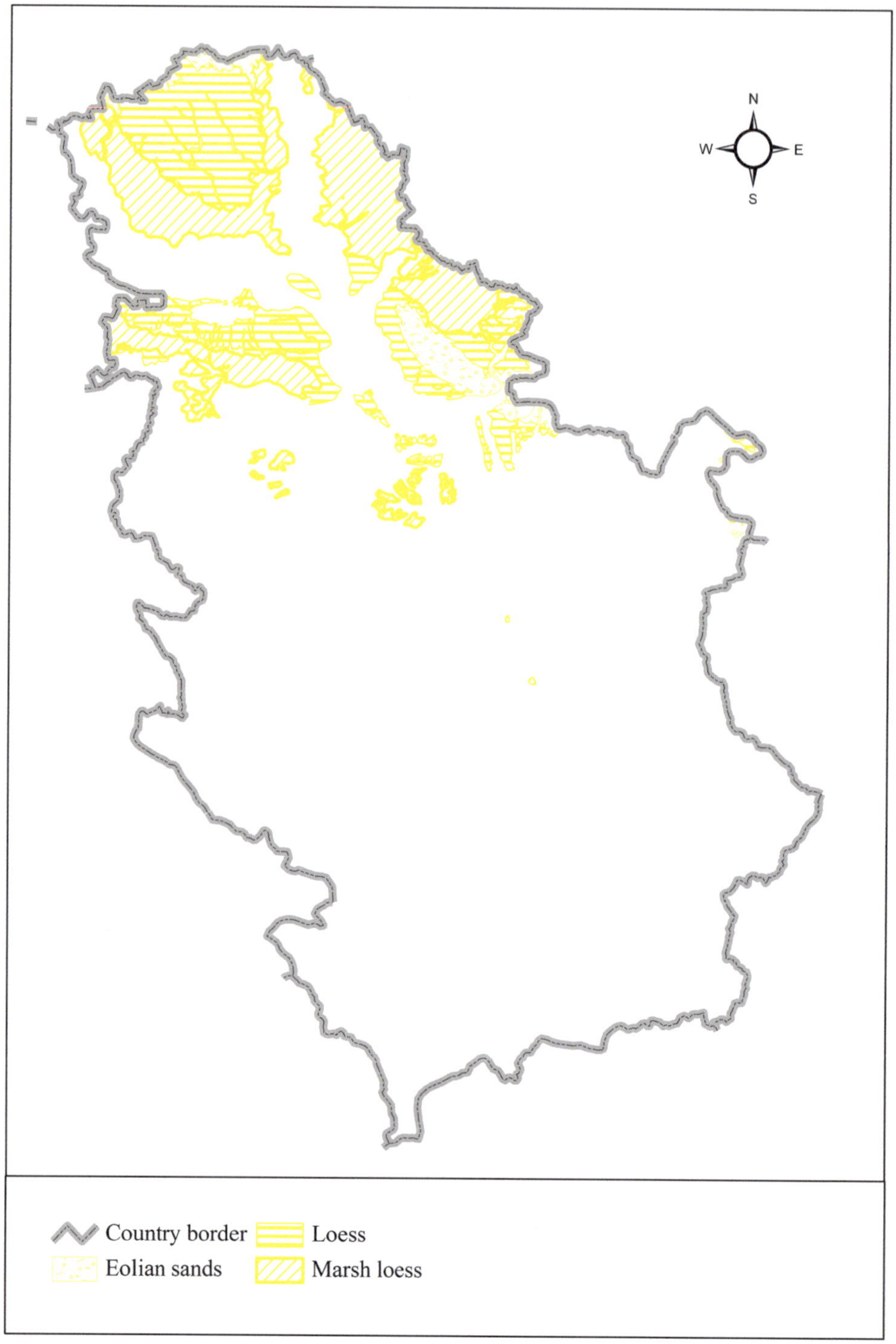

Fig. 5.5 Aeolian deposits

Sea. Thick successions of Neogene sediments consist of coarse clastic deposits (conglomerate, sandstone, shale, clay and reef algal—*Lithothamnium* limestone). In the Dacian depression, there is the Timok graben and Carpathian pediment depressions with marine Miocene outcrops in eastern Serbia (Fig. 5.10).

According to Ćirić (1996), all Tertiary **lake deposits** could be classified as molasse formations which filled the Pannonian and Dacian basins, as well as the adjacent intermountain lakes. The different BGM maps describe the Tertiary deposits as quite diverse in terms of composition and facies. Besides the prevailing clastics of various grain sizes (sand, gravel and conglomerate), there are also marls, limestone, dolomite and magnesite, bituminous shale, and coal, as well as volcanic rocks and their pyroclasts. The thickness of the deposits varies greatly, from 100 to 1000 m.

Fig. 5.6 A river terrace covered with loess near Kovilj, Novi Sad: *1* Danube sands, *2* Loess (*photo* Košćal M., Menković Lj.)

Fig. 5.7 Srem loess plateau with paleosoils near Batajnica (*photo* Košćal M., Menković Lj.)

Fig. 5.8 Srem loess plateau, with an ancient coast line (cliff) near Beška (*photo* Košćal M., Menković Lj.)

It should be noted that hydromica and illite-type clays are common in Posavina, Tamnava and the northern part of the Kolubara basin, while the southern part of the basin also comprises kaolinite, which occurs in large lenses, particularly around the northern fringes of the Bukulja and Brajkovac granites, where fire clays are exploited in the vicinity of Rudovci and Aranđelovac. Further south in Šumadija and around the Velika Morava valley, illite, montmorillonite and beidellite clay types are more common. Even further south, around Leskovac and Niš, as well as in the east Serbian intramountain basins, illite and halloysite are more abundant then kaolinite and smectites. Montmorillonite and bentonite are common in those basins adjoining areas with extensive volcanic activity.

In Šumadija, there are several **lake terraces** descending stepwise from the mountains of Zlatibor and Tara in the south all the way up to the River Sava and Danube in the north. The abrasion terraces are dissected and remodelled by fluvial erosion mass-wasting processes. Lake terraces cut by rivers are also found in eastern Serbia, as well as in the Kosovo and Metohija depression. In southern Serbia, there are several larger lake basins and bays with coal deposits, which were connected to the Aegean Sea during the Tertiary. They are found in Kosovo and Metohija, in the Vranje area and in the Južna Morava valley. The Kosovo and Metohija basins comprise Miocene and Pliocene grey marly and sandy clays interbedded with clastics, marls and carbonaceous clays with several coal seams. Pleistocene marly and ferruginous clays with carbonate concretions are found in Kosovo and Metohija, where they often exhibit horizons of red clays, so-called brand clays of once burned coal and clay (Fig. 5.11).

5.2.2 The Mountainous Region

The mountainous region of Serbia comprises a great variety of sedimentary, igneous and metamorphic rocks.

The formation of Permian red sandstones is one of the most outstanding geological features in eastern Serbia. It extends as two subparallel intermittent zones from Romania across the Danube in the north, along the Carpathian-Balkan Mts. and further on to Bulgaria. The (inner) western zone extends from the Danube valleys along the western margins of the Kučaj Mts. to Mt. Suva planina. It comprises Upper Carboniferous basal conglomerates, sandstone and shale,

Fig. 5.9 Sand dunes in Deliblato Sands near the village of Sušara (*photo* Košćal M., Menković Lj.)

which gradually pass upwards into a Permian varicoloured series of red and white pebbly-quartz sandstones 600–1000 m in thickness, which in turn gradually pass into Triassic white marine sandstones, conglomerates and occasionally lithic arkose and subgreywacke. The eastern zone extends from Veliki Greben to Stara planina. The red sandstone formation on the Stara planina ridge occurs as a gently rolling plate with intercalations of shale and lenses of tuffites and paleobasalt flows.

In western Serbia, siliciclastics are found in an area extending from the Drina in the north to Užice and around Zlatibor and Javor mountains. The succession comprises Carboniferous arenites and greywacke with lenses of argiloschist, which pass into a Permo/Triassic conglomerate, red sandstone and micaceous quartzite. They are overlaid with the Lower Triassic sequence of micaceous and quartz sandstones, about 200 m thick.

On many BGM maps, there are 'undifferentiated' sediments or alternating successions of tightly packed siliciclastic and carbonate beds with relatively rapid vertical and lateral changes of facie and lithology, e.g. as in the case where thin layers of marlstone or shale are interbedded within thick beds of limestone or vice versa—thin layers of limestone within thick marlstone beds. The mixed rocks of this class are found mainly in eastern Serbia along the Bulgarian border, where fine-layered shales alternate with thicker marlstones in beds as thick as 200 m or as thin layers of shale found within thick-bedded limestone, about 600 m in thickness. They also appear on the south-west margin of the Beljanica-Kučaj anticlinorium and then over the eastern slopes of Mt. Rtanj to Stara planina and to the south-east into Bulgaria.

A narrow intermittent zone of shales extends from the Bor pelite in the Timok Eruptive Area, over Metovnica epiclastite formations, south-east of Mt. Rtanj and further south to the border with Bulgaria. This zone comprises bituminous and micaceous shale, silicified pelite and fine tuffite. Several small areas of shale are also found in western Serbia and on the slopes of Fruška Gora. The shales are more frequently found within flysch formations as interbedded with limestone, marlstone and sandstones.

Flysch sediments have specific lithological characteristics as a result of their formation—without much sorting, but

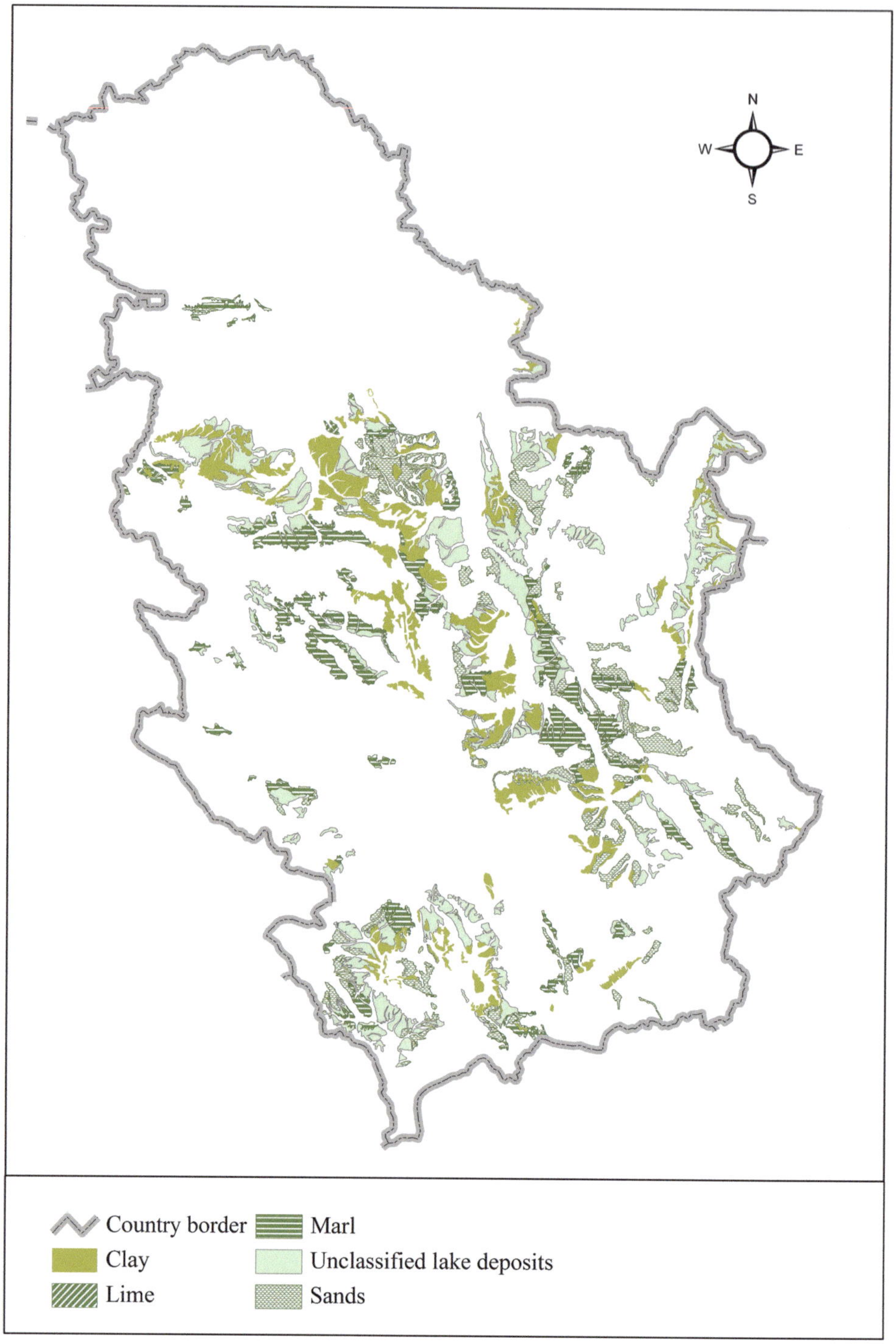

Fig. 5.10 Neogene lake deposits

with graded bedding due to their often fast turbidity deposition in subduction trenches and marine depressions. The sediments comprise psephitic debris flows, polygenic conglomerates, psammitic greywackes with >25% of rock/lithic fragments, marlstones and silt-bearing shales, occasionally with a lot of pyroclastic material. The ages of the flysch formations can vary (Devonian, Jurassic, Cretaceous and Tertiary) and they are found in three distinct zones in Serbia.

The first zone is represented by the large Lower Cretaceous paraflysch of the Central Vardar Subzone, about 250 km long. It extends from Kragujevac and Gledićke Mts. in the north to Kopaonik and further on to Uroševac and Macedonia in the south. The best-studied area of this zone is in the Gledićke Mts., between Kragujevac and Kraljevo. The paraflysch is divided into two mega-sequences and two marginal units. Each mega-sequence is composed of a

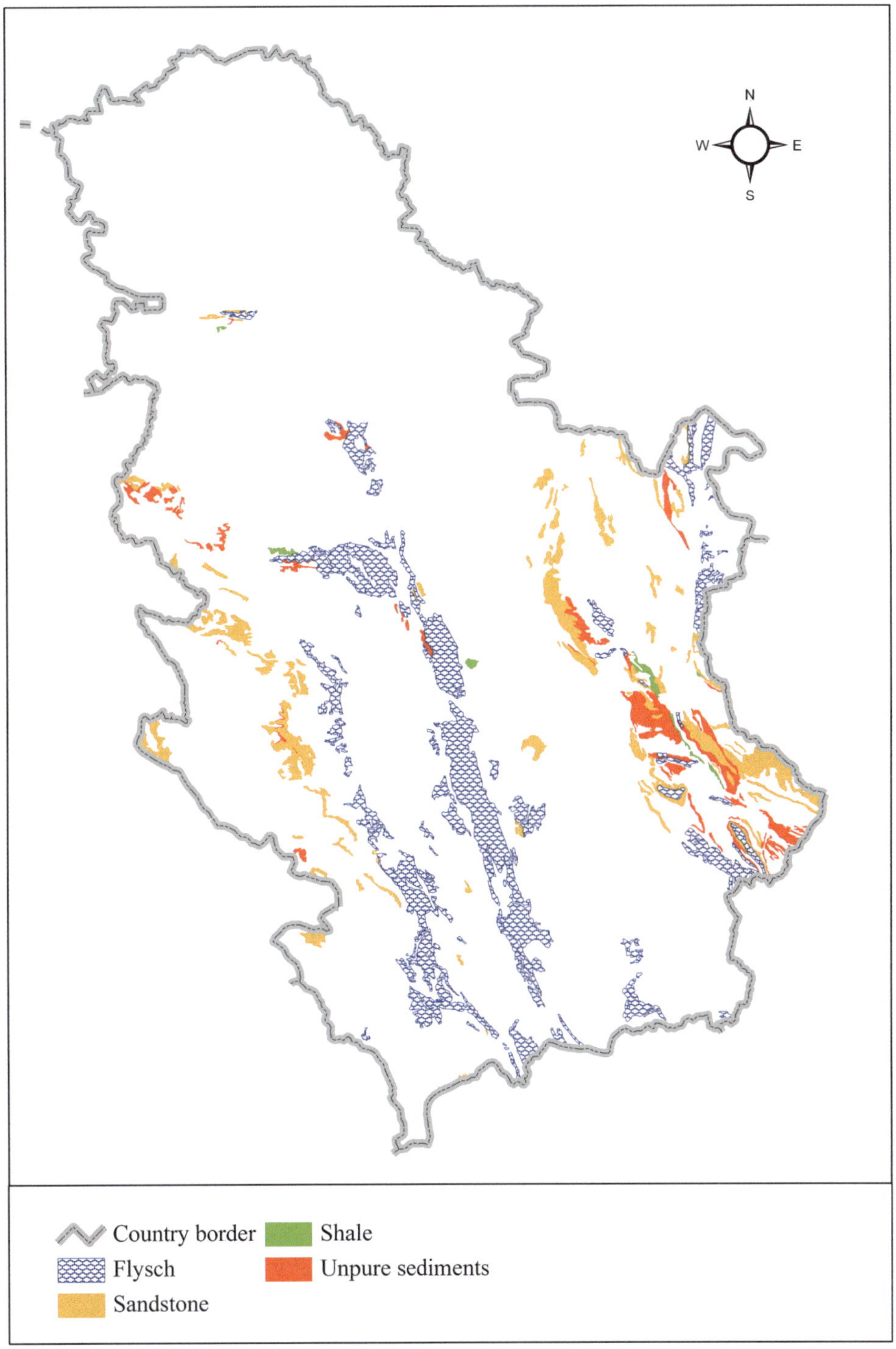

Fig. 5.11 Consolidated clastic sediments

coarse-grained lower and a fine-grained upper unit. The first mega-sequence comprises a coarser-grained unit, while the second one comprises finer-grained units, with an overall thickness of about 1000 m.

The second zone is more or less parallel to the first one and extends from Čačak along the western margin of Kopaonik and further on to Macedonia in the south. The flysch is estimated to be about 1500 m thick, and the depositional basin must have been over 230 km in length.

The third zone—of flysch deposits from the Devonian Age—is found in eastern Serbia. The weakly metamorphosed flysch comprises arenite, siltstone and pelite in two different beds, which are found in the cores of anticlinal or monoclinal forms in Kučaj, Rtanj, Suva planina and Stara planina (Figs. 5.12 and

Fig. 5.12 Veliki Krš, the table mountain (*photo* Košćal M., Menković Lj.)

Fig. 5.13 Rtanj mountain (*photo* Košćal M., Menković Lj.)

5.13). A series of Jurassic/Cretaceous flysch about 2700 m thick appears in south-east Serbia on the slopes of Mt. Ruj. The sediments are layered as follows: conglomerate, sandstone grading to shale, followed by beds of pebbles and a thin layer of black shale, which is overlaid with thick beds of arenites, marlstone and shale, 1000 m thick. In eastern Serbia, there is also Senonian Mokranje flysch, which extends from Romania across the Danube and Mt. Miroč to the plateau of Vrška Čuka and further on into Bulgaria, with an estimated thickness of 250 m near Donji Milanovac and 1400 m near Bor.

Carbonate rocks (limestone and dolomite) are frequently encountered in western and eastern Serbia and can be from any Age, from the Palaeozoic to the Cainozoic. They occur as karstified areas of mero-karst, in the Dinarides and Carpatho-Balkan mountains. The carbonate content in these rocks is generally over 50% and mostly around 80%, but the Mesozoic reef limestone in the Dinarides tends to be the purest carbonate.

Palaeozoic carbonate rocks are present in western Serbia, extending from the Drina and Valjevo in the north, across the Kraljevo area to the Prokletije Mts. in the south. The oldest are the Devonian yellowish cavernous or massive limestones; the Carboniferous is thick-bedded fusulinid limestone, while Triassic carbonates cover large areas in the Dinarides and comprise nodular, flaggy, massive limestone and dolomitic limestone. Jurassic/Cretaceous carbonate rocks are only found as a narrow zone of sandy-limestone and thin-bedded limestone, about 20–30 m thick, in the Dinarides, as well as a series of marly limestone and clayey sandstones, which occur over the Ophiolitic melange.

Unlike the Dinarides, in eastern Serbia, there is complete development of Jurassic sedimentary formations, which starts with calcareous sandstones and sandy-limestone and gradually passes into flaggy and thin-bedded limestone with some dolomite, which extends further south with widespread greyish-green and red ferruginous sandy- and marlylimestones. They occur mostly along the border with Bulgaria from Vrška Čuka to the south-western parts of Stara planina.

The Cretaceous system of the Dinarides and Carpatho-Balkanides exhibits great differences in all its formations, apart from the Senonian deposits. In the Dinarides, the Lower Cretaceous deposits are mainly clastic–psephitic, psammitic and flysch in character, while in eastern Serbia they are dominantly carbonate. The Upper Cretaceous, Senonian formations in western Serbia comprise the flysch sediments, which also occur in eastern Serbia and across the Serbian Massif.

In eastern Serbia, carbonate rocks cover large areas, making limbs of anticlinal forms. They comprise fine- to thick-bedded, rarely massive, crystalline limestone and dolomitic limestone. Sandy-limestones with thin intercalations of marlstone are found more to the west, while to the east there are flaggy and thick-bedded limestones with occasional chert nodules and layers which gradually pass into marlstone with intercalations of shale.

Dolomite and dolomitic limestone are categorized into a separate carbonate class due to their specific type of weathering and their role in soil formation. It appears from the map (Fig. 5.14) that limestone prevails more in the north and dolomite gradually increases to the south of the Dinarides in western Serbia. In south-west Serbia, the large masses of Triassic thick-bedded and massive dolomite and dolomitic limestone, about 1200 m thick, are found on Mokra Gora and further south to the Prokletije Mts., Mt. Koritnik, and on the northern slopes of Mt. Šar. Further north, around the ultrabasites of Mt. Maljen, there is a series of dolomite and dolomitic limestone, about 200 m in thickness. Some of the Triassic limestone intruded with porphyry may contain layers of pyroclastics and is often dolomitized or completely transformed into dolomite.

In eastern Serbia, the dolomite content also increases to the south, with dolomitic limestone and dolomites covering larger areas from the north of Beljanica and Kučaj across Mt. Rtanj and Suva planina and then to the east to Stara planina and into Bulgaria. Bituminous limestone and dolomite, as well as Triassic and Jurassic sandy-limestone and dolomite with interlayers of sandstone and shale, about 250 m thick, are found on Stara planina.

There are several types of mixtures between siliciclastic and carbonate sediments. The most common is marlstone, ideally composed of almost equal portions of clay and carbonate ($CaCO_3$) components. Most of the marlstone occurring in Serbia is bound to the flysch series with generally considerable intercalations of marlstones and limestone. A series of variegated marlstones with intercalations of sandstone with nickeliferous oolitic iron ore, over 1000 m thick, is found along the western margins of Mt. Tara, around Vardište and Mokra Gora. Marlstone is also found in an intermittent zone, parallel to Senonian flysch, from Kosjerić in the north through Požega to Ivanjica and Novi Pazar, and shows variable clastic proportions of predominantly quartz with minor feldspars, micas and chert. Clay particles are predominately illite with minor kaolinite and/or smectites.

In eastern Serbia, marlstone occurs in a few places within the Suprageticum Zone and on Stara planina. The northern limb of the Talambas anticline is overlaid with monotonous Lower Cretaceous marlstone 1000 m or more in thickness. From the Danube to Mt. Miroč, there is a zone of flaggy limestone alternating with marlstone and shale up to 250 m thick, passing upwards into marlstone and shales intercalated with marly limestone, about 350 m in thickness.

Metamorphic and igneous rocks are an important parent material for soil formation since they cover about 28% of the Serbian land mass. The metamorphic rocks are divided into three groups: acid and basic regional metamorphic rocks and weakly metamorphic rocks. They are mainly found in the Serbian Massif and to a lesser extent in the Dinarides and Carpatho-Balkan mountains. From the Vršac Mts. in the north along the Serbian Massif to the southern border with Macedonia and Bulgaria, there is no crystalline schist without granite intrusion, in particular gneiss masses with migmatite where migmatization is not a product of granite intrusions. The Serbian Massif is composed of two crystalline complexes.

The lower complex of Precambrian age, in the west, is composed of acid regional metamorphic rocks of medium- to

Fig. 5.14 Carbonaceous sediments

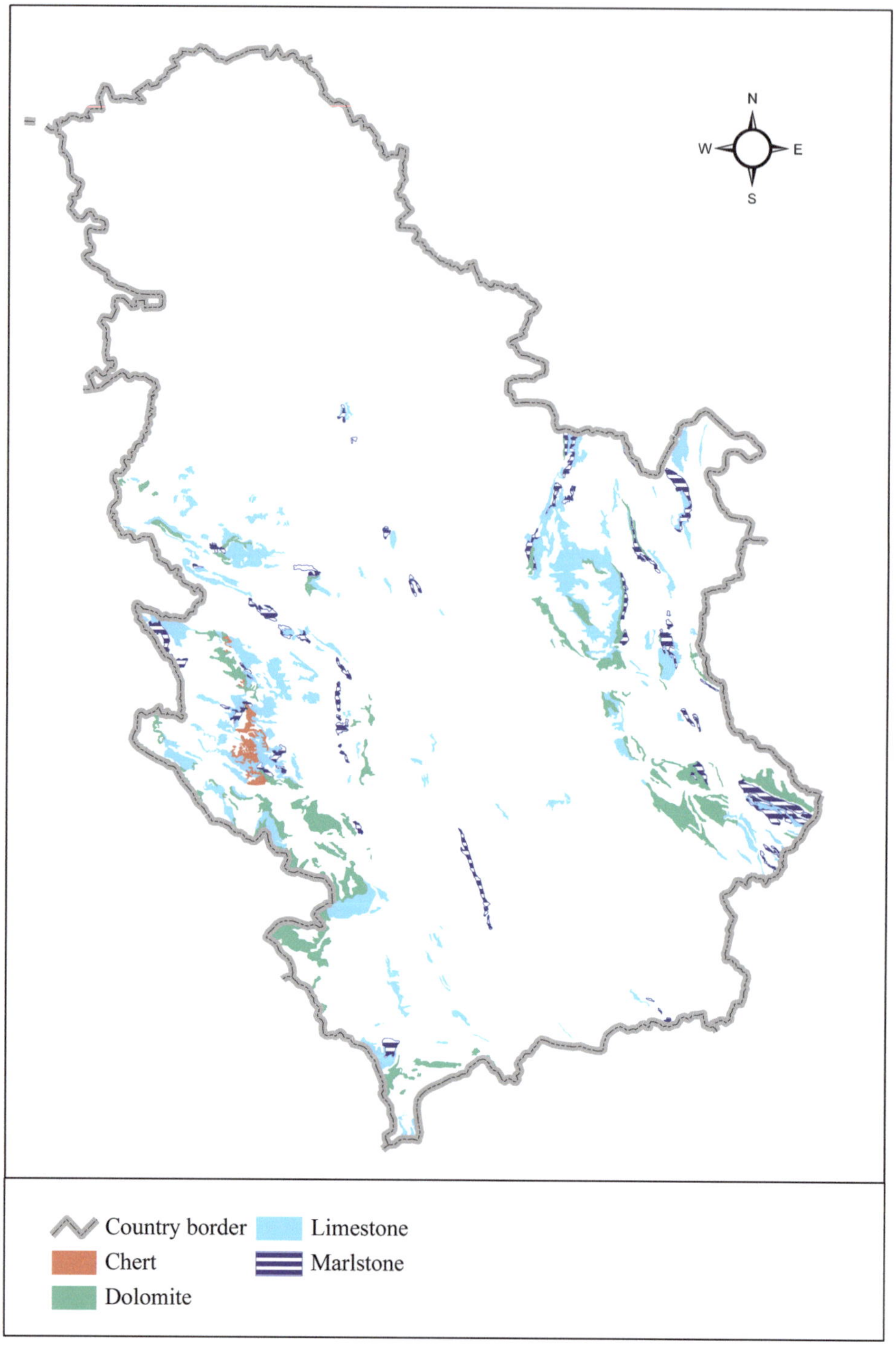

high-grade amphibolite facies. It comprises a series of very thick (over 9 km), fine-grained gneisses and quartzite with numerous pegmatite veins and lenses. This is overlaid with lepthynolithe or feldspathic mica-schists, with intercalation of dolomitic marbles and migmatites, which are found in an area from Crni Vrh in central Šumadija in the north, across mountains near Leskovac, Vranje and Bujanovac in the south, extending over the border with Macedonia to Skopska Crna Gora (Fig. 5.15).

The lower complex is intruded by old Caledonian and Hercynian granites and Alpine granitoids, as well as Tertiary andesite and dacite. The polyphase magmatic activity of the granite pluton progressively affected adjacent crystalline schists, leading to a gradual transition from fine-grained

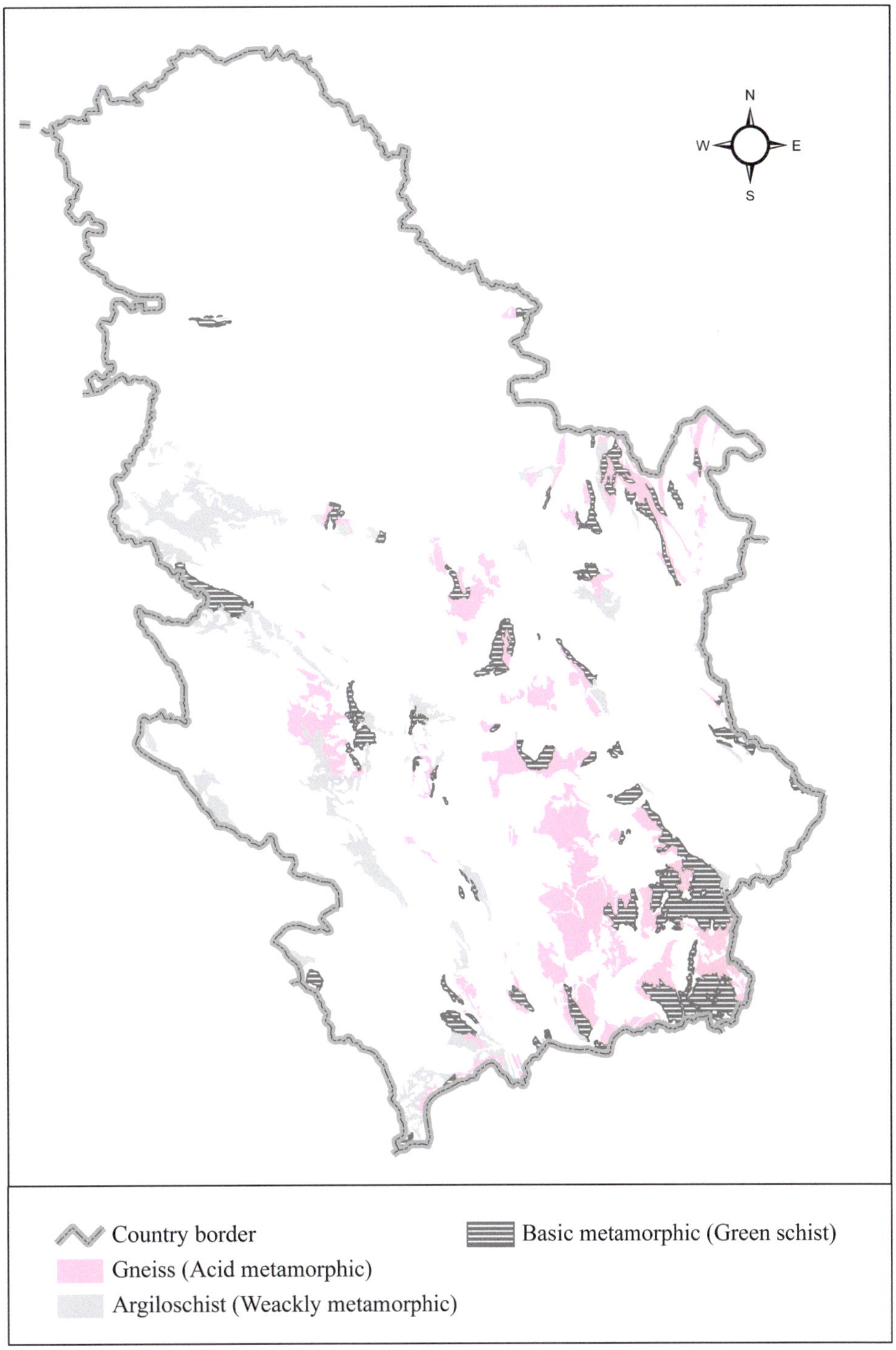

Fig. 5.15 Metamorphic rocks

gneiss to migmatite and later on to the formation of gneiss-granite.

The Upper or Vlasina Complex of the Serbian Massif lies unconformably over the lower complex. It mainly comprises basic regional meso-metamorphic rocks of medium-grade greenschist facies, which consist of greenschists, amphibolite, chlorite-epidote, chlorite-actinolite and amphibole schists, eclogite, calc-schist and talc-schist of Riphean-Cambrian and Palaeozoic age. It is covered by weakly metamorphosed Ordovician to Carboniferous strata and is intruded by the Palaeozoic granitoids and Tertiary granodiorite of Surdulica, which stitch it to the lower complex.

The area which has been studied most lies around Lake Vlasina, which gave the complex its name. The Vlasina Complex consists of greenschists with a mineral composition of chlorite, albite, epidote, actinolite, quartz and sericite.

In the north, part of this complex is metamorphosed to amphibolite facies and includes intercalations of albite-chlorite, albite-actinolite and albite-epidote schists and, rarely, mica-schists. The exposed parts of the geological column are about 5 km thick. Rocks of the greenschist facies are found in the area around Lake Vlasina and Besna Kobila Mts. to the border with Macedonia.

Further north, several keratophyres and quartz keratophyres cut across greenschists (albite-chlorite and chlorite schists associated with meta-psammites), which crop out in several areas separated by Tertiary basins. The majority of Mt. Juhor is composed of two-mica schists, about 2000 m thick, which are invaded by a dense swarm of amphibolite veins with rare marble and quartzite lenses. Further north, similar metamorphic rocks (fine-grained amphibole schist with lenses of tremolite-antigorite and calc-schists) are found on Crni Vrh.

In eastern Serbia, a series of greenschists, about 2100 m thick, occurs in the north Homolje Mts. This area comprises actinolite and actinolite-epidote schists, with lenses of antigorite schists and quartzite with garnets. The actinolite schist passes marginally into greenschists with lenses of meta-sandstone and graphitic schist or chlorite-actinolite schist with intercalations of marble.

Weakly metamorphic rocks are mainly found in Palaeozoic formations comprising argilophyllite in association with phyllite and sericite schist, as well as graphitic schist and mata-sandstones. The rocks were exposed to low-grade anchimetamorphism and often have the prefix *meta-* followed by the name of a protolith, e.g. meta-sandstone. In western Serbia, large masses of such rocks belong to the Drina–Ivanjica Element. It extends intermittently from the River Drina and Valjevo to Čačak and further south across Ivanjica, Pešter and Kosovo to Orahovac and Prizren in Metohija. The formation comprises sericite-phyllite, phyllite, and slates, with intercalations of meta-sandstone and meta-volcanite with lenses of sericite and epidote-actinolite schist and calc-schist of the Late Cambrian to Carboniferous periods, with an overall thickness of about 2700 m. In the Ivanjica region, this formation progressively metamorphosed to gneiss, lepthynolithe, biotite-mica schist, biotitized phyllite, and albite-sericite schist by the shallow granodiorite intrusions on Mt. Golija. During the Miocene, the rocks on Mt. Čemerno were also progressively metamorphosed by the intrusion of Polumir granitoid to lepthynolithe, biotite-mica schist, amphibolite and phyllite. Igneous rocks are often encountered in the mountainous parts of Serbia, particularly on the peaks and along the ridges of the mountains (Fig. 5.16). They are the product of several magmatic phases from the Precambrian, Palaeozoic, Mesozoic and Tertiary periods, with their formation generally linked to the Caledonian, Hercynian and Alpine tectonic epochs.

Acid and intermediate plutonic rocks comprise mainly granites, granodiorite and quartz-diorites (tonalite), but locally also diorites. Most granitoids also contain aplite, pegmatite, porphyrite and lamprophyre dykes and quartz veins, which branch from granite into surrounding rocks. A contact aureole, up to 300 m wide, occurs around all granitoid intrusions, with thermo- and contact metamorphic rocks such as skarn, kornite and spotted phyllite, or hydrothermal alterations such as silification, kaolinization and pyritization. The largest masses of acid and intermediate plutonic rocks are found mainly in the Carpatho-Balkan Mountains of eastern Serbia, the Serbian Massif and the Vardar Zone in western Serbia.

The largest Hercynian granite in Serbia is the Bujanovac pluton, which intrudes the crystalline schist of Mt. Skopska Crna Gora in the south and Jablanica in the north. It is a composite intrusion with a core of granodiorite, which migmatized adjacent rocks, leading to anatexite and granite lenses in the lower complex of the Serbian Massif.

In south-east Serbia, the Surdulica granodiorite pluton stretches along the western slopes of Besna Kobila and Vardenik mountains, with a large contact aureole, which stitches together the Lower and Upper metamorphic complex of the Serbian Massif. This is the largest of all the young (Miocene) intrusive bodies in Serbia. In fractures, the Massif comprises dykes kilometres and decakilometres in size and larger bodies of quartz-latite, as well as smaller dacite and andesite veins.

There is a group of several Hercynian granitoids of eastern Serbia situated in the axial zone of the Dardanian mountain range. The largest among them is the Gornjane granitoid batholith, concordantly intruding surrounding schists along the northern ridge of Mt. Stol. It comprises biotite quartz-monzonite, surrounded by a belt of granodiorite with an extensive contact aureole, which bears lengthy veins of aplite granite and syenite.

The oldest rocks of the Homolje Mts. are Proterozoic gneisses and lepthynolithes, intruded by the aplite two-mica granite of Ziman and the granodiorite to granite-monzonite of Nersnica and Brnjica. The intrusions developed large contact aureoles and migmatized adjacent rocks to banded and augen-plagio-gneiss with ribbons of amphibole gneiss.

In western and central Serbia, there are several younger intrusions from the Oligocene–Miocene epochs which intrude the Vardar Zone. Mt. Cer composite pluton, with quartz-monzonite, granodiorite and tourmaline granites, as well as Mt. Boranja granodiorite, are stitching plutons, linking the Drina–Ivanjica Element to the Vardar Zone. They produced extensive contact aureoles in the Palaeozoic schists, which metamorphosed into spotted schist, kornite and hornfels in a belt up to 400 m in width. Further east, the series of limestone and clastics are intruded by Brajkovac granodiorite and the

Fig. 5.16 Igneous rocks

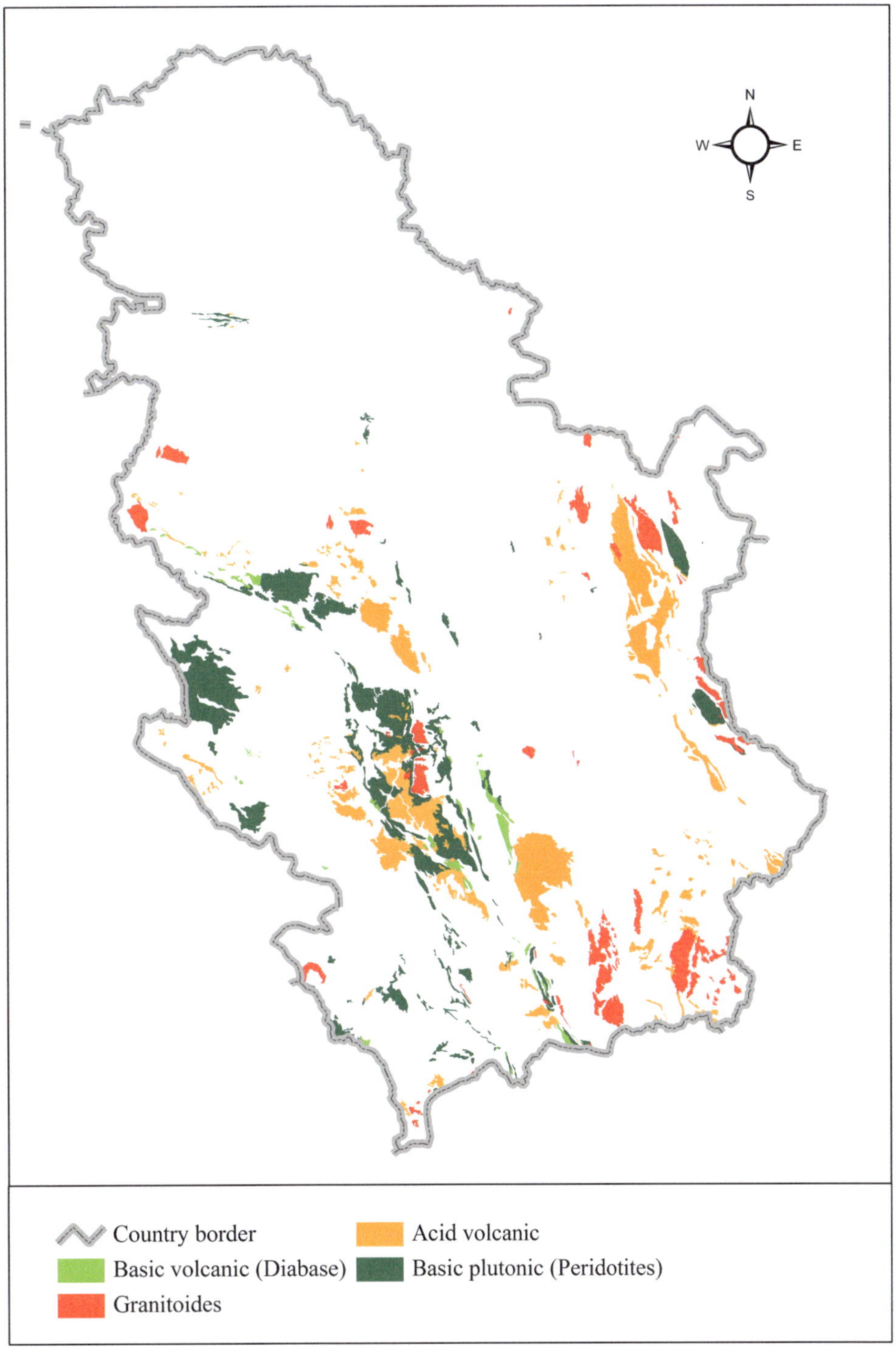

Bukulja composite intrusion of granite, monzonite and granodiorite, which produced contact aureoles with phyllite and large masses of white marble at Venčac. Further south, there are shallow concordant intrusions of Mt. Željin normal-grained granodiorite and Kopaonik granodiorite, quartz-monzonite and granite, which form a dome. The phyllites of the central Kopaonik series are enriched in feldspars around the massif and epidote-actinolite schists are hydrothermally altered.

Peridotites are the most common igneous rocks found in large masses in the Dinarides in western Serbia. They mainly comprise harzburgite and rarely lherzolite and dunite, which are often associated with gabbro, diabase and spilite pillow lavas. They are found in three distinctive generally parallel, but intermittent zones.

The first zone stretches from the Drina and Zlatibor to the Prokletije Mts. in south-west Serbia. It comprises the large serpentinized peridotite massifs of Zlatibor and Ozren mountains and several smaller masses in Kosovo and Metohija and the eastern parts of the Prokletije Mts. The second zone extends from Zvornik on the River Drina to Maljen and Suvobor and then to Stolovi, Goč and Kopaonik and along the Ibar river valley to Brezovica and Raduša on Mt. Šar. The third zone runs from Fruška Gora in the Pannonian basin to Avala near Belgrade and across Stragari near Kragujevac to the Toplica valley and then further south to Macedonia, marking the boundary between the Vardar Zone and the Serbian Massif.

The Zlatibor Ultramafic Massif, the largest in Europe, is composed mainly of harzburgite, subordinate lherzolite and dunite, which are serpentinized mostly on the margins. Along the margins gabbro, olivine gabbro and troctolite (plagioclase containing ultramafite) are also found, as well as diabase and spilite, with abundant amphibolites in the sole of the nappe. The large ophiolitic masses of Maljen and Suvobor represent a complete ophiolite sequence, from layered harzburgite at Divčibare to peridotite cumulates and troctolite and gabbro passing into a sheeted diabase dyke complex with spilite pillow lavas. The peridotite is often serpentinized, while gabbro and diabase are usually transformed to rodingite by metasomatic alteration. Ultramaphites play an important role in the Kopaonik block, where their contact with adjacent rocks is almost exclusively tectonic. The largest is the Ibar Massif (Stolovi and Goč), which lies west of Mt. Željin and stretches to Trepča in the south.

The Ophiolite Belt comprises large masses of serpentinized peridotites in the Dinarides and is closely associated with the Diabase-Chert formation, which comprises a mixture of sedimentary, igneous and metamorphic rocks. Igneous rocks are mainly ophiolitic, and predominantly blocks of serpentinite, diabase and spilite, frequently as pillow lavas and less frequently basalts. Gabbro occurs only as small masses in the mélange and is usually associated with larger ultramafic bodies. Blocks of poorly stratified sandstones are also encountered in many places, as well as inclusions of metamorphic rocks and flysch turbidite and thin-bedded limestone with chert. The largest area of the diabase-chert formation extends from Užice around the Zlatibor Massif across Pešter to the Prokletije Mts. in the south and further on to the Albania Mirdita Zone. Another zone stretches from Zvornik and the Drina, across Povlen, Maljen and Suvobor mountains, and then along the west side of Kopaonik to Kosovo in the south.

In eastern Serbia, the gabbro and serpentinite bodies are located in several places beneath Gethian crystalline schists. Most of the smaller bodies, however, consist of highly schistose serpentinite as thick lenses squeezed within dislocations. The largest of these are the Deli Jovan and Zaglavak gabbros, which also include other rocks: anorthosite, allivalite, dunite, gabbro-pegmatite, diabase, etc.

In present-day Serbia, there are no active volcanos, but the presence of volcanic rocks and their pyroclasts over an area of 3800 km^2 undoubtedly points to their once intensive activity. Cainozoic-type vulcanite comprises dacito-andesite, andesite-basalts and, rarely, basalts and their respective pyroclastics. The volcanic rocks stretch intermittently from Fruška Gora in the north, then wide across Šumadija to Kopaonik, and further on to Mt. Radan and to Vranje and Binička Morava in the south. In eastern Serbia, there is the large Timok Eruptive Area, which extends from Majdanpek in the north to Pirot and further south to Bulgaria, and also a few smaller masses elsewhere (Fig. 5.17).

The Timok Eruptive Area (TEA) is the largest eruptive area in Serbia, covering 780 km^2. It stretches from Majdanpek in the north, across Crni vrh and the Bor area, to Sleme and Knjaževac in the south. In its northern part, the zone is 20–30 km wide, but further to the south, from Knjaževac, the zone forms a relatively narrow strip, in places less than 1 km wide, emerging in the south at the border with Bulgaria, where small andesite bodies are found in the form of scattered volcanic clusters. Volcanic activity started with a strong eruption of andesite, hornblende-biotite andesite or 'timozite', dacite and trachyte with abundant pyroclasts, which produced a series about 200 m in thickness. In the Paleogene, volcanism ended with a very strong eruption, which produced mainly pyroclasts and rare flows of andesite-basalt and finally latite. The estimated thickness of the series is about 3000 m, with about 80% pyroclasts.

The mountain area between Kuršumlija and Leskovac comprises the complex volcanic massif of Mt. Radan, which extends over 600 km^2 (Fig. 5.18). It is composed of large masses (over 75%) of pyroclasts interlayered with hornblende-andesite and pyroxene andesite flows, with rare quartz-latite and dacito-andesite. Post-volcanic activity even today is noted in the form of hot springs, spas and solfataras in Sijarinska Banja.

Neogene polyphase volcanic activity with explosive eruptions in the Rudnik-Borač-Kotlenik area produced an extrusion of quartz-latite, dacite and andesite, as well as trachyte and basalts flows with about 60% stratified pyroclasts across about 350 km^2.

The extensive volcanic activity during the Miocene was connected to the shallow granitoid intrusions at Kopaonik and Golija mountains. Polyphase synchronous explosive volcanic activity comprises dacito-andesite and quartz-latite with coarse-grained crystals of sanidine (5 cm long), which is followed by large masses of pyroclasts along the Ibar valley, on Mt. Rogozna, and in part of the Drina–Ivanjica Element, west of Kopaonik.

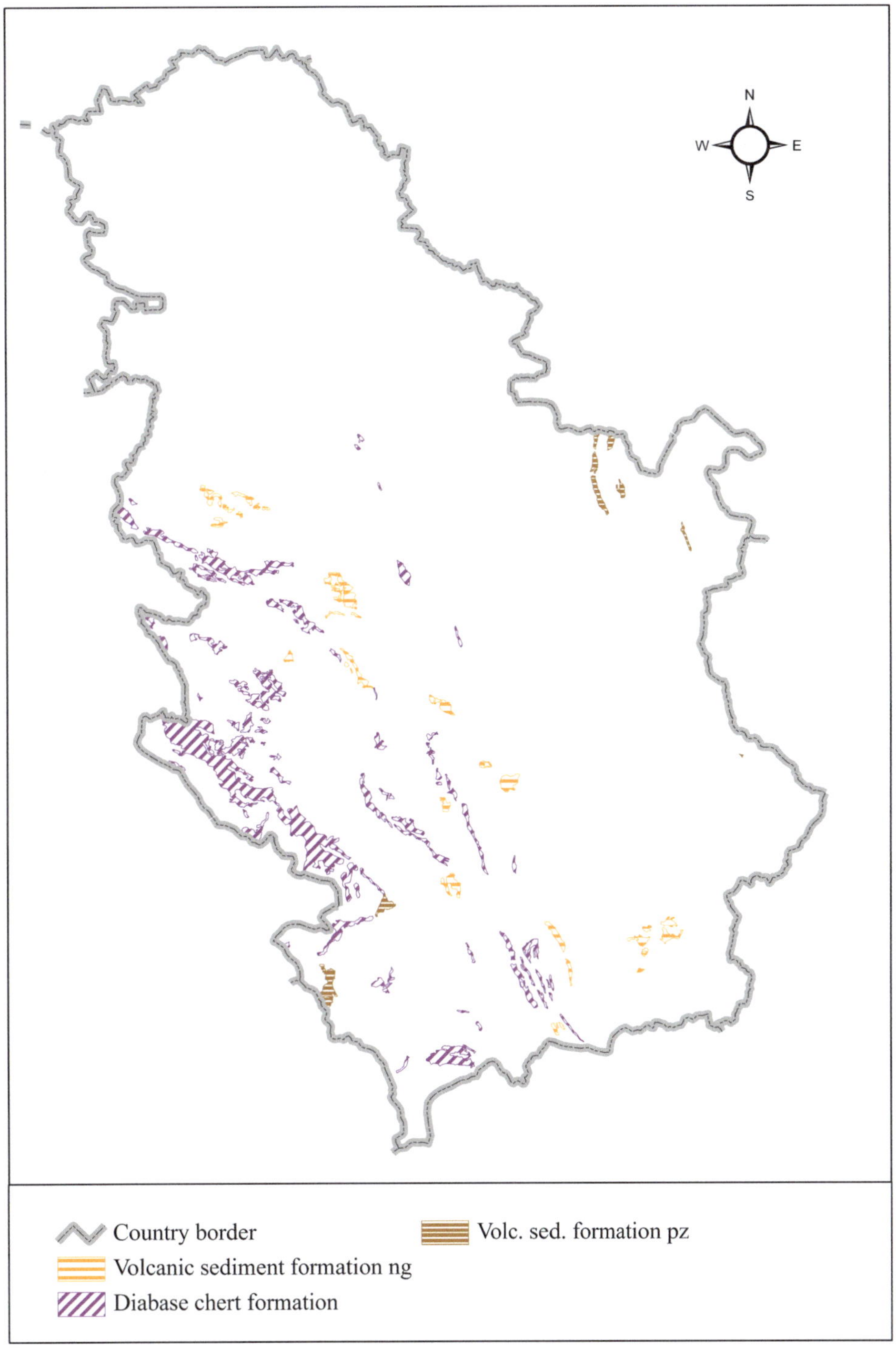

Fig. 5.17 Volcano-sedimentary formations

The class of 'complex lithology' or volcano-sedimentary formations contains intimately mixed rock associations in the inner zones of young and medium-aged orogens. Rocks of this class are present in Serbia and can be divided according to their ages into Palaeozoic, Mesozoic and Tertiary formations.

Palaeozoic volcanogenic-sedimentary formation occurs in eastern Serbia and comprises argiloschists interlayered with meta-conglomerate, meta-sandstone with lenses of sandy-limestone, slates and chert/lydite, which are interbedded with lava flows of quartz-porphyry, diabase, and porphyrite, rhyolite and trachyte and their pyroclasts.

Triassic porphyrite-chert formation is found in the inner zones of the Dinarides, with varying thicknesses of several hundred metres. The area that has been studied most is between the Drina and Valjevo, where the effects of volcanic

Fig. 5.18 'Devil's castle' volcanic tephra of Mt. Radan, near Kuršumlija (*photo* Košćal M., Menković Lj.)

activity on sedimentation consist of the interchange of fine-layered cherty limestone and dolomite associated with submarine flows of porphyrite, andesite and keratophyre, interbedded with volcanic agglomerates, tuffs, red shale, marlstone, variegated chert and green tuffites, so-called pietra verdes.

Tertiary volcano-sedimentary formation is found in the Dinarides and the Serbian Massif. The series comprises vari-grained clastics, dolomitic limestone, dolomite and magnesite, which are associated with volcanic flows of dacito-andesite, rare trachyte and trachybasalt. The formation also comprises alterations of fine-layered shale and marlstone with green tuffite and shale; red, green, grey and white chert, up to several hundred metres thick.

The volcano-sedimentary formation of the Vranje basin comprises sand, bentonite clay and marlstone with intercalations of andesite, altered dacite and pyroclastics. The bentonite clays are alteration products and devitrification of volcanic ash in subaquatic conditions. The formation comprises several occurrences and the deposit of bentonite, over 20 m thick.

5.3 Soil Mineralogy

During large-scale sample collection financed by the Ministry of Science and Technology, an orthogonal regular 10 × 10 km grid was used to avoid bias at the site location. From a set of 4950 samples, a selection of samples from Vojvodina and central Serbia was taken to represent the most important soils. The mineralogy of the clay fraction and bulk soil was obtained by microscopic image and X-ray diffraction

Table 5.2 Average mineral composition of the clay fraction of various soils

Soil type	Illite	M-L-S (10–14)	Chlorite	M-L-S (14–14)	Vermiculite	Smectite	Kaolinite	Goethite	Talc
Pannonian basin									
ARS	67.80	7.25	1.83	1.38	2.87	3.45	4.90	0.46	–
CHZ	60.44	6.83	5.05	1.22	4.32	8.65	5.63	0.64	–
SEG	51.10	7.31	6.27	2.58	4.71	17.70	5.97	0.86	–
HUG	48.73	10.18	7.92	1.14	4.19	21.04	5.07	1.43	–
HAL	61.55	8.85	7.00	2.03	3.20	4.86	5.02	0.75	–
FLU	46.43	5.49	7.48	1.32	4.26	23.66	4.60	0.68	–
EUG	24.89	4.85	8.66	0.24	2.43	57.73	4.41	1.33	–
Peripannonian region									
HUG	41.83	10.77	6.13	1.70	7.35	25.68	5.40	1.58	–
CHZ	56.89	6.32	9.91	1.08	5.05	14.12	5.44	0.27	–
SEG	44.82	5.35	5.47	1.87	5.81	22.69	6.59	0.39	0.18
FLU	42.62	9.12	6.51	1.32	5.06	24.48	8.39	0.53	0.33
PSG	26.72	6.81	10.79	3.57	6.84	10.87	8.04	0.97	–
VRT	32.16	10.15	5.29	2.12	5.70	28.14	13.21	0.81	0.15
ECS	40.09	9.66	4.75	2.17	7.32	19.30	13.01	0.94	0.26
LUV	54.60	3.43	4.09	4.10	9.44	10.85	10.93	2.18	0.68
DCS	31.24	14.16	4.16	1.17	10.91	11.14	7.70	2.34	1.03
RNK	14.13	3.57	6.04	2.19	12.26	10.54	18.71	3.17	5.20
RDZ	58.44	6.42	7.42	1.42	6.30	9.72	7.97	2.01	0.61

analyses, respectively. The obtained data were statistically analysed, and raster maps were drawn up with five classes showing mineral distribution.

The Pannonian basin comprises loess and loessoid deposits, which form loess plateaus and alluvial plains and terraces. In the Peripannonian region, most of the area is covered by Neogene lake deposits, which are partially covered with loess, as well as alluvial plains, while the periphery is composed of a variety of rocks.

A mineralogical study of the soils in Serbia has led to the conclusion that most of the minerals are inherited from the parent material or geological substratum, which predetermines the composition of soils, in both the initial and mature phases of development.

Bulk soil comprises dominant quartz and micas, associated with minor amounts of, in descending order: feldspars (plagioclase and orthoclase), biotite, chlorite, carbonates (calcite and dolomite) and clay minerals (smectite, kaolinite, vermiculite and mixed-layer-silicates). In the clay fraction of soils, illite, smectite and kaolinite are predominant in association with chlorite, vermiculite, mixed-layer silicates and minor amounts of goethite and talc.

The climatic conditions of the Pannonian basin and Peripannonian Serbia, with a temperate climate and relatively low precipitation, are advantageous for the biogeochemical alteration of primary minerals (mica, chlorite and feldspar) through their transformation to clay minerals and the formation of illite, smectite and kaolinite and to a lesser extent vermiculite during the calco-, humo-, hallo- and ferro-forms of the bisiallitisation processes of weathering. From the data presented in Table 5.2, it can be seen that the quartz content varies between 33 and 72% in bulk soils, showing its geomorphological position in the relief. It is the most abundant mineral (72%) in a arenosol developed on the aeolian sands of Subotica and Deliblato Sands, as well as luvisol and cambisols developed on Mesozoic flysch sediments and Palaeozoic schists. Chernozem developed on loess, plateaus and terraces around the sand plains contains on average 52–56 quartz. Eugley (WRB, Gleysol), humogley (WRB, Gleysol, humic) and humofluvisol (WRB, Fluvisol, humic) developed on the alluvial plains of Vojvodina, as well as the vertisol on Neogene lake deposits in Šumadija, contain 45–50% quartz. The lowest amounts of quartz (less than 45%) are found in the eugleys (WRB, Gleysol) from the marshes of Bačka and in a few leptosols developed on metamorphic rocks and limestone along the Peripannonian rim.

The second most abundant soil minerals are micas, mainly muscovite and rarely biotite, which occur within the range of

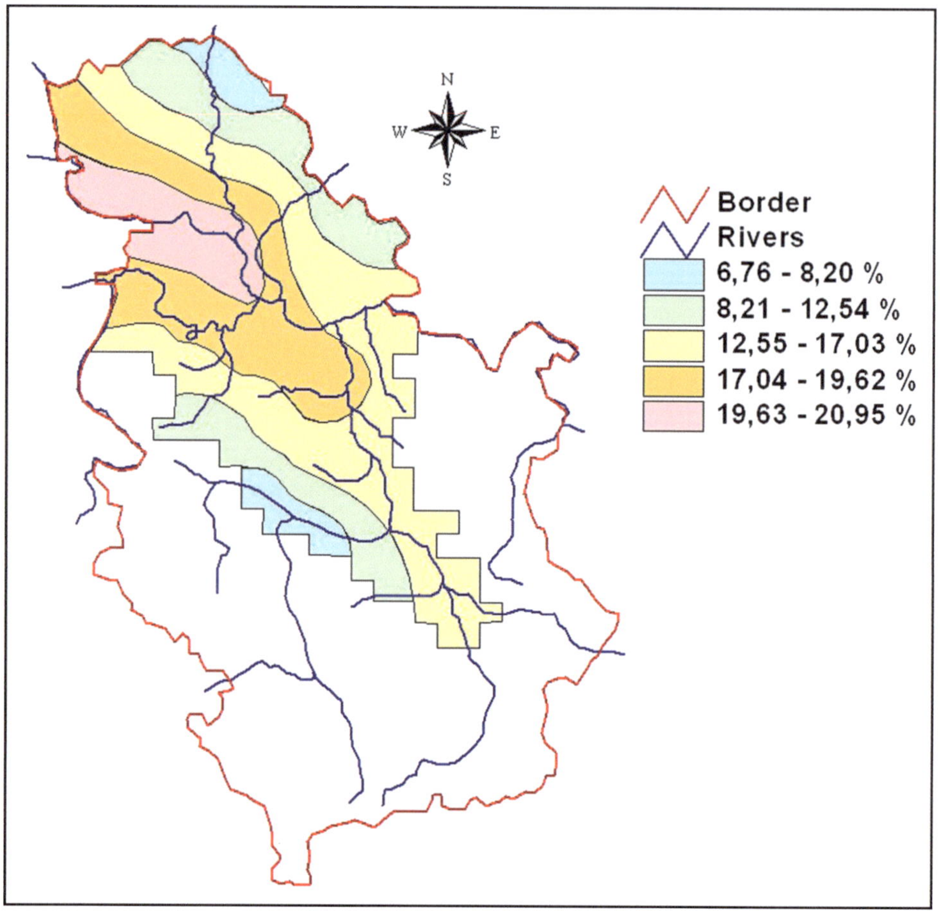

Fig. 5.19 Distribution of micas in bulk soil

7–21%. They show a negative correlation to quartz—an increase in micas in soils always causes a decrease in quartz. The map of the spatial distribution of micas (Fig. 5.19) shows that the highest values are in Bačka and Srem around the Danube and this gradually decreases towards the periphery of the Pannonian basin. Unlike quartz, micas are most abundant in the hydromorphic soils of fluvisol and eugley (WRB, Gleysol) and gradually decrease in amount through gleys, humofluvisol (WRB, Fluvisol, humic) and chernozem to arenosol in the north of Vojvodina.

Micas gradually decrease with distance south, from Srem, through Šumadija, to the Čačak-Kraljevo and Aleksinac-Niš depressions. It should also be noted that minor amounts (up to 5%) of glauconite mica and talc are found in rankers (WRB, Leptosol) and cambisols developed on Mesozoic sediments between Topola and Kragujevac.

Feldspars are found in lesser amounts (2–17%) and are mainly represented by Na–Ca plagioclase of albite-oligoclase composition. The K-feldspars (orthoclase and microcline) are more susceptible to weathering and only minor amounts (0.4–1.5%) occur in soils.

Carbonate minerals (calcite and dolomite) (Fig. 5.20) occur within the range of 0–22%. They exhibit an interesting distribution spatially, as well as within the soil types. Dolomite is dominant in north-west Vojvodina, particularly in Bačka, where calcareous chernozem and humofluvisol (WRB, Fluvisol, humic) occur with some scattered spots of solonetz and rarely solonchak. From an average of 5.5% in arenosol and humofluvisol (WRB, Fluvisol, humic) in Bačka, dolomite gradually decreases towards the south-east to about 1% in eugley (WRB, Gleysol), which occurs in the marshes of the Sava and Danube, and vertisol in Šumadija.

Unlike dolomite, calcite is found in more or less equal concentrations of about 3–6% in the majority of the soils. However, the amount of calcite in arenosol increases up to 7%, while in vertisol and solonetz, it decreases to about 1% or even less. In north Bačka, in some calcareous soils such as chernozem, humofluvisol (WRB, Fluvisol, humic) and humogley (WRB, Gleysol, humic), amounts of calcite and dolomite can reach concentrations as high as 15%.

The average composition of the clay fraction (−0.002 mm) is presented in Table 5.2, which shows that the soils consist

Fig. 5.20 Distribution of carbonates in bulk soil

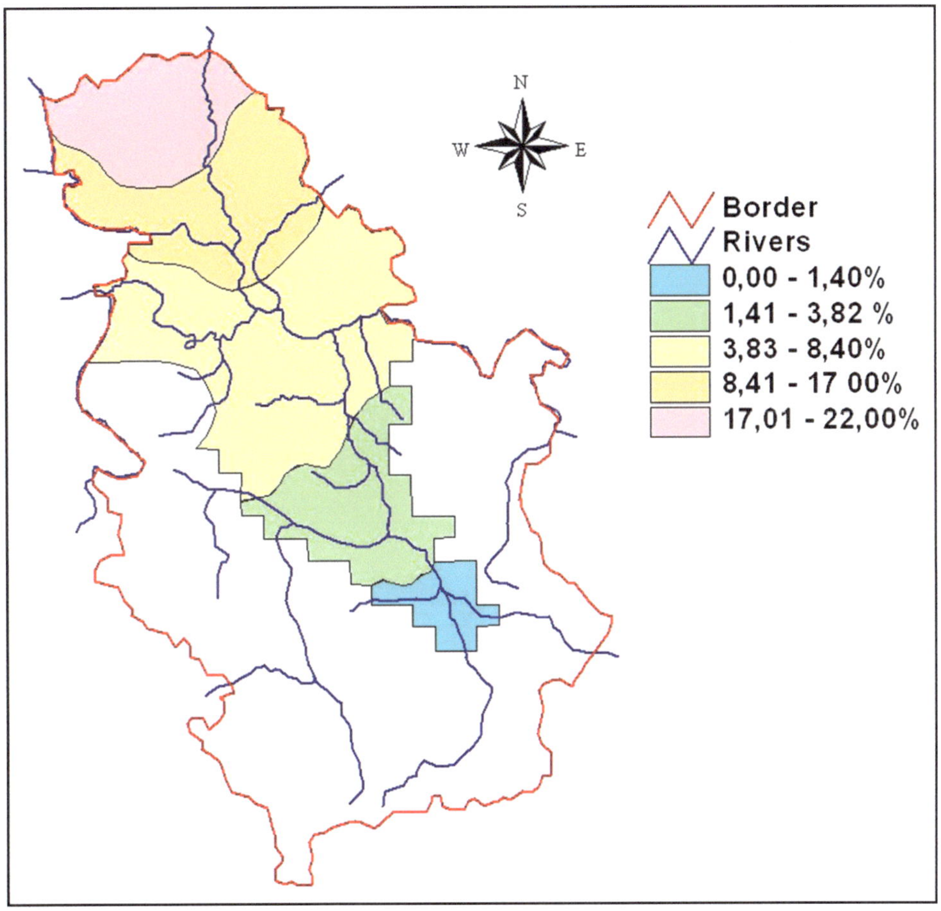

predominantly of illite, smectite and kaolinite, associated with chlorite, illite-smectite (10–14) and chlorite-vermiculite (14–14) type mixed-layer silicates (M-L-S), vermiculite, and minor amounts of fine-grained talc and goethite-limonite. In this fraction trace amounts of quartz, feldspars, calcite and dolomite are also found.

Illite is the most common and most abundant mineral in the clay fraction of the soils. From the map presented in Fig. 5.21, it can be seen that the spatial distribution of illite varies between 14 and 71% in the soils. Illite is most abundant in chernozem and arenosol developed on loess plateaus and terraces in Bačka and Banat. Under the influence of moisture in wetlands, the amount of illite gradually decreases from about 60% in chernozem, through 46–49% in humofluvisol (WRB, Fluvisol, humic), humogley (WRB, Gleysol, humic) and fluvisol, to about 25% in eugley (WRB, Gleysol) developed in marshy environments.

South of the Sava and Danube rivers, in Šumadija, the amount of illite varies between 14% in some rankers (WRB, Leptosol) developed on flysch, 56% in chernozem, luvisol on Neogene sandy-silty clays and 58% in some rendzina (WRB, Leptosol, calcaric) on marlstone. All other soils in Šumadija contain the average 32–55% of illite with the exception of pseudogley (WRB, Stagnosol) developed on Neogene lake deposits in western Serbia. Fluvisol along the River Morava and its tributaries comprises an average 43% illite.

The second most abundant clay mineral group is smectite (Fig. 5.22), mainly montmorillonite and beidellite and rarely ferro-beidellite and nontronite, which occur in the range of 4–42%. In Vojvodina, unlike illite, the content of smectites increases from 3.5 to 8.6% in arenosol and chernozem, respectively, through 18% in slightly wet humofluvisol (WRB, Fluvisol, humic), and 21% in humogley (WRB, Gleysol, humic) to 24% in wet fluvisol and 58% in eugley (WRB, Gleysol) developed in the marshy environments of Banat.

In Šumadija, these soils contain slightly higher amounts of smectite, but instead of eugley (WRB, Gleysol), Vertisol is developed on Neogene lake clay deposits and can have an average of 28% smectite. Pseudogley (WRB, Stagnosol), luvisol, dystric cambisol, and ranker (WRB, Leptosol,

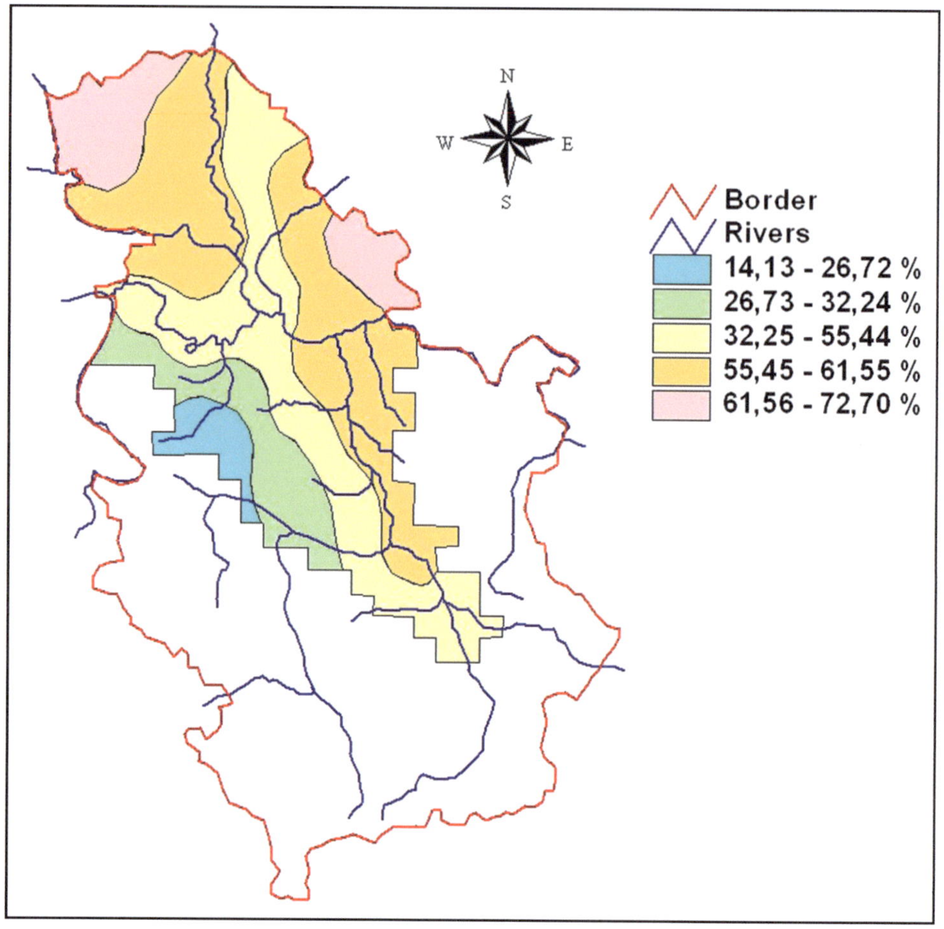

Fig. 5.21 Distribution of illite in clays

eutric) and rendzina (WRB, Leptosol, calcaric) contain less than the average amount of smectite due to their either slightly acid or well-drained soil environment (Dugalić et al. 2004).

The kaolinite and vermiculite (Figs. 5.23 and 5.24) in the clay fraction of the soils of Vojvodina are present in almost equal amounts (about 5 and 4%, respectively), with the exception of the content of kaolinite in vertisol in Banat, which is developed on Miocene/Pliocene clays, with 7.6% kaolinite inherited from the parent material.

In the soils of central Serbia, however, the content of kaolinite and vermiculite gradually increases from about 5–8% in chernozem, eutric cambisol and rendzina (WRB, Leptosol, calcaric) to 9–14% in dystric cambisol, ranker (WRB, Leptosol) and luvisol following an increase in the annual rainfall in the area, advantageous for mineral alterations in the slightly acidic soil environment. It should be noted that soils in Šumadija developed on marlstone comprise illite and glauconite associated with halloysite instead of kaolinite. The same is found in the fluvisol of the Kolubara area, where halloysite dominates over kaolinite in the clay fraction. Some relict terra rossa (WRB, no adequate name) from Gruža near Kragujevac and the southern slopes of Mt. Ozren near Aleksinac, as well as all rhodic cambisol in eastern Serbia, contains the dominant halloysite and rarely kaolinite in the clay fraction.

Mixed-layer silicates (M-L-S) can be considered a by-product of mica, illite and chlorite transformations to smectite and vermiculite. The illite-smectite (10–14) type of M-L-S is the most common in the soils, with the content varying between 1.6 and 12%. It is most abundant in dystric cambisol, where slightly acid conditions are favourable for its development. The (14–14) types of M-L-S or chlorite-vermiculite and rarely chlorite-smectite are present in minor amounts, on average between 0.5 and 4.6%. It is most abundant in luvisol and pseudogley (WRB, Stagnosol) with an average content of 4.10 and 3.57%, respectively. The rare corrensite, an ordered M-L-S of the illite-chlorite

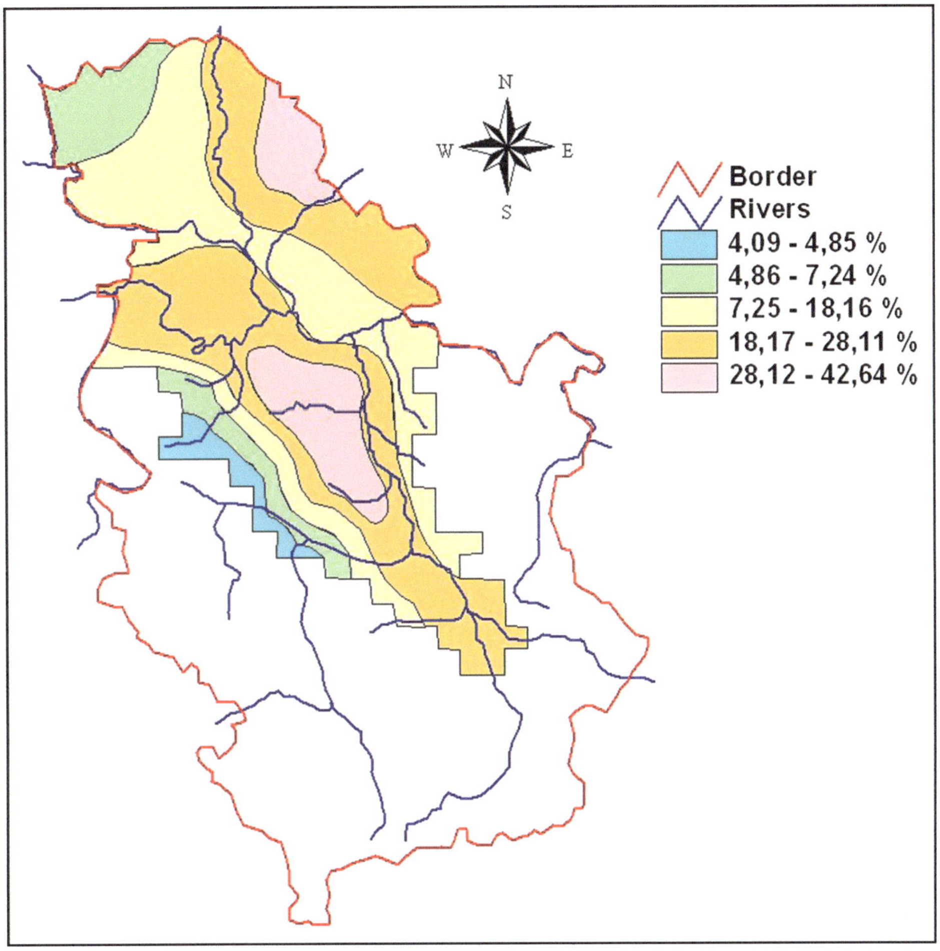

Fig. 5.22 Distribution of smectites in clays

(10–14) type, is found in eutric cambisol developed on chlorite-sericite-actinolite schists near Leskovac, where it is associated with other (14–14) types: chlorite-vermiculite and chlorite-smectite and halloysite/kaolinite.

The hilly and mountainous area of Serbia comprises up to 14% igneous rocks, approximately 14% metamorphic rocks and approximately 20% hard consolidated sedimentary rocks. However, for the mountainous regions of Serbia, only sporadic mineralogical analyses of soils have been carried out and only then upon request of the mapping teams as support for the soil classification of specific areas at a scale of 1:50,000.

The largest ultramafic massifs in Europe are located in Serbia. They cover large areas of western Serbia and comprise mainly harzburgite and serpentinized peridotite. Rankers (WRB, Leptosol eutric) from Kopaonik, Zlatibor, Maljen and Fruska gora are developed on serpentinized peridotite residue. The clay fraction of the soils comprises serpentine-antigorite, chlorite, talc, the chlorite-vermiculite (14–14) type of M-L-S, nontronite and minor amounts of disordered kaolinite (Kostić et al. 1998b). Acid and intermediate volcanic rocks in Serbia are present across approximately 3800 km^2. Ranker (WRB, Leptosol) is developed on generally fresh rocks, while andosol is often developed on volcanic tephra (pyroclasts).

The first discovery of andosol in Serbia was on Mt. Radan near Leskovac. The soil developed on pyroclasts of labrador andesite with about 40% volcanic glass comprises also halloysite/kaolinite and cristobalite–tridymite (Opal-CT). It is associated with minor amounts of illite, the illite-smectite (10–14) and chlorite-vermiculite (14–14) types of M-L-S, allophane and volcanic glass tarnished by goethite (Topalović et al. 2004). Further studies of soils on quartz-latite and dacito-andesite pyroclasts have shown a similar composition of the clay fraction with the only difference being in the amount of allophane and volcanic glass. The soils comprise 12.5–25% volcanic glass and 10–15% allophane. According to the WRB classification, these andosols were classified as Vitri–andic and Sil–andic andosols, respectively (Topalović et al. 2004; Kostić et al. 2005).

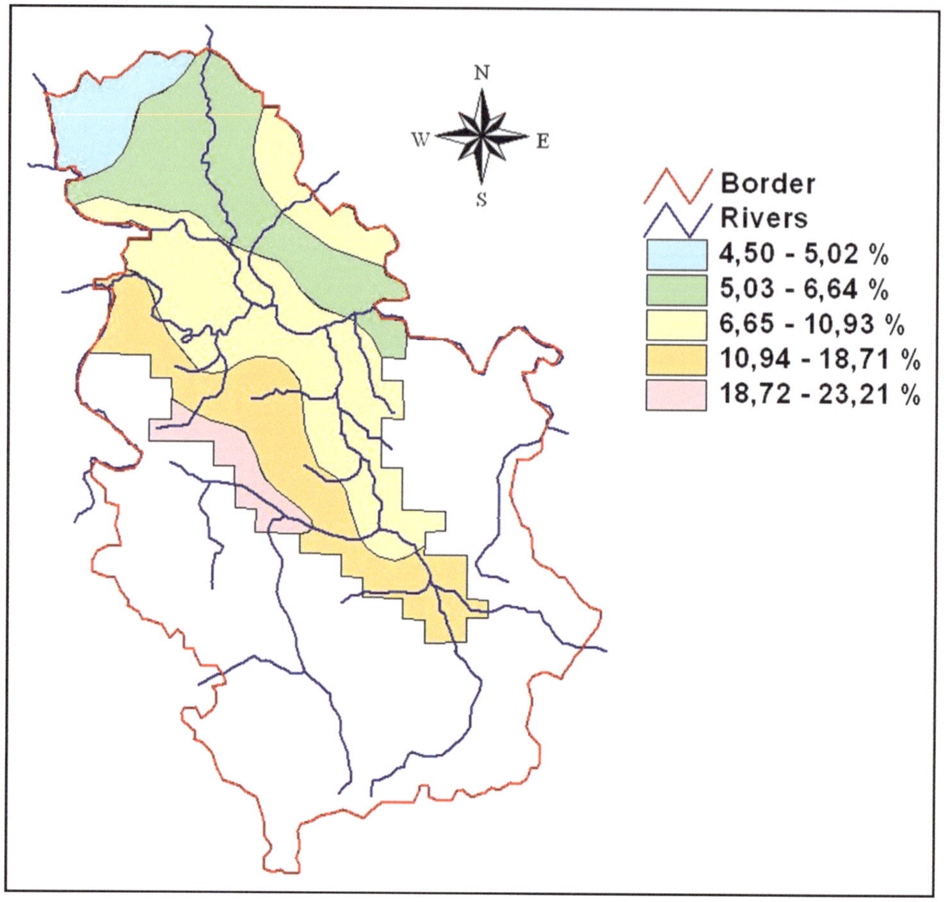

Fig. 5.23 Distribution of kaolinite in clays

Mineralogical analyses of several rankers (WRB, Leptosols) developed on quartz-latite, andesite, latite and dacite have shown that illite is the predominant clay mineral in the soils (50–60%) and is associated with chlorite, vermiculite, and minor amounts of disordered kaolinite/halloysite, or illite associated with kaolinite and beidellite, or disordered kaolinite or illite associated with ferro-beidellite. Halloysite and disordered kaolinite gradually replace each other in the soils. Rankers (WRB, Leptosols) developed on the western slopes of Mt. Besna Kobila comprise 4–6% rare todorokite (MnO_2), manganese being inherited from dacite (0.43% MnO) (Kostić 1998a).

Metamorphic rocks cover over 14% of the Serbian land mass, mainly in central and western Serbia. Eutric cambisol and dystric cambisol are soils developed on argiloschists in Fruška Gora, phyllite near Kragujevac, and gneiss-granite on Crni vrh and Jastrebac. The clay fraction of the soils investigated comprises mainly illite and/or chlorite associated with the illite-smectite (10–14), illite-vermiculite (10–14) and chlorite-vermiculite (14–14) types of mixed-layer silicates (M-L-S), with minor vermiculite, smectite and kaolinite.

In the kalkocambisols (WRB, no adequate name) and eutric cambisol developed on carbonate rocks on Rtanj, Jablanik and Tara mountains, the clay fraction of the soils comprises up to 40% illite and about 20% smectite associated with kaolinite, the illite-smectite and illite-vermiculite (10–14) types of M-L-S, and minor chlorite and vermiculite, as well as 8–12% of goethite and haematite (Kostić 1990). The relict terra rossa (WRB, no adequate name) is developed on the residue from the weathering of Triassic and Cretaceous pure limestones in eastern and western Serbia. The soil comprises disordered kaolinite/halloysite, illite, gibbsite-hydrargilite, goethite and haematite with minor quartz, chalcedony, smectite, the kaolinite-smectite (7–14) type of M-L-S and lepidocrocite. However, the terra rossa (WRB, no adequate name) in eastern Serbia comprises more halloysite than kaolinite, while similar soils found on the mountains of Šar and Koritnik contain more goethite and

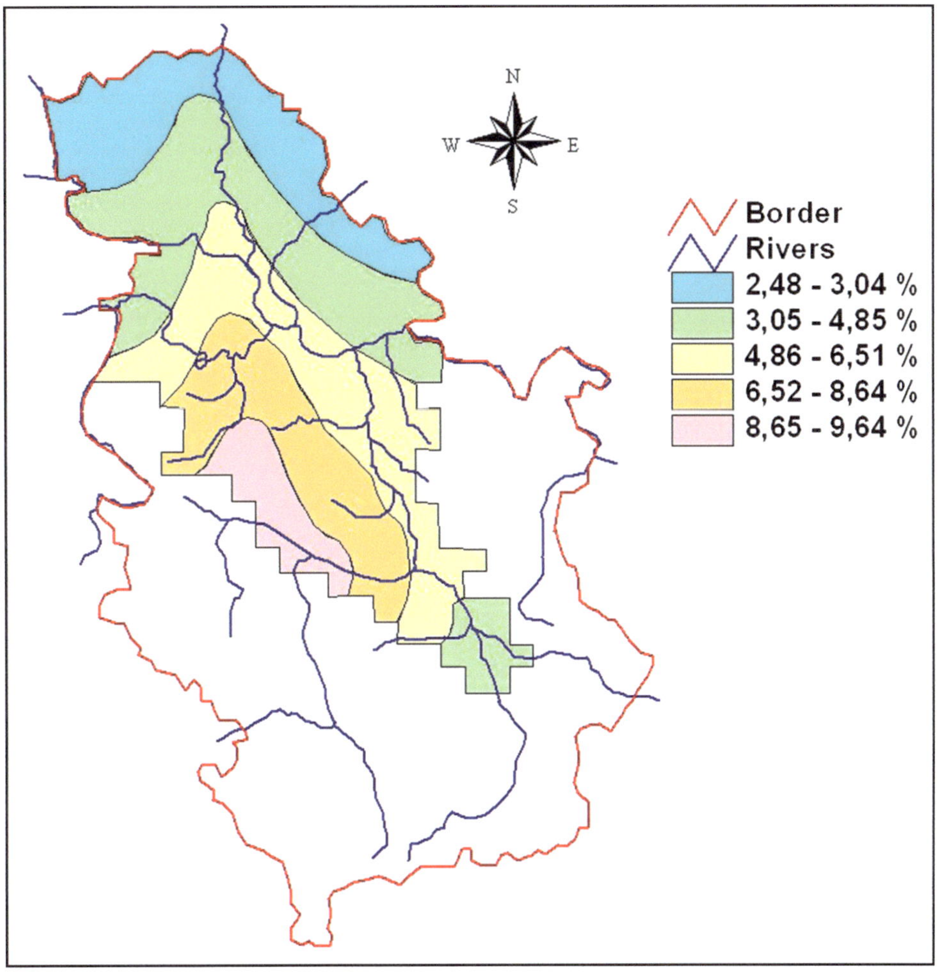

Fig. 5.24 Distribution of vermiculite in clays

haematite, which can amount to 15% (Kostić 1990; Topalović et al. 1996).

References

Aleksandrović D (1956) An attempt to classify loess deposits in Podunavlje and Pomoravlje. Soil Plant Belgrade 3:1–3 (in Serbian)

Ćirić B (1996) Geology of Serbia. Geokarta, Belgrade (in Serbian)

Dimitrijević MD (1997a) Geology of Serbia. Geokarta, Institute of Geology, Special Publications, Belgrade (in Serbian)

Dimitrijević MD (1997b) Geology of Yugoslavia. Special Publication, Geological Institute Gemini, Belgrade, p 187 (in Serbian)

Dugalić G, Kostić N, Živković M, Jovšić D (2004) Mineralogical composition of pseudogley of the experimental station, College of Agronomy, Kraljevo. Soil Plant Belgrade 53(1):11–20 (in Serbian)

Federal Geological Survey (1970) Geological map SFRJ in scale 1:500,000. Federal Geological Institute of SFRY, Belgrade (in Serbian)

Finke P, Hartwich R, Dudal R, Ibàñez J, Jamagne M, King D, Montanarella L, Yassoglou N (2001) Georeferenced soil database for Europe: manual of procedures version 1.1. European Soil Bureau, Scientific Committee, Office for Official Publications of the European Communities, Luxembourg, p 184

Geological Survey of Serbia (1968–1995) Basic Geological Maps (BGM) 1:100 000, 76 sheets, with the corresponding explanation books (in Serbian)

Klute A (ed) (1986) Methods in soil analysis. I. Physical and mineralogical methods, 2nd edn. Agronomy no. 9, American Society of Agronomy, Medison, WI

Koščal M, Lj Menković (2005) Explanation for the geomorphological map of Vojvodina. Geozavod-Gemini, Belgrade

Kostić N (1990) An investigation on clay mineral formation in brown soil profiles in Central Serbia, Yugoslavia. Sci Geol Mem France 85 (1):155–166

Kostić N (1999) Pedomineralogy and geochemistry of soils of Vojvodina. Mineralogija Godisnjak JAM II(2):16–28 (in Serbian)

Kostić N (2001) Pedomineralogy and geochemistry of some important soils in Serbia. In: Proceedings of 10th congress of Yugoslav Soil Science Society (JDPZ), Vrnjačka Banja, pp 134–148 (in Serbian)

Kostić N (2012) Lithological map of Serbia, in scale 1:500 000 (in Serbian)

Kostić N, Protić N (2000) Pedology and mineralogy of loess profiles at Kapela-Batajnica and Stalać, Serbia. Catena 41:217–227

Kostić N, Hadžić V, Protić JN (1998a) Manganese geochemistry of soils of Vojvodina, Sumadija and Northern Pomoravlje. In: Proceedings of 2nd Balkan symposium of field crops, vol 2. Ecology and physiology, pp 9–21

Kostić N, Wilson JM, Živković M, Bain D (1998b) Mineralogy and geochemical speciation of heavy metals in some serpentine soils of

Serbia. In: Proceedings of 16th world congress of soil science, vol 1. Elsevier, Amsterdam, pp 444–455

Kostić N, Jakovljević M, Hadžić V, Protić N (2001) Mineralogy and agrochemistry of magnesium in soils Vojvodina, Šumadija and Northern Pomoravlje. Proc Nat Sci Matica Srpska Novi Sad 101:37–48 (in Serbian)

Kostić N, Dugalić G, Tomić Z, Jovšić D (2005) Mineral and chemical composition of soils developed on tephras of Radočelo, Igrošoa and Takovo. In: Proceedings of XI congress of DPZ-SCG (Society of Soil Science of Serbia and Montenegro), Budva, pp 126–137

Topalović M, Kostić N, Miletić Z (1996) Two-layered profiles of Terra Rossa in the surroundings of Prizren, Šara Mt. Soil Plant Belgrade 45(2):79–90 (in Serbian)

Topalović M, Kostić N, Pavlović P (2004) Andosol from Radan Mt., Serbia. Soil Plant Belgrade 53(2):12–26 (in Serbian)

Yugoslav Geological Survey (1970) Geological map of Yugoslavia, in scale 1:500,000 (in Serbian)

6 Soil Classification

6.1 A Historical Overview of Soil Classification

The initial stage in the development of soil classification in Serbia occurred at the beginning of the twentieth century, and in particular during the 1920s when the Russian genetic school was introduced into the first soil classification, which had a marked impact. The period from 1926 to 1956 was characterized by the creation of a comprehensive regional soil classification, which absorbed some of the European experience, but also took account of the specific conditions of the country.

This important period for the development of soil science in the Kingdom of Yugoslavia including Serbia was characterized by the simultaneous use of genetic classifications (both Russian and Central European) and the creation of the first national classifications. The first national soil classification, set out by Stebutt (1926) in his work entitled 'Our main agricultural regions', was simple and involved dividing soils into four basic classes: a class of undeveloped soils, a class of soils characterized by the formation of aluaminosilicates, a class of soils characterized by degradation processes and a class of soils with destructive pedogenesis. In addition, Stebut (1930) published two general soil classifications which had a major impact outside Europe as well. Unlike the Russian genetic schools of Dokuchaev, Sibirtsev and Glinka, Stebut's classification was not based on the principle of zonality, but on typical pedogenetic processes which result in the formation of different types of soil (Neugebauer et al. 1963). The second classification by Gračanin (1950) was based on the degree of genetic development and the soils were categorized into two large groups: undeveloped or atypical soils and developed typical ones. For higher units, the classification criteria were pedogenetic, while morphological characteristics of the profile, texture, chemical composition and geological character of the parent material were used for lower units (Neugebauer et al. 1963). Later, Kovačević (1956) suggested a soil classification based on the pedogenetic processes but with the introduction of notions from the American classification, and later Živković (1961) highlighted some of the problems connected to classification.

In the 1930s, the first classification system was developed based on the Russian principles and adjusted to local soils. This was modified by Gračanin (1950) and then again in 1963, leading to the first official version of the *Classification of Yugoslav Soils* (CYS), which was presented internationally in 1964 at the eighth International Congress of Soil Science (Filipovski et al. 1964). This version was also influenced by ideas from Kubiëna (1953) and the German system (Mückenhausen and Vogel 1962). The CYS was revised three more times: in 1964, 1973 and 1985 (Škoric et al. 1985).

From 1960 to 1985, detailed research was undertaken into different soil types, on improving their classification. To this end, the following congresses were important: Ohrid (1963), Belgrade (1972) and Novi Sad (1980), with new findings not only in the world but also in particular those at the national level reviewed (including checking the first versions of the classification). In between these congresses, symposiums, working groups, meetings of the Commission on Soil Genesis, Taxonomy and Mapping of what was then known as the Yugoslav Society of Soil Science, work on the FAO project, the publishing of textbooks and extensive soil mapping all took place. During this period, three versions of soil classification were published:

- the first based on the genetic principle (Neugebauer et al. 1963) was adopted at the second Congress of Pedology in Ohrid (1963) and was used across the whole of what was then the Socialist Federal Republic of Yugoslavia (SFRY);
- the second was promoted at the eighth International Congress on Soil Science in Bucharest (Filipovski et al. 1964);
- the third, which was based on the same principle as the first, was adapted to the international classification valid

P. Pavlović et al., *The Soils of Serbia*, World Soils Book Series,
DOI 10.1007/978-94-017-8660-7_6

at that time in Europe and was used up until 1985 (Škorić et. al. 1973) and
- the fourth, which is still used in Serbia, but which, unfortunately, does not entirely correspond to the WRB's criteria (Škorić et al. 1985).

The most recent version of the CYS (Škorić et al. 1985) is still in use in all the Republics of the former Yugoslavia, although there have been proposals to abandon it in Croatia (Bogunović and Racz 2001) or modify it in Serbia and Montenegro (Antonović and Protić 1997), and in Serbia (Antonović et al. 2010, 2011).

6.2 Soil Map Research

Research for the soil map of Serbia also began between the two world wars, under the influence of the Russian school of soil genesis. The first soil map of the Kingdom of Yugoslavia was published by Stebut in the scale of 1:3,500,000 and 1:1,200,000 (Stebut 1926, 1931). After World War Two, Neugebauer et al. (1961) published the Basic Soil Map of Yugoslavia (1:1,000,000) and later a map of Vojvodina (1:50,000), Neugebauer et al. 1971a, b). The Basic Soil Map of Yugoslavia (1:50,000) was produced by the Institute for Agricultural Research in Novi Sad, while a Soil Map of Kosovo and Metohija was completed by Pavićević et al. (1974) and published by the 'Jaroslav Černi' Institute for the Development of Water Resources, Belgrade. The mapping of soils in central Serbia was undertaken by scientists from the Institute of Soil Science (Belgrade) in several phases between 1958 and 1982, leading to the publication of various 1:50,000 maps: a map of the river basins of the Velika Morava and Mlava (1958), a map of western and north-western Serbia (1963), a map of eastern Serbia (1970-76) (Antonović 1972), and a map of part of southern Serbia (1979 and 1982) (Protić et al. 2005). More recently, the south-eastern part of Serbia (about 700,000 ha) was mapped in the period 2007 2011 (Mrvić et al. 2013). Between 2009 and 2011, the Institute of Soil Science undertook field and laboratory studies, mapping and digitizing, leading to the publication of 19 maps as part of the project entitled 'The development of a national soil map of parts of central and southern Serbia' with Dr. E. Saljnikov as project leader; Dr. G. Antonović, Dr. V. Mrvić, Dr. D. Čakmak, Dr. S. Maksimović, Mr. B. Brebanović, ing. M. Nikoloski, MSci.V. Perović, MSci. D. Jaramaz and MSci. Lj. Kostić-Kravljanac as authors and contributors (Mrvić et al. 2013). At the same time, several important monographies were published: 'Soils of the Velika Morava and Mlava' (Tanasijević and Pešić 1965, 1967), 'The pedological cover of western and north-western Serbia' (Tanasijević et al. 1966), 'The soils of the Braničevo-Zvižd district and Homolje' (Antonović et al. 1975), 'The soils of the Timok river basin' (Vojinović and Jeremić 1974), 'The soils of Stari Vlah and Raška' (Pavićević et al. 1968) and 'The soils of the Nišava river basin' (Antonović and Mrvić 2008).

Meanwhile, researchers were working on a pedogeochemical map, as part of a project entitled 'The control of fertility and levels of harmful and hazardous substances in the soils of Serbia', funded by the Republic Fund for the Protection, Use, Improvement and Development of Soil under Agriculture and by the Serbian Ministry of Agriculture, Forestry and Water Management. The project was carried out in Vojvodina from 1992 to 1997 by the Institute of Field and Vegetable Crops and the Faculty of Agriculture, Novi Sad, with Prof. Vladimir Hadžić as project leader, and in central Serbia by the Institute of Soil Science in Belgrade, in cooperation with the University of Belgrade's Faculty of Agriculture. The research was divided into two cycles: the first from 1993 to 2007 under the leadership of Dr. Nebojša Protić and Dr. Ljubiša Martinović; and the second from 2008 under the leadership of Dr. Vesna Mrvić (Mrvić et al. 2013).

In addition, Antonović et al. (1997) started preparing a number of manuals for the devising and maintenance of a strategic plan for the protection, regulation and use of agricultural land. The plan was designed for municipalities in the scale 1:25,000, but the concept could be extended to farm level. Its methodology and information system began with the concept of open systems whereby individual problems, e.g. soil conservation, were considered within the scope of conditions and processes in natural and socio-economic environments (Protić et al. 2005).

6.3 Actual Soil Classification

The Classification of Yugoslav Soils (Škorić et al. 1985) is the official soil classification which is used today in the Republic of Serbia (Table 6.1). Table 6.1 gives possible names for soil types according to the WRB classification (World Reference Base 2006). It has six categories of taxons (orders, classes, types, subtypes, varieties and forms) connected hierarchically based on the logical principles. The basic principles of this classification can be summarized as follows:

1. The classification is based on the characteristics of soils, rather than the external conditions affecting their formation. The soil properties taken as the basis for categorizing soils are morphologically visible and easily measurable. Quantitative indicators of these properties are used to a great extent. Horizons are also precisely defined.

2. The classification is morphological and genetic and takes into account soil evolution. When classifying into types and higher categories, the sequence and character of genetic horizons are used. Grouping soil types into higher categories is done on a genetic and evolutionary basis.
3. The central unit of this classification is soil type, and the categorization into types and higher categories is based on the common criteria. The central unit of our classification is soil type and categorisation into types and higher taxonomic levels is based on common criteria. However, this is not the case when categorising soils into lower units (subtypes, varieties and forms)—for each soil type, those characteristics that result in the greatest variability are chosen as criteria. This means that variations between soils in a higher taxon are greater than for those soils in a lower category.
4. The classification was to serve as the basis for an assessment of a soil's production and ecological value and for the creation of various applicative classifications. When dividing soil types into lower categories, characteristics were taken into account which, besides genetic significance, also had ecological and productive importance. These criteria are as follows: the evolution to other soil types, depth of the solum and upper horizons, texture and skeleton content, character of the parent substrate, lithic and regolithic characteristics, carbonate content, salinity and alkalization, vertic, eutric and/or dystric qualifiers, depth of the salts accumulations, the character of the humic horizon, etc.
5. Given the ever greater anthropogenization of soils in the former Yugoslavia, the highest systematic units (orders) of this classification included special classes for anthropogenic soils, which were then further categorized into types.

6.3.1 Definitions and Labelling of Genetic Soil Horizons

Genetic horizons are labelled with upper-case letters. A supplemental label specifying the nature of the horizon or subhorizons is in lower-case letters.

O—(Organic) organic surface horizon, which lies on the mineral soil in mainly aerobic conditions. The subhorizons of this horizon are denoted according to the degree of decomposition of organic matter:

Ol—(Litter) is a subhorizon of organic material in which plant parts can be clearly differentiated, connected by fungal mycelia. It is characteristic of forest soils.
Of—(Fermentation) is a subhorizon of half-decomposed organic material in which it is difficult to identify macroscopically the primary structure of the plant remains.
Oh—(Humus) is a humified organic subhorizon in which the structure of the plant matter cannot be identified, with sporadic admixture of mineral grains.

(A)—(Accumulation) the initial poorly developed horizon, which is the biologically active surface layer of soil where the majority of plant roots develop. Here, structure aggregates begin to be formed. It differs slightly from the C horizon in terms of colour.

A—typical humus-accumulative horizon in which well-humified colloidal organic matter is mixed with mineral soil as an organo-mineral complex. Subhorizons are denoted according to the character of the humus.

Amo—(*L. mollis*—soft, gentle) a dark, deep subhorizon, with a marked structure and base saturation of over 50%.
Aum—(*L. umbra*—shade) in terms of colour and depth is similar to the mollic subhorizon, but has a lower degree of base saturation (below 50%) and a poorly developed structure.
Aoh—(Gr. ochros—pale) a light-coloured subhorizon, not as deep as the mollic and umbric subhorizons. Its structure is poorly developed.

E—(eluviation–flushing) the eluvial horizon, found beneath the O or A horizon. Compared to the horizon below it, it contains lower amounts of one or more of the following components: clay, humus and/or sesquioxides. It is usually lighter in colour than the horizons immediately above and below it.

B—the illuvial horizon, lying beneath the E horizon. In comparison, it contains higher amounts of one or more of the following components: clay, humus, and/or sesquioxides, resulting in several variations.

Bt—(translocated clay) the argillic subhorizon (L. argilla—clay and L. luvi—rinse), which lies beneath the E horizon and is an accumulation zone for clay leached from the E horizon. Colloidal membranes are macroscopically visible around larger grains and structural aggregates or on the walls of larger pores.
Bh—the humic-illuvial subhorizon, which lies beneath the E horizon and is the zone where humus leached from the O and E horizons mainly accumulates in acidic conditions. It is dark brown to black in colour and has a sandy texture.
Bfe—(ferrous) a subhorizon situated beneath the E horizon or the Bh and is the zone where sesquioxides leached from the E horizon accumulate in acidic conditions.

(B)—the cambic horizon of in situ metamorphosis found between the O or A horizons and the C or R horizons. It differs from them in terms of its yellow or red hue, altered primary structure of parent material, and frequently higher clay content.

(B)v—a modification of the cambic horizon, which originates from the oxidative and hydrolytic decomposition of primary minerals in situ and argillosynthesis, and is characteristic for silicate substrates.
(B)rz—a modification of the cambic horizon originating mainly from residual accumulation arising when calcareous rocks dissolve.

G—the gley horizon, which exhibits signs of reduction and secondary oxidation in permanent or temporary anaerobic conditions and is characterized as being blue, green or grey in colour, with rusty mottling on the surface of aggregates.

Gso—a subhorizon formed from the secondary oxidation of part of the G horizon. It is characterized by dominant rusty mottling on the grey-green and blue reduced microzones. Rusty-coloured specks are found on the surface of the structural aggregates.
Gr—a subhorizon with permanently stagnant water and hence dominated by reduced grey-green and blue microzones.
g—a horizon occurring as the result of stagnating surface water and the alternation of wet and dry periods. It is characterized by marked marbling—streaks of rusty and whitish grey specks, with the grey zones dominating in the cracks and on the surface of the aggregates, and the rusty mottling dominating within the aggregates. This is the only horizon in this classification which is not denoted with a capital letter.

T—peat horizon. Non-decomposed organic material which accumulates in anaerobic conditions.

T_{fi}—the fibric subhorizon, fibrous, poorly humified organic material (humification <30% of the organic mass)
T_{he}—the hemic subhorizon (hemisis–half), moderately decomposed (humification 30–60% of the organic mass)
T_s—the sapric subhorizon (sapros–rotten) well-humified organic mass (humification >60% of the organic mass)

C—the loose part of the parent substrate, virtually unaffected by the pedogenetic processes characteristic of the other horizons.
R—hard rock.
P—a horizon created through multiple natural horizons or layers being ploughed, mixed and homogenized, and enriched anthropologically with organic or mineral material.

In addition to these horizons and subhorizons, this classification allowed for the addition of other lower-case letters, separated with a comma, denoting the presence of specific materials which are not included in the mandatory properties for each individual horizon, but may instead appear in several different horizons. These are the following symbols:

ca—accumulation of carbonates
sa—accumulation of soluble salts, apart from gypsum in quantities of 0.25% for chloride and sulphate salinization and 0.15% for sodium salinization
cs—accumulation of gypsum
na—Na accumulation in the adsorption complex over 7% of '*T*' (e.g. in solonetz: Bt, na)
cn—accumulation of concretions enriched in sesquioxides and manganese (e.g. g, cn)
m—high-level cementation of horizon
b—buried horizon
vt—(Latin verto) verticity (cracks and slip planes)
p—ploughed horizon
a—hydromorphic humus

6.3.2 Soil Classification

According to this classification, soils in Serbia are divided into four orders depending on the character of soil wetting and the composition of the water which wets the soil:

1. Automorphic soils, which are wetted only by precipitation. Water percolates freely without prolonged retention on impermeable horizons.
2. Hydromorphic soils, which are characterized by the temporary or permanent saturation of all soil pores with water which is neither saline nor alkaline. This leads to reduction processes which cause processes of gleization.
3. Halomorphic soils are exposed to additional wetting by saline and alkaline ground (and also less commonly, surface) water. They have one or both of the following characteristics: (1) up to a depth of 125 cm in one part of the profile, they contain at least 1% salts (for chloride and sulphate salinization) and at least 0.7% salt for sodium salinization; (2) the A and B horizons of the profile clearly differentiate in terms of texture and other properties, with the B horizon having a Na^+ concentration of more than 15% of the total capacity for cations and a columnar or prismatic structure.
4. Subaqueous (subhydric) soils, which develop under water (lakes, ponds). Pedogenic processes often combine with sedimentation processes.

Table 6.1 Classification of Serbian soils (Škorić et al. 1985)

Class	Type	Subtype	Variety	Form	
Order of automorphic soils					
I (A)-C or (A)-R (undeveloped)	1. Lithosol WRB name: leptosol	1.1 On acidic rocks 1.2 On neutral and basic rocks 1.3 On peridotite 1.4 On limestone and dolomite			
	2. Regosol (sirozem on loose substrates) WRB name: Leptosol	2.1 Siliceous	2.1.1 Dystric 2.1.2 Eutric	According to texture (for both varieties): 1. Sandy 2. Clayey 3. Loamy	According to skelet content (for both varieties): 1. Low skelet content (<25%) 2. Moderate skelet content (25–50%) 3. High skelet content (>50%)
		2.2 Calcaro-siliceous	2.2.1 On loess and loess-like sediments	2.2.1.1 Loamy 2.2.1.2 Clayey	
			2.2.2 On marl, marly clay and harder marly limestone 2.2.3 On calcareous sandstone		
		2.3 Sandy-dolomite	2.3.1 Shallow (up to 40 cm) 2.3.2 Deep (>40 cm)		
	3. Arenosol (aeolian sand) WRB name: Arenosol	3.1 Quartz 3.2 Siliceous 3.3 Calcaro-siliceous	3. (1–3) 1. Continental 2. Marine		
	4. Colluvium (colluvial soils) WRB name: Colluvial soil	4.1 Eutric-siliceous 4.2 Dystric-siliceous 4.3 Calcareous 4.4 With fossil soil	For all subtypes: 1. With a predominance of detritus rock, non-gleyed 2. With a predominance of detritus rock, gleyed 3. With a predominance of soil material, non-gleyed 4. With a predominance of soil material, gleyed 5. Alluvial-colluvial, non-gleyed 6. Alluvial-colluvial, gleyed	According to texture (for all varieties): 1. Sandy 2. Clayey 3. Loamy According to skelet content (for all varieties): 1. Low skelet content (<25%) 2. Moderate skelet content (25–50%) 3. High skelet content (>50%)	

(continued)

Table 6.1 (continued)

Class	Type	Subtype	Variety	Form
II A-C or A-R (Humus-accumulative soils)	1. Kalkomelanosol (Limestone–dolomite black soil) WRB name: Leptosol	1.1 Organogenic	1.1.1 Lithic 1.1.2 Skeletal colluvial	According to the type of humus horizon: 1. With a mollic horizon 2. With an ochric horizon
		1.2 Organo-mineral	1.2.1 Lithic 1.2.2 Colluvial	
		1.3 Brownized		
		1.4 Rubificated		
	2. Rendzina WRB name: Leptosol, calcaric	2.1 On marl, marly and soft limestone	For all subtypes: 1. Calcareous 2. Leached 3. Brownized 4. Colluvial	According to texture for 2. (1–3) and all varieties: 1. Sandy 2. Clayey 3. Loamy According to skelet content for 2.(1–3) and all varieties: 1. Low skelet content (<25%) 2. Moderate skelet content (25–50%) 3. High skelet content (>50%) According to depth for 2.(1–6): 1. Shallow (up to 20 cm) 2. Moderately deep (20–40 cm) 3. Deep (>40 cm)
		2.2 On loess and loess-like sediments		
		2.3 On moraine		
		2.4 On dolomite gruss		
		2.5 On calcareous sand		
		2.6 On calcareous gravel		
	3. Ranker (Humus-siliceous soil) WRB name: Leptosol eutric, dystric	3.1 Eutric	3.1.1 Lithic 3.1.2 Regolithic 3.1.3 Brownized lithic 3.1.4 Brownized regolithic 3.1.5 Colluvial	According to texture for all varieties: 1. Sandy 2. Clayey 3. Loamy According to skelet content for all varieties: 1. Low skelet content (<25%) 2. Moderate skelet content (25–50%) 3. High skelet content (>50%)
		3.2 Dystric	3.2.1 Lithic 3.2.2 Regolithic 3.2.3 Brownized lithic 3.2.4 Brownized regolithic 3.2.5 Podzolized lithic 3.2.6 Podzolized regolithic 3.2.7 Colluvial	
	4. Chernozem WRB name: Chernozem	4.1 On loess and loess-like sediments	4.1.1 Calcareous 4.1.2 Leached 4.1.3 Brownized 4.1.4 Calcareous gleyed 4.1.5 Leached gleyed 4.1.6 Brownized gleyed 4.1.7 Salinized and alkalized	According to A horizon depth for all varieties: 1. Shallow (<40 cm) 2. Moderately deep (40–80 cm) 3. Deep (>80 cm)
		4.2 On calcareous aeolian sand	4. (2–3) 1. Calcareous 2. Leached 3. Calcareous gleyed 4. Leached gleyed	
		4.3 On alluvial deposits		
	5. Vertisol (Smonitza) WRB name: Vertisol	5.1 Calcareous	5.1.1 On marl 5.1.2 On clayey sediments	According to solum depth for variety 5.3.2, and according to depth of the humus horizon for all other varieties: 1. Shallow (<40 cm) 2. Moderately deep (40–60 cm) 3. Deep (>60 cm)
		5.2 Non-calcareous	5. (2–3) 1. On clayey sediments 2. On basic and ultrabasic rocks	
		5.3 Brownized		

(continued)

Table 6.1 (continued)

Class	Type	Subtype	Variety	Form
III A-(B)-C or A-(B)-R Cambic soils	1. Eutric cambisol (Eutric brown soil) WRB name: Cambisol, eutric	1.1 On loess and loess-like sediments	1.1.1 Typical 1.1.2 Luvic 1.1.3 Typical gleyed 1.1.4 Luvic gleyed	According to texture for all varieties: 1. Sandy 2. Clayey 3. Loamy According to skelet content for all varieties: 1. Low skelet content (<25%) 2. Moderate skelet content (25–50%) 3. High skelet content (>50%)
		1.2 On basic and neutral eruptive rocks	1.2.1 Lithic 1.2.2 Regolithic 1.2.3 Vertic	
		1.3 On peridotite and serpentinite	1.3.1 Lithic 1.3.2 Vertic	
		1.4 On lacustrine sediments	1.4.1 Typical 1.4.2 Luvic 1.4.3 Vertic 1.4.4 Pseudogleyic	
		1.5 On alluvial, colluvial and aeolian deposits	1.5.1 Typical 1.5.2 Luvic 1.5.3 Vertic 1.5.4 Typical gleyed 1.5.5 Luvic gleyed 1.5.6 Vertic gleyed	
		1.6 On shale	1.6.1 Typical 1.6.2 Luvic 1.6.3 Vertic	
		1.7 On amphibolitic schists	1.7.1 Lithic 1.7.2 Regolithic 1.7.3 Luvic	
	2. Dystric cambisol (Dystric brown or brown acid soil) WRB name: Cambisol, dystric	2.1 Typical 2.2 Humus rich 2.3 Luvic 2.4 Psuedogleyic	For all subtypes: 1. On breccia 2. On sandstone 3. On shale 4. On sand 5. On clay 6. On chert 7. On acidic eruptive rocks 8. On neutral and basic eruptive rocks 9. On crystalline schists	According to solum depth for all varieties: 1. Shallow (<40 cm) 2. Moderately deep (40–70 cm) 3. Deep (>70 cm)
		2.5 Podzolic	2.5.1 On acidic silicate substrates 2.5.2 On quartz substrates	
	3. Kalkocambisol (Brown soil on limestone and dolomite) WRB name: no adequate name	3.1 Typical 3.2 Luvic	According to solum depth for all subtypes: 1. Shallow (<35 cm) 2. Moderately deep (35–50 cm) 3. Deep (>50 cm)	According to texture for all varieties: 1. Loamy 2. Clayey
	4. Terra rossa (Red soil) WRB name: no adequate name	4.1 Typical 4.2 Luvic	According to solum depth: 1. Shallow (<40 cm) 2. Moderately deep (40–70 cm) 3. Deep (>70 cm)	
		4.3 Ferralic		

(continued)

Table 6.1 (continued)

Class	Type	Subtype	Variety	Form
IV A-E-B-C or A-E-B-R Eluvial–Illuvial	1. Luvisol (Luvic or lessive) WRB name: Luvisol	1.1 On siliceous and calcaro-siliceous substrates	1.1.1 Typical 1.1.2 Pseudogleyic 1.1.3 Typical gleyed 1.1.4 Pseudogleyic gleyed 1.1.5 Podzolized	According to texture of A + E or P horizon: 1. Sandy 2. Clayey 3. Loamy According to skelet content: 1. Low skelet content (<25%) 2. Moderate skelet content (25–50%) 3. High skelet content (>50%)
		1.2 On pure limestone and dolomite	1.2.1 Typical 1.2.2 Acric (heather)	1.2.2.1 Outside karst sinkholes 1.2.2.2 In karst sinkholes
	2. Podzol WRB name: Podzol	2.1 Ferrous	According to depth of E horizon: 1. Weak (0–10 cm) 2. Moderate (10–20 cm) 3. Strong (>20 cm)	According to parent substrate for all varieties: 1. On quartz sandstone 2. On quartzite 3. On chert 4. On acidic eruptive rocks 5. On phyllite 6. On sand
		2.2 Humus-ferrous		
	3. Brunipodzol (Brown podzolic soil) WRB name: Podzol, humic	3.1 On quartz sandstone	According to lithic contact: 1. Lithic 2. Regolithic	According to texture for all varieties: 1. Sandy 2. Loamy According to skelet content: 1. Low skelet content (<25%) 2. Moderate skelet content (25–50%) 3. High skelet content (>50%)
		3.2 On quartzite		
		3.3 On chert		
		3.4 On acidic eruptive rocks		
		3.5 On phyllite		
		3.6 On sandstone		
V P-C Anthropogenic	1.Rigosol (Deeply ploughed soil) WRB name: Anthrosol	1.1 Vineyard soil (Vitisol)		
		1.2 Intensive orchard soil		
		1.3 Crop soil		
	2. Hortisol (Vegetable garden soil) WRB name: Anthrosol, hortic			
VI Technogenic soils I, II, III	1. Deposol (soil of disposal sites) WRB name: Anthrosol	1.1 According to material: soil, ore, sludge		
	2. Flotisol (Flotation material, deposits from waste water) WRB name: Anthrosol	1. According to depth and type of deposit		
	3. Aeroprecipitates (atmospheric deposition) WRB name: no adequate name			

(continued)

Table 6.1 (continued)

Class	Type	Subtype	Variety	Form
Order of hydromorphic soils				
I A-E/g-Bg-C Pseudo-gleyed	1. Pseudogley WRB name: Stagnosol	1.1 Plateau (on level ground) 1.2 Slope (on slopes)	According to depth: 1. Shallow (<25 cm) 2. Moderately deep (25–50 cm) 3. Deep (>50 cm)	For all varieties: 1. Eutric 2. Dystric
II Layers or (A)-G or (A)-C (Undeveloped)	2. Fluvisol (fluvial or alluvial) WRB name: Fluvisol	1.1 Calcareous 1.2 Calcareous gleyed 1.3 Calcareous gleyed both salinized and alkalinized 1.4 Non-calcareous 1.5 Non-calcareous gleyed 1.6 Non-calcareous both salinized and alkalinized	For all subtypes: 1. Shallow (<40 cm) 2. Moderately deep (40–80 cm) 3. Deep (80–120 cm) 4. Very deep (>120 cm) 5. Alluvial with fossil soil	According to texture for all varieties: 1. Sandy 2. Loamy 3. Clayey According to skelet content: 1. Low skelet content (<25%) 2. Moderate skelet content (25–50%) 3. High skelet content (>50%)
III A-C-G Semi-gleyed	1. Humofluvisol (Fluvial meadow) WRB name: Fluvisol, humic	1.1 Gleyed at a shallow depth (G between 100–150 cm) 1.2 Gleyed at a moderate depth (G between 150–200 cm) 1.3 Gleyed at a considerable depth ($G > 200$ cm)	For all subtypes: 1. Calcareous 2. Calcareous salinized and alkaline 3. Non-calcareous 4. Non-calcareous salinized or alkaline	According to texture for all varieties: 1. Sandy 2. Loamy 3. Clayey
IV A-G Gleyed	1. Pseudogley-gley WRB name: Stagnosol, gleyic			
	2. Humogley (Wetland black soil) WRB name: Gleysol, humic	2.1 Calcareous 2.2 Non-calcareous 2.3 Calcareous vertic 2.4 Non-calcareous vertic	According to degree of salinization for all subtypes: 1. Slightly salinized (0.25–0.50% salts) 2. Moderately salinized (0.50–0.70% salts) 3. Highly salinized (0.70–1.00% soli) According to degree of alkalinity for all subtypes: 1. Slightly alkaline (7–15% adsorbed Na) 2. Moderately alkaline (15–40% adsorbed Na) 3. Strongly alkaline (>40% adsorbed Na)	According to texture for all varieties of calcareous and non-calcareous subtypes: 1. Sandy 2. Loamy 3. Clayey

(continued)

Table 6.1 (continued)

Class	Type	Subtype	Variety	Form
	3. Eugley (wetland gley) WRB name: Gleysol	3.1 Hypogley	For all subtypes: 1. Mineral 2. Humic 3. Peat-gleyed (T up to 30 cm)	For all varieties of 3.1 (1–3): 1. Calcareous 2. Non-calcareous 3. Calcareous salinized and alkaline 4. Non-calcareous salinized and alkaline
		3.2 Epigley		For all varieties of 3.2 (1–3): 1. Calcareous 2. Calcareous vertic 3. Non-calcareous 4. Non-calcareous vertic
		3.3 Amphigley		For all varieties of 3.3 (1–3): 1. Calcareous 2. Calcareous vertic 3. Non-calcareous 4. Non-calcareous vertic
V T-G Peat	1. High moor peat WRB name: Histosol	1.1 Shallow (30–50 cm)	For all subtypes: 1. With a T_{fi} peat subhorizon 2. With a T_{he} peat subhorizon 3. With a T_s peat subhorizon	
		1.2 Moderately deep (50–100 cm)		
		1.3 Deep (>100 cm)		
	2. Transient peat WRB name: Histosol	2.1 Shallow (30–50 cm)	For all subtypes: 1. With a T_{fi} peat subhorizon 2. With a T_{he} peat subhorizon 3. With a T_s peat subhorizon	For all varieties: 1. Oligotrophic 2. Mesotrophic
		2.2 Moderately deep (50–100 cm		
		2.3 Deep (>100 cm)		
	3. Planohistol (Low moor peat) WRB name: Histosol	3.1 Shallow (30–50 cm)	For all subtypes: 1. With a T_{fi} peat subhorizon 2. With a T_{he} peat subhorizon 3. With a T_s peat subhorizon	For all varieties: 1. Oligotrophic 2. Mesotrophic
		3.2 Moderately deep (50–100 cm)		
		3.3 Deep (>100 cm)		
VI P-G Anthropogenic	1. Deep-ploughed peat WRB name: no adequate name			
	2. Hydroameliorated WRB name: no adequate name	According to categorization of parent soil		

(continued)

Table 6.1 (continued)

Class	Type	Subtype	Variety	Form
Order of halomorphic soils				
I Asa-G or Asa-CG (acutely salty)	1. Solonchak WRB name: Solonchak	1.1 Soda	For all subtypes: 1. Surface salinization (0–10 cm) 2. Moderately deep salinization (30–80 cm) 3. Deep salinization (80–125 cm)	For all varieties: 1. Ochric 2. Mollic 3. Vertic
		1.2 Sulphate		
		1.3 Chloride		
		1.4 Soda-sulphate		
		1.5 Sulphate-soda		
		1.6 Soda-chloride		
		1.7 Chloride-soda		
		1.8 Sulphate-chloride		
		1.9 Chloride-sulphate		
		1.10 Mixed, with three salts		
II A/E-Bt, on–C	1. Solonetz WRB name: Solonetz	1.1 Solonetz-solonchak	For subtypes: 1. Non-saline 2. Soda 3. Sulphate-soda 4. Chloride-soda 5. Chloride-sulphate 6. Sulphate-chloride 7. Sulphate 8. Chloride	According to depth of $B_{t,na}$ horizon for all varieties: 1. Shallow (up to 7 cm) 2. Moderately deep (7–15 cm) 3. Deep (>15 cm)
		1.2 Typical		
		1.3 Mollic		
		1.4 Luvic		
		1.5 Pseudogleyic	1.5.1 Typical 1.5.2 Glossic	
Order of subaquatic soils				
I (A)-CG	1. Protopedon WRB name: no adequate name			
II A-C or A-G	1. Gyttja WRB name: no adequate name			
	2. Dy WRB name: no adequate name			
	3. Sapropel WRB name: no adequate name			
III P-C or P-G Anthropogenic	1. Drained gyttja WRB name: no adequate name			
	2. Drained sapropel WRB name: no adequate name			

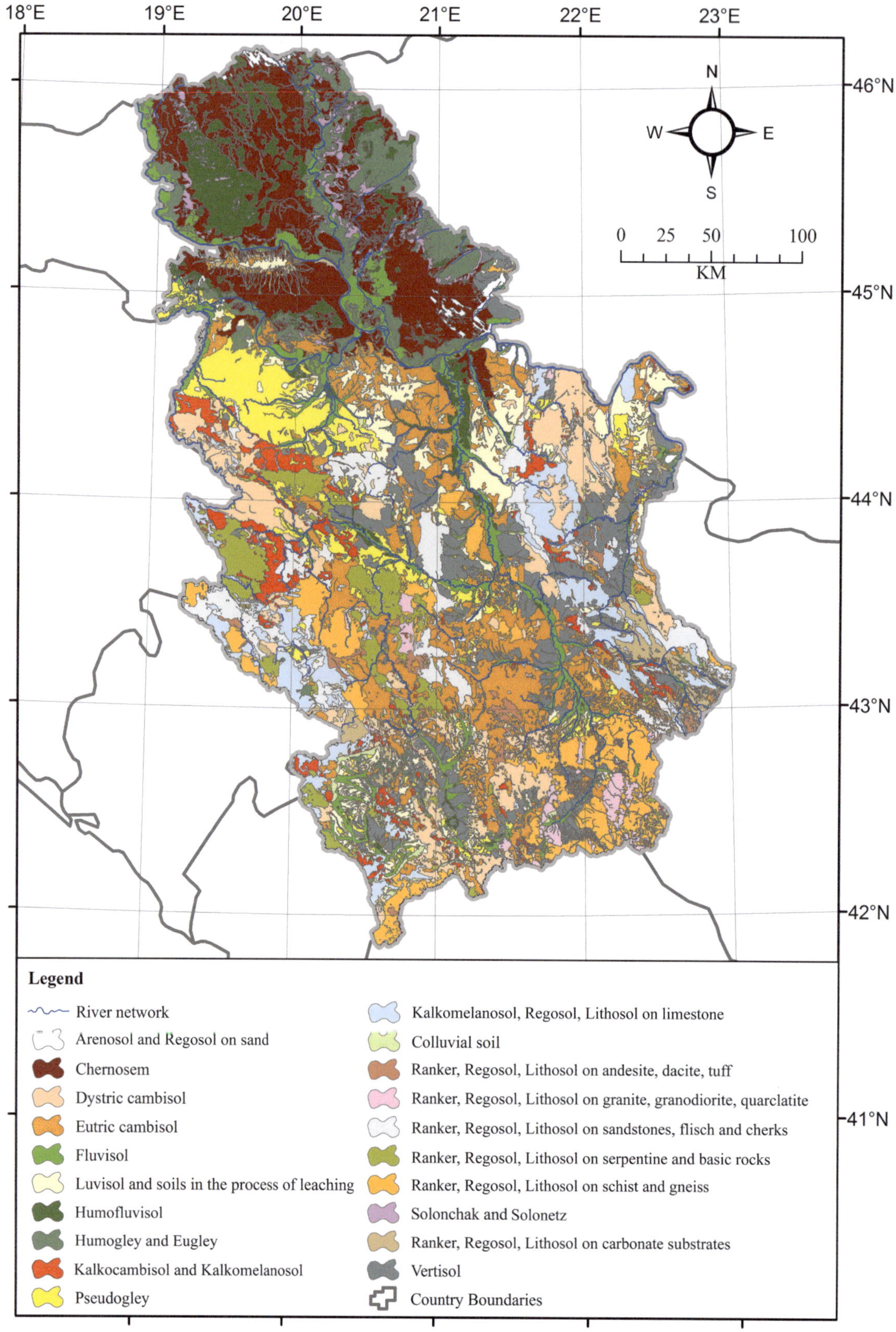

Fig. 6.1 Soil map (Mrvić et al. 2013)

The pedological map of Serbia has been created on the basis of existing, digitalized pedological maps at a scale of 1:50,000, produced by the Institute of Soil Science (Belgrade), the Institute of Field and Vegetable Crops (Novi Sad), the 'Jaroslav Černi' Institute for the Development of Water Resources (Belgrade), including data from the Environmental Protection Agency of the Serbian Ministry of Agriculture and Environmental Protection. For analysing spatial data, the ArcGIS software package was used. In a vector format with ESRI shapefiles, the soil map had an attributes table complemented with a new column with the text of the revised classification. Thereafter, we merged the data using Dissolve (Arc Toolbox/Data Management). The newly created multipart shapefiles could then be converted to singlepart, creating a feature class containing singlepart features generated by separating multipart input features (Fig. 6.1).

References

Antonović G (1972) Soil map of SFRY, sheet Pirot 1, 1:50,000. Institute of Soil Science, Belgrade (in Serbian)

Antonović G, Mrvić V (eds) (2008) Soils of the Nišava river basin. Institute of Soil Science, Belgrade (in Serbian)

Antonović G, Protić N (1997) Proposal for the classification of the soils of Yugoslavia. Acta Biol Yugoslavica—Ser A: Soil Plant 46(2):97–106

Antonović G, Nikodijević V, Tanasijević Đ, Vojinović Lj (1975) Soils of the Braničevo-Zvižd district and Homolje, vol 3, no 126. Institute of Soil Science, Belgrade (in Serbian)

Antonović G, Milivojević J, Miladinović M, Ševarlić M, Lješević M, Protić N, Đorović M, Radojević R, Stojanovic S, Rudić D, Spasov P, Andrić S, Kuburić J (1997) Methodology for devising a plan for the protection, regulation and use of agricultural land. Yugoslav Authors Agency, S-25/97, Belgrade (in Serbian), p 470

Antonović G, Mrvić V, Saljnikov E, Perović V, Kostić Kravljanac Lj, Nikoloski M (2010) Pedological Cartographie and Soil Classification in Serbia. Soil Plant, Belgrade 59(3):139–157

Antonović G, Mrvić V, Saljnikov E, Perović V, Nikoloski M, Jaramaz, D (2011) Proposal for the new Serbian soil classification. Soil Plant, Belgrade 60(3):93–101

Filipovski G, Neugebauer V, Ćirić M, Škorić A, Živković M (1964) Soil classification in Yugoslavia. 8th international congress of soil science, Bucharest

Gračanin M (1950) A supplement to the genetic classification of soils, p 280. Rad JAZU (Yugoslav Society of Soil Science), Zagreb (in Serbo-Croatian)

Kovačević P (1956) Problems of classifying and mapping soils and their uses in practice. Agron J, Zagreb 4

Kovačević P, Kalinić M, Pavlić V (1967) Detailed soil classification and the production of a soil map for Croatian Posavina at a scale of 1:50,000. Soil Plant (Belgrade) 16(1–3):247–259

Kubiëna W (1953) Bestimmungsbuch und Systematik der Böden Europas. Stuttgart

Mrvić V, Antonović G, Čakmak D, Perović V, Maksimović S, Saljnikov E, Nikoloski M (2013) Pedological and pedogeochemical map of Serbia. In: Proceedings of the 1st international Congress on Soil Science, XIII National Congress in Soil Science: Soil-Water-Plant, 23–26 Sept, Belgrade, Serbia, pp 93–105

Mückenhausen E, Vogel F (1962) Entstehung, Eigenschaften und Systematik der Böden der Bundesrepublik Deutschland. Deutsche Bodenkundliche Gesellschaft, Frankfurt am Main, p 148

Neugebauer V, Ćirić M, Živković M (1961) Soil map of Yugoslavia (scale of 1:1,000,000), Yugoslav Society of Soil Science, p 107

Neugebauer V, Ćirić M, Filipovski G, Škorić A, Živković M (1963) Classification of Yugoslav soils (Yugoslav Society of Soil Science). Soil Plant (Belgrade) XII: 1–3 (in Serbian)

Neugebauer V, Živković B, Tanasijević MĐ, Miljković N (1971a) Soil map of Vojvodina 1:50.000, sheet Sremski Karlovci 2. Institute for Agricultural Research, Novi Sad

Neugebauer V, Živković B, Tanasijević Đ, Miljković N (1971b) Soil map of Vojvodina scale 1:50,000, sheet Sremski Karlovci 2. Institute for Agricultural Research, Novi Sad

Pavićević N, Antonović G, Nikodijević M, Tanasijević Đ (1968) Soils of Stari Vlah and Raška. Institute of Soil Science, Belgrade (in Serbian)

Pavićević N, Grujić L, Ljubomir M, Petar K, Gradimik V (1974) Soil map of the socialist autonomous province of Kosovo, SR Serbia, scale: 1:50,000. The Jaroslav Černi Institute for the Development of Water Resources, Belgrade

Protić N, Martinović Lj, Miličić B, Stevanović D, Mojasevic M (2005) The status of soil surveys in Serbia and Montenegro. In Jones RJA, Houšková B, Bullock P, Montanarella L (eds) Soil resources of Europe, second edition, Research report no. 9. European Soil Bureau, Luxembourg, pp 297–315

First YSSS Assembly (1954) Conclusions from the first YSSS assembly, 8–11 Nov 1953. YSSS Bulletin, No. 2, Belgrade

Stebutt A (1926) Our main agricultural regions. 'St. Sava' Press, Belgrade (in Serbian)

Stebut A (1930) Lehrbuch der allgemeinen Bodenkunde, Berlin

Stebut A (1931) Pedološka karta Kraljevine Jugoslavije: pregled tipova zemljišta. Geografsko društvo, Beograd, Štamparija "Davidović, Pavlovića i druga", p 15

Škorić A, Filipovski G, Ćirić M (1973) classification of Yugoslav Soils. A basic soil map of Bosnia and Herzegovina 1:50,000. Faculty of Forestry, Sarajevo (in Serbian)

Škorić A, Filipovski G, Ciric M (1985) Classification of Yugoslav Soils. Academy of Sciences and Arts of Bosnia and Herzegovina, Sarajevo, Interacademic Board for Soil Science, Book 13, pp 71 and 72 (in Serbian)

Tanasijević Đ, Pešić BR (1965) Soils of the Velika Morava and Mlava rivers. Union ofAgricultural Engineers and Technicians of Yugoslavia, Belgrade (in Serbian)

Tanasijević Đ, Pešić R (1967) Soils of the Velika Morava and Mlava basins. Springfield, Clearinghouse for Scientific and Technical Information, VA, USA

Tanasijević Đ, Antonović G, Aleksić Ž, Pavićević N, Filipović Đ, Spasojević M (1966) Pedologic cover of western and north-western Serbia. Institute of Soil Science, Belgrade, p 382 (in Serbian)

Vojinović Lj, Jeremić M (1974) Soils of the Timok river basin. Institute of Soil Science, Belgrade (in Serbian)

World Reference Base for Soil Resources (2006) FAO, ISRIC and ISSS, Rome

Živković M (1961) Problems with local soil classification. Contemporary Agriculture Novi Sad 10 (in Serbian)

Order of Automorphic Soils

7

Automorphic soils predominate in Serbia, covering approximately 66,323.58 km^2 or 74.9% of the land area. This order is divided into four classes according to the degree of profile development.

7.1 Class I: Undeveloped Soils with (A)-C or (A)-R Soil Horizon Sequence

Soil types in this class have a humus horizon in the initial phase of development. It has not formed an organo-mineral complex with the mineral part of the soil, nor impacted on its evolution in any way. On compact rocks, this horizon forms amongst moss and lichens, while on loose substrates, it develops under the influence of higher plants that take root easily. The parent substrate has a major impact on the physical and chemical characteristics of these soils. Extreme climatic conditions and water or wind erosion can mean that undeveloped soils remain at this stage almost permanently. However, in the climatic conditions that prevail in Serbia, they evolve relatively quickly. In Serbia, these undeveloped soils are called sierozems (from the Russian meaning *crude earth*).

7.1.1 Lithosol

The name of the soil type. Its name comes from the word *lithos* (rock) because rock completely dominates. The subtypes of this soil according to the classification system used in Serbia are as follows: lithosol on acidic rocks, lithosol on neutral and basic rocks, lithosol on peridotite, and lithosol on limestone and dolomite (Škorić et al. 1985). In Serbia, these soils are known as sierozems on compact rocks.

The WRB name. Leptosol (Lithic, Calcaric, Dystric, Eutric, Haplic).

Genesis and sequence of horizons. They are formed on rocks which produce rocky detritus during the phase of physical disintegration (peridotite and serpentinite rocks, quartzite, quartz-porphyrite and chert), Fig. 7.1. Pure Mesozoic limestone does not usually disintegrate mechanically, but brecciated and karstified limestone or limestone from fault zones can produce detritus, a condition for the occurrence of lithosol. This detritus only remains in situ on flattened terrains, while on slopes it is transported by gravity in the form of scree.

Climatic regions with marked temperature oscillations, which accelerate mechanical weathering, are favourable for the formation of this soil type. These soils can be divided into groups according to the character of physical disintegration: lithosols on igneous and metamorphic rocks, and lithosols on limestone and dolomite, where chemical weathering dominates. The formation of lithosols on acidic igneous rocks (granite and diorite) progresses considerably more quickly than on neutral rocks, and even more so than on basic rocks (gabbro, diabase, and basalt). The pedogenesis of lithosols on basic rocks is slow, and they remain in the initial stages of development for a long time. The soil forms in situ through the physical disintegration and erosion of fine particles. In the extreme ecological conditions in which these soils originate, only meagre, specific vegetation grows, providing only a small amount of organic residue. This is easily washed out through the large pores of the rock detritus, which is why there is little and only sporadic accumulation of humus here. The morphology of the profile of this soil type is (A)-R or (A)-C-R.

Position in an evolutionary series. Lithosols are the first evolutionary phase on physically altered hard rocks, and the pedogenesis of these soils is slow due to the young age of the soil, severe erosion or unfavourable climatic conditions. They evolve in the direction of kalkomelanosol or dystric cambisol on acidic parent substrates, eutric cambisol on rocks rich in bases or carbonates and kalkocambisol on limestone and dolomite.

Distribution. These are the soils of mountainous regions where low temperatures are the main factors in the

P. Pavlović et al., *The Soils of Serbia*, World Soils Book Series,
DOI 10.1007/978-94-017-8660-7_7

Fig. 7.1 Lithosol (*photo* Mrvić, V.)

disintegration of rocks. They form on slopes in hilly and mountain areas, on different geological substrates, and also on level to partly sloping land and on the gentle slopes of rocky massifs. In Serbia, they are found at high altitudes (on hard limestone above 1800 m), where extremely unfavourable conditions for the development of vegetation predominate. They have been recorded on Šara Mt., at an altitude of 2300 m a.s.l., on the blocks of acid silicate rocks of gneiss and granite, and on blocks of serpentinite (Jović et al. 1976), as well as on limestone and dolomite in *Pinus heldreichii* forest regions (Jović 1969).

Physical and chemical properties. Lithosol can be up to 20 cm deep, and beneath it there is compact or poorly fragmented rock. All its properties depend on the characteristics of the substrate on which it forms. Horizon (A) is characterized by the fragmentary accumulation of humus in some large pores, and due to the ease with which it can fall through these, isolated clumps of humus can also be found at greater depths. The domination of the rock fraction is the major factor in determining the physical properties of lithosol. It is characterized by its extreme permeability to water and the absence of water retention properties. Due to minimal contact with the liquid phase, the chemical processes connected to the solid phase are drastically reduced, and hence the differences between the mineral and chemical composition of the substrate are minimal. Because of the small adsorption area and the reduced liquid phase, lithosols are poor in plant-available nutrients. Soil depth varies depending on where it forms. On flattened terrains, they are shallower (10–20 cm), while on slopes where material is redistributed due to gravity, deeper layers may form (scree). The solum can be virtually unified or can show signs of layering, when, in addition to the gravitational redistribution of dry material, surface water also moves material. The major process is the physical disintegration of rock, with particles <2 mm predominating. They can be found among the larger particles, most frequently beneath the surface layer, on the hard parent rock, marking the initial phase of the (A) horizon. In this zone, the accumulation of humic compounds occurs.

Land use. Lithosols are extremely dry habitats exposed to marked temperature increases and are unfavourable for the development of plant root systems. In these habitats, adapted xerothermic plant species are found. On scree, plants are exposed to mechanical damage due to the mobility of the substrate. They are used as low-quality pasture.

Guidelines for management. Afforestation with oligotrophic plant species adapted to the xerothermic conditions of the habitat is possible. In this way, these soils are protected from water and wind erosion, facilitating their evolution towards more complex pedological stages.

7.1.2 Regosol

The name of the soil type. Regosols (*rhegos*—blanket) are undeveloped or poorly developed soils on loose substrates and substrates which break up easily (the exception being alluvial, diluvial, and aeolian deposits and skeletal detritus). They occur through the erosion of previously formed soils and the initial processes of pedogenesis, which has not resulted in the formation of humus A horizon due to youth, erosion or anthropogenic influences. The subtypes of this soil according to the classification system used in Serbia are as follows: siliceous regosol, calcaro-siliceous regosol and sandy-dolomite regosol. In Serbia, these soils are known as sierozems on loose substrates.

The WRB name. Regosol (Lithic, Calcaric, Dystric, Eutric, Haplic).

Genesis and sequence of horizons. The parent substrate on which this soil forms breaks up easily and gives regosol its fine granulometric composition. Such substrates include igneous rocks with a granular structure, some crystalline schists and sandstones, marlstone and marly limestones, and crystalline dolomites. Vegetation contributes to the formation of this soil type as the roots accelerate weathering, and through the accumulation of organic residue and its humification, this soil evolves into the later developmental stages (Fig. 7.2). Climate has an indirect impact on the formation of this soil by causing the erosion and denudation of rocks, while it directly affects mechanical weathering, meaning areas with intensive frosts or high temperatures are more prone to disintegration and the formation of regosol. These soils represent an initial phase in soil evolution. As a rule, there are as many regosols as there are parent substrates: calcareous, clayey, sandy, siliceous and calcaro-siliceous. All regosols are characterized by pronounced petrographic material, and hence their characteristics boil down to the characteristics of the parent substrate. They all have a very weak, thin humus-accumulative horizon with organic matter only mechanically mixed with the mineral component. Humus consists of almost pure organic matter affected by the transformation process to varying degrees. The morphology of the profile of this soil is (A)-C, and the pattern of humus in the surface (A) horizon is fragmented. Regosols on loess or glacial deposits are always deep because their C horizon correlates with the thickness of the layers of these sediments. The C horizon, which makes up the main part of the profile, can be homogenous (e.g. on loess) or can inherit a degree of layering from the substrate (marlstone or redeposited regolith).

Position in an evolutionary series. Regosols originate through the erosion of previously formed soils and through initial processes of pedogenesis, which has not resulted in the creation of humus A horizon (mollic) due to youth or the impact of anthropogenic factors. The initial character of the regosol stage represents the weakest evolution of the organic and mineral components. This soil evolves into soils with an A-C or A-R sequence of horizons. The wide variations which exist between different types of regosol point to the very varied conditions for their formation, not only during the initial stages, but also for the further members of the evolutionary series. In many cases, it is actually the characteristics of regosol which determine how pedogenesis progresses. Classification at the level of varieties and forms is made according to the criteria set out in Table 6.1 (Chap. 6 Soil Classification).

Distribution. In Serbia, the relatively small amount of regosol that there is appears on smaller areas of exposed marl, where erosion has removed the soil material. It is mostly found in lowland regions and on plains, mainly on loose, sedimentary rocks or diluvial and alluvial deposits. Most commonly it forms on loess, loess-like material, sands and clays. In mountain regions, it occurs locally on igneous rocks, crystalline dolomites and schists. It has been found on Fruška Gora and in Šumadija.

Physical and chemical properties. The depth of these soils depends on the rock type; hence, regosols on marl are usually deeper than on marly limestones and igneous rocks, unless there is a colluvial accumulation of their regolith. Their physiological depth is often greater than the physical depth of the soil profile (plant root systems can also develop in loose substrate). Regosols already have some marked physical and chemical characteristics. The water-holding capacity of certain regosols such as sandy, clayey and those on loess can already meet the needs of higher vegetation. Hence, regosols do not go through the stage characterized by

Fig. 7.2 Regosol (*photo* Mrvić, V.)

pioneer species, so soil development progresses considerably more quickly, which is why regosols as an initial phase are far less widespread than lithosols.

The water-holding capacity of these soils depends on the mechanical composition. Regosols on marlstone are clayey, on loess are loamy, and on crystalline dolomite and schists are sandy. Regosols that form on gruss of igneous rocks are dominated by fractions of coarse sand and fine gravel. Those on loess have the most favourable water–air properties.

The chemical characteristics depend entirely on the mineralogical properties of the parent substrate. The greatest difference exists between soils formed on carbonate substrates (on marlstone, loess, calcareous sandstones) and siliceous substrates (regolith of igneous rocks and crystalline schists). The presence of carbonate gives a weak alkaline reaction (pH = 7.8–8.5), with the domination of Ca and Mg ions in the adsorption complex. Regosols on marlstone have greatest adsorption capacity because the clay fraction is dominated by montmorillonite clay. Other regosols, apart from that on loess, have low adsorption capacity. Siliceous regosols on basic eruptive rocks have neutral chemistry and a base saturation >50%. Regosols on schists and sandstones are usually acidic and have a base saturation <50%. The humus content is usually low (<1%), but regosols have more humus than lithosols on solid rocks. They have a low content of readily available nutrients.

Land use. There are significant differences between some forms of regosol: regosols on loess and marlstone have better air–water properties and a high base saturation and as such are suitable for agricultural production (orchards and vineyards).

Guidelines for management. The most important ameliorative measures are halting erosion processes, thereby allowing humus accumulation and an increase in fertility, and compensating for the lack of phosphorus and nitrogen through the use of mineral fertilizers. Siliceous regosols occupy only small areas in mountain regions. For the most part, they are coarse in texture and hence these are dry habitats, particularly on southern slopes where as a rule they are found. Hence, they are prone to more pronounced erosion. Afforestation prevents erosion, allows an increase in fertility, and facilitates the evolution of regosols to the next developmental stage.

7.1.3 Arenosol

The name of the soil type. This is a markedly sandy soil (Latin *arena*—sand), formed through aeolian processes. Local names for it are *living sand* and *flying sand.* Subtypes of this soil according to the classification system used in Serbia are as follows: quartz sands (>95% quartz), siliceous (>95% quartz) and calcaro-siliceous (>95% SiO_2 + $CaCO_3$) (Škorić et al. 1985).

The WRB name. Arenosol.

Genesis and sequence of horizons. Due to the surface layers being continually moved by winds, a true humus A horizon does not form, meaning the soil can remain an undeveloped soil with an (A)-C horizon sequence permanently. In the (A) horizon, the partial accumulation of decomposed organic matter can be noted in the form of isolated spots or veins.

Position in an evolutionary series. With the development of vegetation, it gradually evolves into a soil with an A-C soil horizon sequence. Classification at the level of varieties and forms is made according to the criteria set out in Table 6.1 (Chap. 6 Soil Classification).

Distribution. Arenosols in Serbia are mostly found on the Pannonian Plain (Deliblato and Subotica Sands) and on the right Danube terrace upstream from Đerdap (Ram, Gradište and Golubac Sands), Figs. 7.3 and 7.4.

The majority of these soils have been 'calmed' by the planting of vegetation, and only a small proportion have a surface layer of mobile sand with undeveloped soil. Arenosols characteristically form on valley-like terrain. Aeolian deposits can even be loamy and clayey. In Serbia, arenosols cover approximately 0.6% of the land mass, i.e. about 571.23 km^2.

Physical and chemical properties. Arenosols are particularly dry and warm soils, well-aerated, with a low water holding capacity. The depth of the aeolian sandy deposit can be between 50 and 200 cm. This layer is homogenous throughout the whole depth, and beneath it various sediment or fossil soils can be found. In view of its aeolian origin, the granulometric composition of the soil is dominated by a fraction of fine sand (90–95%), while the proportion of coarse sand can be up to 2%. Quartz is the dominant mineral in arenosol. In the aeolian sands of Vojvodina, it accounts for more than 70% of the total minerals, followed by feldspars (18–30%), while the heavy mineral fraction makes up only 3–5%.

In the A horizon, only fragmentary humus accumulation can be noted. The humus content in loose sand is lower than 1 and 0.5% in bound sands. Some sands contain carbonates (calcite and dolomite), while others are non-calcareous or siliceous. Their adsorption capacity is low, and they have N and P deficiencies.

Land use. Arenosols have low production potential. The natural vegetation is xerophytes. At Deliblato Sands, these are mainly semi-desert plant communities and less commonly grass-steppe communities. Crops on these soils suffer from drought.

Guidelines for management. The most effective ameliorative measure is afforestation with xerothermic tree species.

Fig. 7.3 Arenosol (*photo* Mrvić, V.)

Black pine, black locust, manna ash and Austrian oak are most commonly used for this. Other measures include adding clay, organic fertilizers, and irrigation.

7.1.4 Colluvial Soils (Colluvium)

The name of the soil type. The name derives from the Latin word *coluo*—to wash out, to rinse, highlighting the way these soils originate through the washing out and deposition of soil material. The term diluvial soil was used in the past. Subtypes of this soil according to the classification system used in Serbia are as follows: eutric-siliceous colluvial soil, dystric-siliceous colluvial soil, calcareous colluvial soil and colluvial soil with fossil soil.

The WRB name. Colluvial soil.

Genesis and sequence of horizons. Colluvial deposits form at the base of slopes where eroded soil and rock material from higher reaches accumulates. These soils are characterized by sharp-edged skeletal material. On the upper slopes, they consist mainly of sand and coarser skeletal material, while in lower reaches they comprise finer (colloid) material. Colluvial soils do not usually exhibit the layering characteristic of alluvial soils. As colluvial deposits are one result of soil erosion, the formation of colluvial soils is affected by the same factors that influence erosion (the destruction of vegetation, heavy rain, the level of drainage and binding of the substrate, and inadequate soil management).

The only exception is the influence of the terrain. Here, it is a factor in material accumulating rather than being washed away with the terrain characteristic for colluvial soils being level to gently sloping land or plains abutting slopes. Colluvial soils are usually deep and, besides an (A) horizon with an insignificant amount of humus, may have a slight degree of layering. In the deeper layers, fossil soil horizons can be found, Fig. 7.5. Classification at the level of varieties and forms is made according to the criteria set out in Table 6.1 (Chap. 6 Soil Classification).

Distribution. The largest expanses are found in the Vranje valley and Kosovo and Metohija regions, while it occurs only locally in other hilly and mountain regions (Fig. 7.6). In Serbia, colluvial soil covers approximately 1.4% of the land mass, i.e. about 1279.35 km^2.

Physical and chemical properties. In terms of texture, colluvial soils are very heterogeneous (both horizontally and vertically). Deposition does not lead to any significant granulometric sorting so there is no obvious layering. They range from clayey to sandy. Colluvial soil from acid rocks has many coarse fragments and a little clay, while that from base rocks is richer in clay. Almost without exception, these soils contain coarse fragments, making them well-drained, aerated and warm. They have a low water holding capacity, meaning specific xerothermic vegetation grows on them.

Physical and chemical characteristics depend on the soils and rocks the colluvial soils originated from through erosion. As these factors can vary greatly, it is impossible to talk about the characteristics colluvial soils share other than a

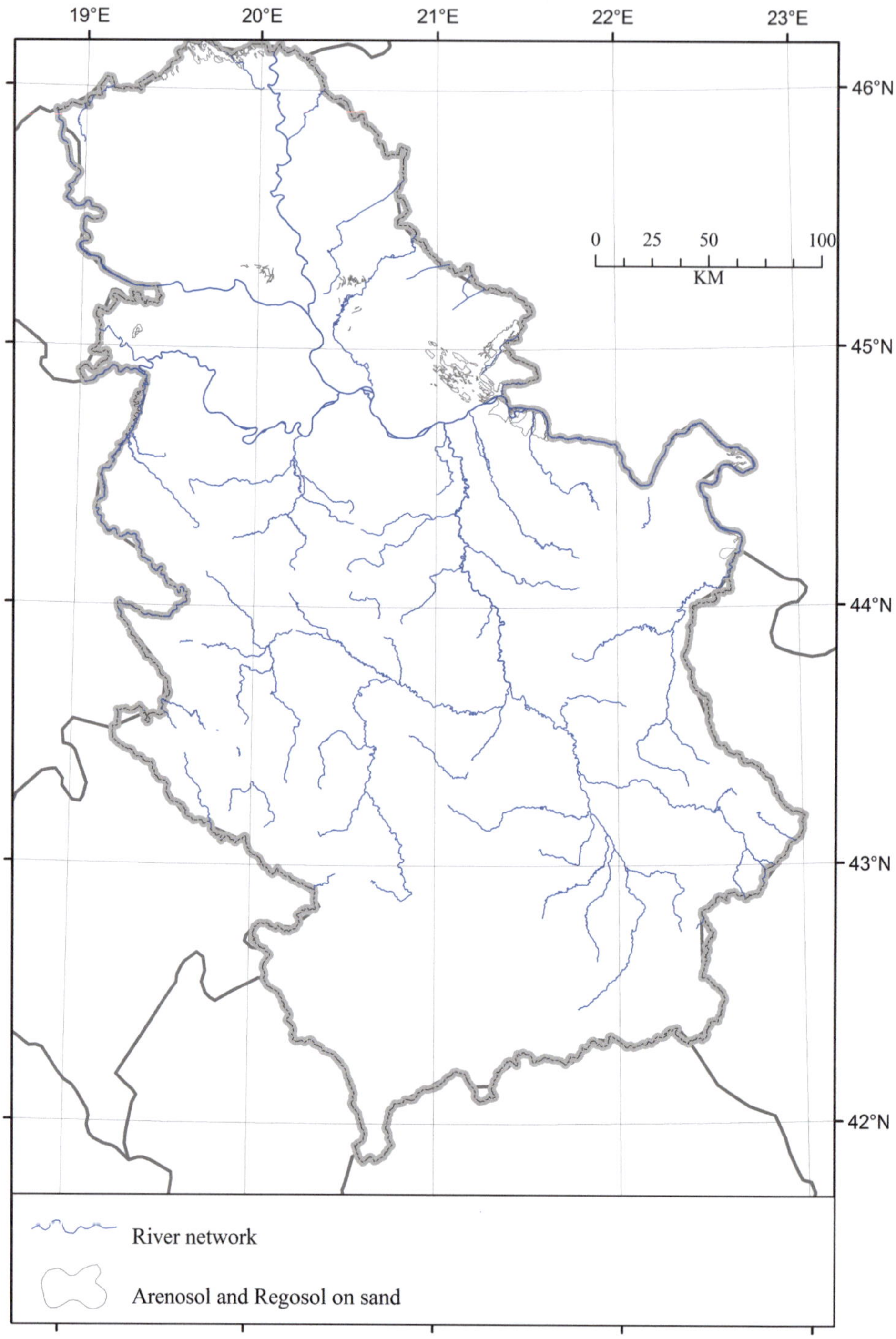

Fig. 7.4 Distribution of arenosol and regosol on sand

lack of humus and a marked soil skeleton. Greater humus content can be found in colluvial soils which have a higher proportion of soil material. However, this is not humus that forms in the soil itself, but that which has been transported with the eroded material. They can be calcareous or non-calcareous. The pH of non-calcareous colluvial soils is neutral to slightly acidic, while calcareous colluvial soils are neutral to slightly alkaline. The adsorption capacity depends on the clay particle content, while the soil chemistry affects the degree of base saturation.

Land use. The ecological and productive features of colluvial soils vary and depend on the physical and chemical characteristics of the soils and rocks they originated from. The production value and usefulness of these soils is largely determined by their texture. The least favourable colluvial soils are those which have a lot of coarse fragments while

Fig. 7.5 Colluvial soil (*photo* Čakmak, D.)

those with fewer such fragments and where ground water is deep are more favourable. They are used as arable land, orchards, pastures or forest land.

Guidelines for management. The management of these soils involves erosion protection, humization, fertilization, deep ploughing and irrigation. Afforestation of colluvial soils is necessary in order to combat erosive processes and to halt the influx of new material.

7.2 Class II: Humus-Accumulative Soils with A-C, A-R or A-C-R Soil Horizon Sequence

Soils of this class are characterized by the presence of a well-developed humus-accumulative horizon as the only genetic horizon that can be considered diagnostic. If they form on compact rocks, the humus horizon lies directly on them (lithic contact) with an A-R soil profile. If, however, the humus horizon lies on loose substrate, an A-C-R soil profile forms with a frequent transitional AC horizon. This class encompasses a large number of soil types with different conditions for pedogenesis. In Serbia, the name black soil is used for the whole class.

7.2.1 Ranker

The name of the soil type. With its name derived from the word *rank*—slope (Kubiena 1948, 1953), this soil forms on steep slopes. Like its substrates, it is non-calcareous and can be neutral, moderately acidic or highly acidic depending on its natural substrate and altitude. Subtypes of this soil according to the classification system used in Serbia are as follows: eutric ranker, which has a mollic or ochric humus horizon and forms on neutral or basic siliceous rocks, and dystric ranker, which has a limited O horizon and/or umbric horizon and forms on acidic siliceous and quartz rocks (Škorić et al. 1985). In Serbia, these soils are also called humus-siliceous soils, as they are rich in humus and primary silicates.

The WRB name. Leptosol eutric, dystric.

Genesis and sequence of horizons. Ranker is a soil which has a mollic, umbric or ochric horizon, which most

Fig. 7.6 Distribution of colluvial soil

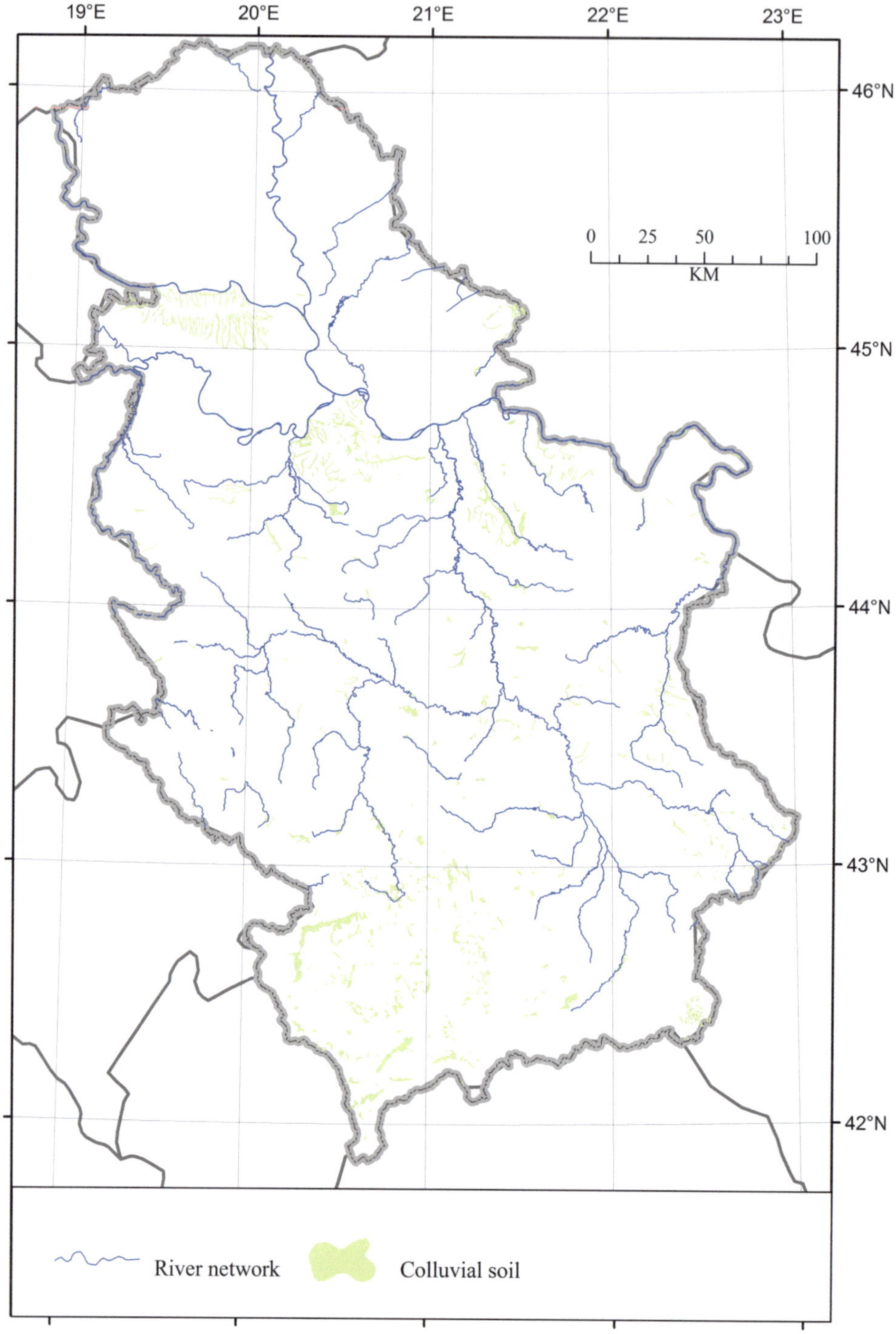

frequently lies directly on hard rock and more rarely on the products from the mechanical weathering of the rock (e.g. gruss of eruptive rocks). Depending on this, an A-C-R soil profile (lithic) or A-AC-C-R (regolithic) horizon sequence forms (Fig. 7.7). Shallow rankers, formed on compact rocks, do not have a transitional AC horizon, but those formed on a loose substrate (diluvial deposits) or a substrate which quickly breaks up into a loose mass can have one. Some subtypes of ranker which are overgrown with dense forest or dense grass vegetation also have a well-expressed organogenic horizon (O). Those subtypes in a stage of transition towards other soil types have a cambic horizon. The depth of the whole profile varies from just a few centimetres to 60 cm. It is characteristic for sharp-edged skeletal particles to occur throughout the whole depth of the profile.

The mull stage of ranker is most commonly found on serpentinite, at different altitudes, where it marks a transitional stage. All the evolutionary phases within the A-C

Fig. 7.7 Distric ranker and eutric ranker (*photo* Čakmak, D.)

stage occur, depending on the incline of the terrain. For example, protoranker soil occurs on north-facing slopes of 45° populated by a hornbeam and sessile oak (*Ostryeto—Quercetum sessiliflorae serpentinicum* subass. *seslerietosum* Vukićević 1965) association. The woody species of this association grow between the large blocks of rock where organic matter has collected. On more gentle slopes, with less humus and a greater soil depth, sessile oaks (*Quercetum montanum serpentinicum* Jovanović 1959) grow. Mull-moder ranker occurs as a partially developed A-C stage on igneous and metamorphic rocks (granite, gneiss, phyllite, etc.), but erosion processes inhibit the further development of this stage.

In places where erosion is less pronounced, leaching is more intensive and the occurrence of clay minerals is greater, with the transition to the next stage easier. The presence of moder ranker to moder-mor ranker in Serbia is linked to an altitude >1500 m, but can even be found up to 1900 m a.s.l. (Jović 1965), where woody vegetation [Norway spruce (*Picetum excelsae serbicum silicicolum*)] and shrub vegetation [European blueberry, juniper berry and spruce (*Vaccinieto–Junipereto–Picetum*)] grow (Fig. 7.8).

Position in an evolutionary series. Lithic rankers form directly on hard (compact) siliceous and quartz rocks as well as the eluvium from these rocks. Regolithic rankers form on substrates which break up easily. The A-C stage is brief and, as humus accumulates, a cambic (B) horizon also begins to form, i.e. brownized soil begins to form. Rankers form on non-calcareous rocks, on highly acidic, quartz-rich rocks (quartzite/quartz latite, sandstone and chert), and on basic rocks (gabbro, basalt, diabase and serpentinite). They most often occur on siliceous acidic substrates (acidic igneous and metamorphic rocks: granite, gneiss, mica-schist, phyllite, shale, sandstone, etc.). The formation of rankers is characterized by processes of humus accumulation, intensive physical and weak chemical weathering of rocks, intensive base leaching from the whole profile and intensive erosion. Humus accumulation arises from unfavourable water and temperature extremes which decelerate the mineralization of humus. Such conditions also slow chemical weathering of rocks, meaning rankers contain a lot of coarse fragments. During their development, rankers on basic rocks evolve into eutric cambic soils, while rankers on siliceous rocks become dystric cambisol, brown podzolic soils and podzol. On the soil map of Serbia, the same colour is used to denote undeveloped soils (regosol and lithosol) and humus-accumulative soils (ranker) on the same parent substrates (Fig. 6.1, Chap. 6 Soil Classification). In this way, it is clearly highlighted that these form evolutionary series on the same substrates (andesite, dacite, tuff, Fig. 7.9a; granite, granodiorite, quartz latite, Fig. 7.9b; sandstone, flysch and chert, Fig. 7.9c; serpentinite and basic rocks, Fig. 7.9d;

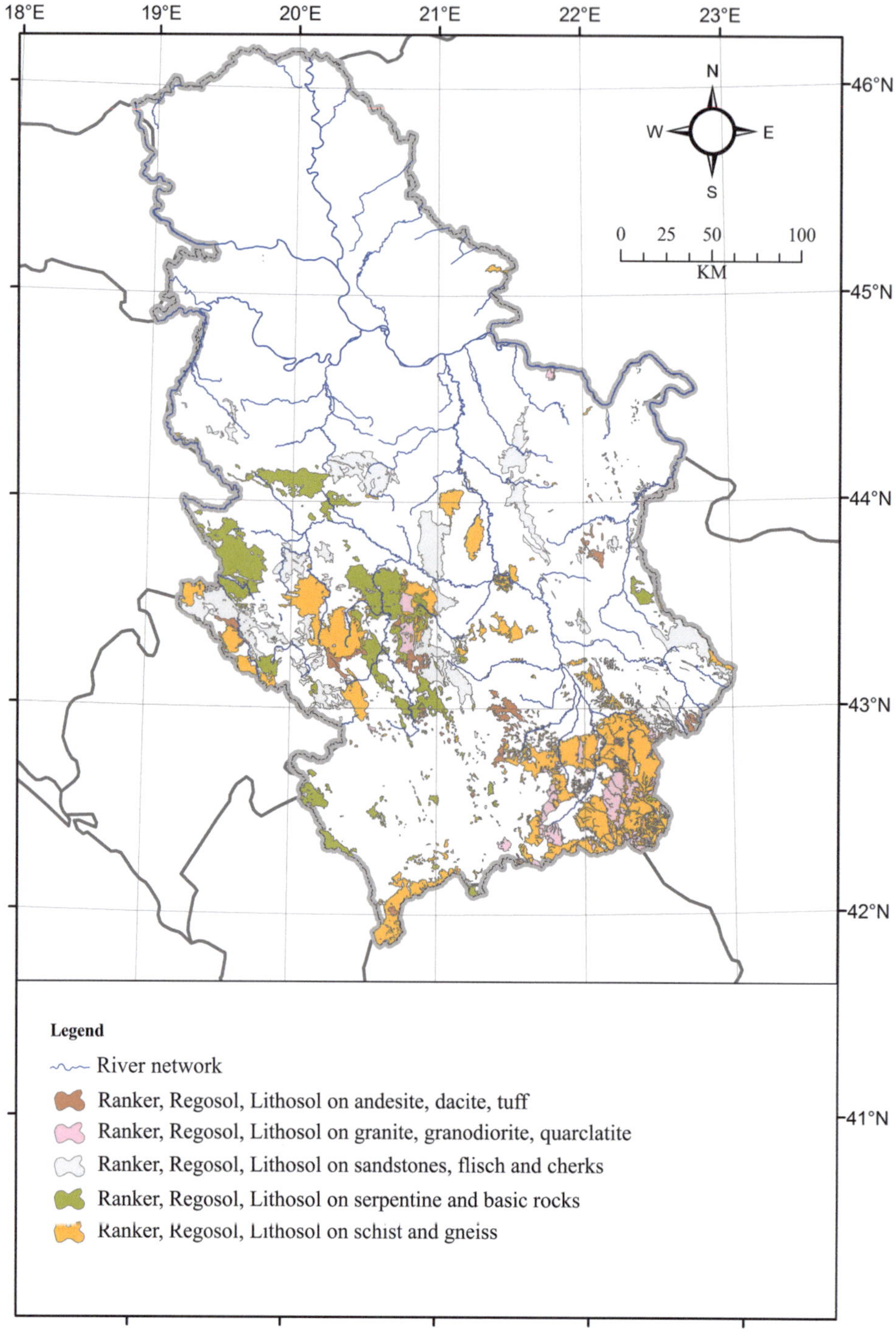

Fig. 7.8 Distribution of ranker, regosol and lithosol on different parent substrates

schist and gneiss, Fig. 7.9e). In Serbia, these soils (ranker, regosol and lithosol) cover approximately 16.4% of the terrain, i.e. approximately 14,541.69 km^2. Classification at the level of varieties and forms is made according to the criteria set out in Table 6.1 (Chap. 6 Soil Classification).

Distribution. In Serbia, rankers are found at altitudes of 150–2000 m, but mainly between 800 and 1600 m a.s.l. (Živković 1965). Wide expanses are found on the following mountains: Tara, Zlatibor, Rudnik, Kopaonik, Radan, Maljen, Suvobor, Jelica, Golija, Rudnik, Miroč and Deli-Jovan (Fig. 7.8).

Physical and chemical properties. One characteristic of the mechanical composition of rankers is the large proportion of skeletal particles, the low proportion of clay and the absence of textural differentiation. As a rule, skelet content increases with depth (Table 7.1). Clay content is low and is never over 25% (Tables 7.1, 7.2, 7.3, 7.4 and 7.5). Micro- and macroaggregates are very stable, which can be explained

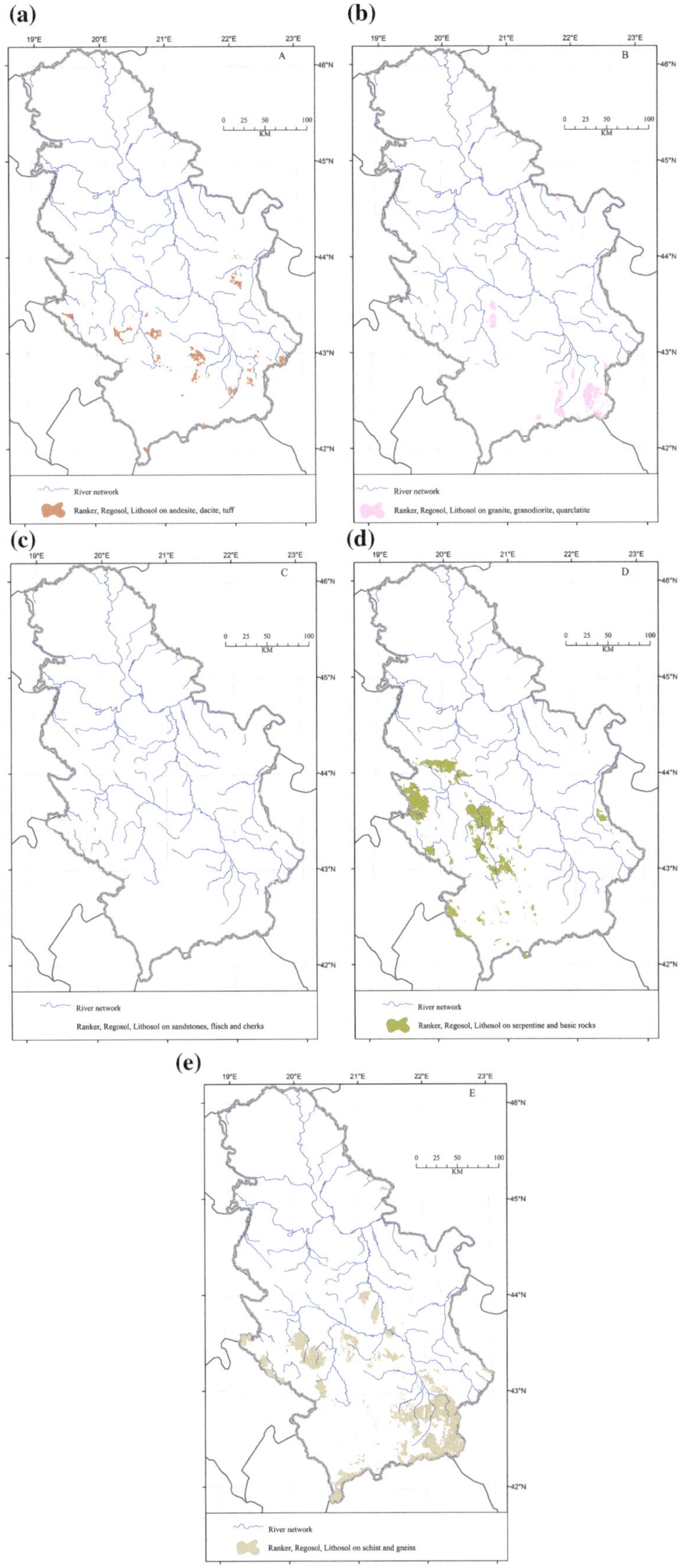

Fig. 7.9 Distribution of ranker, regosol and lithosol on andesite, dacite, tuff (**a**); granite, granodiorite, quartz latite (**b**); sandstone, flysch and chert (**c**); serpentinite and basic rocks (**d**); and schist and gneiss (**e**)

Table 7.1 Physical properties of rankers on serpentinite under forests of *Pinetum nigrae-sylvestris* on Mt. Maljen

Horizon	Depth (cm)	Hydros. humidity (%)	Granulomet ric composition in %								
			>2.0 mm	2.0–0.2 mm	0.2–0.06 mm	0.06–0.02 mm	0.02–0.006 mm	0.006–0.002 mm	<0.002 mm	Σ Clay + silt	Σ Total + silt
Olf	0–5	–	–	–	–	–	–	–	–	–	–
A	5–28	6.05	4.68	14.72	16.76	18.25	22.58	11.19	16.50	50.27	49.73
A/C	28–40	5.14	20.92	16.78	10.31	11.94	30.18	12.09	18.70	60.97	39.03

Source Pavlović (1998)

Table 7.2 Chemical properties of rankers on serpentinite under forests of *Pinetum nigrae-sylvestris* on Mt. Maljen

Horizon	Depth (cm)	pH		Y_1 (ccm)	Adsorption complex				Humus (%)	C (%)	N (%)	C:N	P_2O_5 (mg/100 g)	K_2O (mg/100 g)
		H_2O	KCl		T-S (mg ekv/100 g)	S (mg ekv/100 g)	T (mg ekv/100 g)	V (%)						
Olf	0–5	5.07	4.00	–	–	–	–	–	–	36.19	1.00	36.19	–	–
A	5–28	6.29	5.30	16.19	4.86	37.86	42.72	88.62	12.25	7.11	0.65	10.93	0.10	10.80
A/(C)	28–40	6.45	5.50	7.96	2.39	54.84	57.23	95.82	4.38	2.52	0.30	8.4	0.10	12.10

Source Pavlović (1998)

Table 7.3 Physical characteristics of eutric ranker

Horizon	Hydroscopic water, humidity (%)	Granulometric composition (%)							
		2.0–0.2 mm	0.2–0.06 mm	0.06–0.02 mm	0.02–0.006 mm	0.006–0.002 mm	<0.002 mm	Total sand	Clay + silt
A									
X	7.01	12.55	23.80	17.30	26.10	8.40	11.85	53.65	46.35
Min	5.42	5.40	22.20	16.10	17.70	8.30	10.20	43.70	36.40
Max	8.60	19.70	25.40	18.50	34.50	8.50	13.50	63.60	56.30
A/C									
X	6.34	22.54	17.73	9.80	19.40	10.17	20.37	50.07	49.93
Min	5.98	7.57	12.75	5.30	13.70	7.30	14.30	42.50	35.30
Max	6.70	39.50	20.53	14.90	24.40	12.90	27.10	64.70	57.50

Source Košanin and Gajić (2008) and Knežević and Košanin (2009)

Table 7.4 Chemical characteristics of eutric ranker

Horizon	pH		Y1 ccm n/10 NaOH	Adsorption complex				Humus (%)	C (%)	N (%)	C/N	P_2O_5 (mg/100 g)	K_2O (mg/100 g)
	H_2O	$CaCl_2$/KCl		T-S (cmol kg^{-1})	S (cmol kg^{-1})	T (%)	V (%)						
A													
Average	7.17	6.47	8.08	5.11	65.73	70.84	92.54	12.39	7.19	0.64	9.85	0.80	10.10
Min	7.00	6.20	8.08	4.96	53.26	58.51	91.03	3.80	2.20	0.30	7.30	0.80	5.50
Max	7.34	6.73	8.08	5.25	78.20	83.16	94.04	20.98	12.17	0.98	12.40	0.80	14.70
A/C													
Average	7.22	6.49	11.59	6.32	57.88	64.20	89.83	6.00	3.67	0.46	10.05		8.17
Min	6.7	6.17	7.54	3.9	48.6	52.5	84.46	2.29	1.9	0.27	9.9		6.8
Max	7.57	6.80	15.63	10.16	64.90	75.06	92.57	10.92	6.34	0.64	10.20		10.00

Source Košanin and Gajić (2008) and Knežević and Košanin (2009)

Table 7.5 Physical characteristics of dystric ranker

Horizon	Hydroscopic water, humidity (%)	Granulometric composition (%)							
		2.0–0.2 mm	0.2–0.06 mm	0.06–0.02 mm	0.02–0.006 mm	0.006–0.002 mm	<0.002 mm	Total sand	Clay + silt
A									
X	4.21	11.18	28.12	14.65	24.02	10.42	11.63	53.94	46.06
Min	3.40	5.54	22.62	7.80	19.50	7.64	9.40	50.70	43.48
Max	4.72	20.82	33.29	17.80	28.70	13.43	14.85	56.52	49.30
(B)/C									
X	2.80	9.32	27.22	13.24	22.35	12.13	15.76	49.77	50.23
Min	2.71	6.17	25.94	10.70	21.20	11.60	15.71	49.10	49.56
Max	2.88	12.46	28.50	15.77	23.50	12.65	15.80	50.44	50.90

Source Kostić (2007)

by the high content of humus and its properties. The humus content varies from 3.80 to 21%, depending on the various stages of development and the ranker subtype in question, but on average it is 10–12% in the A horizon. The humus content is greatest in the earlier stages of evolution, on steep terrain and at higher altitudes (unfavourable conditions for mineralization). In terms of pH, the soil ranges from strongly acidic to neutral (pH = 4–7), depending on the properties of the rock on which ranker forms and the climatic conditions and vegetation. Rankers at higher altitudes are strongly acidic regardless of the substrate.

At lower altitudes, in warm and dry climatic conditions, rankers are slightly acidic (on acidic rocks) or neutral to slightly acidic (on basic rocks). Base saturation varies greatly, from 24 to 95%, depending on the parent substrate and the bioclimatic conditions in which they are formed (Tables 7.2, 7.4 and 7.6).

Land use. Rankers are porous soils, well-aerated, with good water permeability. Although the field water capacity of the humus horizon is high, only a small amount of water is retained in rankers due to them being shallow and during the summer they quickly become dry following evapotranspiration. They are dominated by relatively arid pedoclimatic conditions. Rankers are characterized by poor cohesion and low adhesion, and they give little resistance to any tillage treatment whatever the moisture level. They are very susceptible to wind and water erosion. In natural conditions, they are populated with forest communities of black pine and Scots pine, mountain beech, beech and fir, spruce, and undergrowth of juniper and bilberries. Due to the high content of trace elements (Ni, Hr and Co), which can be toxic for many plants, specific vegetation grows (serpentinophytes), often including representatives of endemic flora.

Guidelines for management. The most important productivity limitations of rankers on compact rocks are the small reserves of available water due to their shallowness. The more prone to physical disintegration the rock is, the more favourable the properties of ranker are. Although it contains low amounts of total N (0.3–2.3%) and a narrow C/N ratio (9–12), there is not enough readily available N for plants due to the soil's poor mineralization and the probable cyclical forms of fixation. They are poor in P and moderately supplied with K.

The production value of rankers is linked to their depth and the properties of the parent substrate. Rankers formed on serpentinite and sandstones are least fertile. Those on serpentinite contain a large amount of Mg and some trace elements (Mn, Cr and Mi), which can be present in toxic concentrations. Rankers are most often used as forest habitats or as pastures and meadows, and also partly as arable land for the growing of potatoes, oats and barley.

7.2.2 Rendzina

The name of the soil type. The name 'rendzina' is of Polish origin (*rzedzic*—the rustling sound when a plough touches this soil). Rendzinas, or calcimorphic soils as termed by Duchaufour (1965), are soils with an A-C profile on limestone and dolomite. The term was first introduced by Sibirtsev (1895, 1896, 1953). This soil type includes nearly all stages of development up to a mature A-C profile, often very varied in terms of the character of the humus. Subtypes according to the classification system used in Serbia are as follows: rendzinas on marl, marly limestone and soft limestone; rendzinas on loess and loess-like sediments; rendzinas on moraine; rendzinas on dolomite gruss; rendzinas on calcareous sand; and rendzinas on calcareous gravel. Rendzinas can also be categorized according to humus character and the place where they form, into steppe (mull rendzinas), mountain (mull rendzinas) and alpine (moder-mor rendzinas) (Antić et al. 1982).

Table 7.6 Chemical characteristics of dystric ranker

Horizon	pH		Y1 ccm n/10 NaOH	Adsorption complex				Humus (%)	C (%)	N (%)	C/N	P_2O_5 (mg/100 g)	K_2O (mg/100 g)
	H_2O	$CaCl_2$/KCl		T-S (cmol kg^{-1})	S (cmol kg^{-1})	T (%)	V (%)						
A													
Average	4.72	3.65	49.57	32.22	13.65	45.87	30.32	9.25	5.37	0.45	11.78	1.98	16.33
Min	4.48	3.52	35.64	23.17	12.00	35.17	24.46	5.63	3.27	0.30	10.90	0.80	8.50
Max	4.89	3.82	67.26	43.72	15.94	57.88	34.12	13.27	7.70	0.61	12.62	3.30	26.00
(B)/C													
Average	5.71	3.92	16.57	10.77	14.07	24.84	56.52	1.42	0.82	0.08	10.32	0.50	3.75
Min	5.55	3.79	14.56	9.46	12.20	24.27	50.27	1.32	0.76	0.07	9.78	0.40	3.00
Max	5.87	4.04	18.57	12.07	15.94	25.40	62.76	1.52	0.88	0.09	10.86	0.60	4.50

Source Kostić (2007) and Dinić et al. (2001)

Fig. 7.10 Rendzina on calcareous gravel, brownized rendzina on carbonate and rendzina on scree (*photo* Čakmak, D.)

The WRB name. Leptosol, calcaric.

Genesis and sequence of horizons. The soil profile has a mollic horizon with a gradual transition to a loose calcareous C horizon. Most frequently, a transitional AC horizon can be identified, resulting in an Amo-AC-C-R profile (Fig. 7.10). The substrate has a high carbonate content (>20%), and the whole profile is calcareous except for those varieties known as leached and brownized rendzina. The substrate on which rendzina forms contains more than 10% $CaCO_3$ and is loose as it is prone to disintegration. The soil forms on loose carbonate sediments and soft limestone which contains a large amount of siliceous residue. The $CaCO_3$ content is often less than the amount of siliceous material.

Position in an evolutionary series. Rendzinas which form on regolith have a relatively deep solum (>50 cm) and an A-AC-C-R horizon sequence, while rendzinas on marly limestone can be shallow soils with an A-R profile. The humus-accumulative A horizon is usually deeper than 25 cm and is granular in structure. In the transitional AC horizon, there may be signs of $CaCO_3$ accumulation. Rendzinas are often preceded by regosol. From the start, they are populated by higher vegetation. As the substrate most often contains free clay, clay–mineral complexes form during the initial phase of soil creation. If regosols are populated by dense natural vegetation, rendzinas quickly form due to intense physical disintegration and accelerated biogenesis. The first phase in the evolution of rendzinas is the leaching of carbonate and the formation of leached (non-calcareous) rendzinas.

Further development leads to the beginnings of the formation of a (B) horizon, which marks the evolution of rendzina into brown soil (brownization)—brownized soils. Rendzinas are short-term phases and quickly evolve to the next stage, most often with an A-(B)-C profile. This occurs on substrates such as marl, marly limestone, gypsum, calcareous sandstone, loess and calcareous glacial deposits, which most often contain a lot of clay, hold water and enable the formation of these soils to a great depth. Due to the favourable physical properties (a good air–water regime), microbiological processes of humus mineralization are more intensive than with gley, and due to the presence of clay in the substrate, a clay–humus complex forms. On the soil map of Serbia, the same colour is used to denote undeveloped soils (regosol and lithosol) and humus-accumulative soils (rendzina). In this way, it is clearly highlighted that these form an evolutionary series on a calcareous substrate. In Serbia, these soils (rendzina, regosol and lithosol) cover approximately 1.8% of the terrain, i.e. approximately 1621.97 km^2. Classification at the level of varieties and forms is made according to the criteria set out in Table 6.1 (Chap. 6 Soil Classification).

Distribution. In Serbia, rendzinas are found on the mountains of Miroč, Beljanica and Malinik, on the Svrljig Mts., the Balkan Mts., Vidlič, Giljeva Mt., Jadovnik, Velika Ninaja, the northern side of Kopaonik, Jabuka and Radočelo (Fig. 7.11). In Serbia, the presence of shallow steppe rendzinas, poor in humus, has been recorded in the Đerdap region, from Golubac to Pesača, on Triassic marly limestone (Antić and Mišić 1971). The soil is populated with

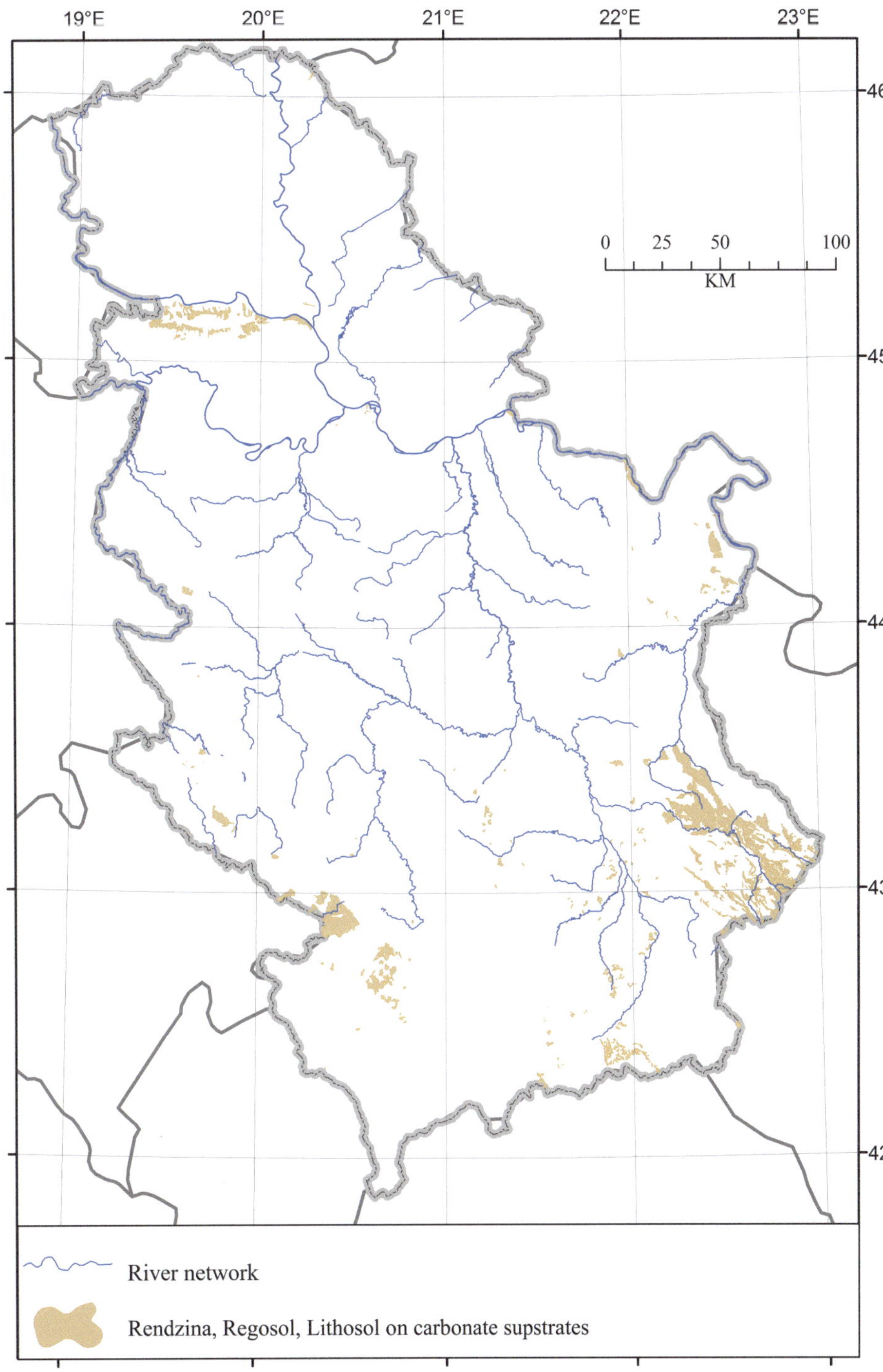

Fig. 7.11 Distribution of rendzina, regosol and lithosol on carbonate supstrates

xerothermic relic communities of lilac, Montpelier maple and oaks (*Syringo–monspessulo–Quercetum mixtum calcicolum*, Mišić 1971).

Physical and chemical properties. The content of clay in rendzina depends on its levels in the substrate on which the soil forms (Table 7.7). They can be heavy clay soils (if they form on clay-rich marl), loamy sand (on dolomite), very skeletal (on moraine), or have the most favourable texture (on loess). Other physical and mechanical characteristics depend on this. They are characterized by good retention properties. The available water capacity is favourable (high content of capillary pores), and they have a granular structure.

The carbonate content depends on its levels in the substrate and on the developmental stage and intensity of leaching, which is higher in wetter regions (Table 7.8). Younger soils formed on a substrate with high carbonate levels have 10–50% $CaCO_3$ in the A horizon. Older rendzinas and those which form on substrates poor in carbonates, in wetter regions, can have a non-calcareous A horizon. The pH of the soil, which can be neutral to slightly alkaline (pH 7–8), depends on the carbonate content. Deeper rendzinas which are brownized or leached are slightly acidic and the base saturation is low.

Rendzinas with natural vegetation have 5–10% humus in low-lying regions and up to 20% in higher altitude zones (dolomite rendzina). The concentration of nutrients (K, P, N) is moderate to high, but their total amount is sometimes restricted for several reasons: the dryness of the environment and reduced biological activity (weak mineralization), the stabilizing influence of $CaCO_3$ on the mineralization of humus, the reduced release of nitrogen, the immobilization of Fe, Mn and B compounds in an alkaline environment, the retrogradation of P into poorly soluble phosphorite (rock)/ phosphate, and the antagonistic activity of Ca ions on the uptake of K and Mg ions by plants.

The soil of steppe rendzinas is very shallow, only up to 20 cm deep. The humus-accumulative layer is compact and is clearly separated from laminated, marly limestone.

Table 7.7 Physical characteristics of rendzina

Horizon	Hydroscopic water, humidity (%)	Granulometric composition (%)							
		2.0–0.2 mm	0.2–0.06 mm	0.06–0.02 Mm	0.02–0.006 mm	0.006–0.002 mm	<0.002 mm	Total sand	Clay + silt
A									
X	6.60	2.21	14.74	15.82	26.02	62.38	28.75	32.63	67.37
Min	3.55	0.10	4.00	7.00	17.10	3.00	19.90	19.20	54.00
Max	8.36	11.70	38.31	25.60	32.30	805.00	39.60	46.00	80.80
AC									
X	6.38	5.29	11.91	14.36	25.27	13.07	29.96	31.56	68.44
Min	2.44	1.80	4.10	11.50	20.30	9.60	19.60	21.90	50.50
Max	8.19	11.09	28.30	19.40	34.00	17.40	38.60	49.50	78.10

Source Cvjetićanin and Knežević (2000), Knežević and Košanin (2004, 2009, 2010) and Košanin and Knežević (2007)

Table 7.8 Chemical characteristics of rendzina

Horizon	pH		Y1 ccm n/10 NaOH	Adsorption complex				Humus (%)	C (%)	N (%)	C/N	$CaCO_3$ (%)	P_2O_5 (mg/100 g)	K_2O (mg/100 g)
	H_2O	$CaCl_2$/ KCl		T-S (cmol kg^{-1})	S (cmol kg^{-1})	T (%)	V (%)							
A														
Average	7.06	6.42	17.05	12.04	58.79	70.83	81.80	12.17	7.06	0.54	12.46	13.85	1.44	18.63
Min	5.37	4.40	2.27	2.28	26.60	55.04	48.33	3.00	1.74	0.28	6.20	1.20	0.30	11.50
Max	8.03	7.43	43.76	28.44	81.60	89.55	96.97	20.20	11.72	0.77	17.50	38.00	2.80	34.20
AC														
Average	7.72	6.99	5.89	3.83	73.00	76.83	95.01	8.46	4.91	0.51	11.00	9.10	1.68	14.97
Min	7.06	5.98	5.89	3.83	73.00	76.83	95.01	1.68	0.98	0.33	9.10	1.67	0.50	3.00
Max	8.66	7.80	5.89	3.83	73.00	76.83	95.01	15.33	8.89	0.79	12.70	24.25	4.50	33.60

Source Cvjetićanin and Knežević (2000), Knežević and Košanin (2004, 2009, 2010) and Košanin and Knežević (2007)

The colour is dark brown, and its structural aggregates are tiny. The colloidal fraction content is extremely low, which is connected to the character of the limestone on which this type of soil forms. The humus form is mull (Amo), and its content in the soil is <10%. The soil is rich in potassium, but is poor in phosphorus, which is a general characteristic of rendzinas. Steppe rendzinas are characterized by a favourable ratio of mineral to organic components and usually take the form of mull rendzinas. Triassic and to a large extent Jurassic limestones are richer in the clay component (from 1–2 up to 5%).

As opposed to hard limestone, marly limestone is prone to physical disintegration, which allows for accelerated chemical weathering, leading to the occurrence of mesophilic beech forests on it. The significant presence of skeletal material in the humus-accumulative horizon ensures the continuous buffering of organic acids and the predominant creation of potassium humate, and with it a stable organo-mineral complex. Its pH is neutral or even alkaline with the relatively intensive release of nitrogen and to a lesser degree other nutrients (Antić et al. 1982).

Mountain rendzinas are found on limestone dating from the Palaeozoic era (Devonian, Carboniferous Periods) and crystalline Mesozoic and Tertiary limestone, at altitudes >900 m. In such conditions, the limited presence of the mineral component in limestone cannot mobilize all the organic matter, which results in its accumulation (10–30% humus). Base saturation is also high (V% = 70–80) due to the high humus content. In terms of pH, the soil is slightly acidic to neutral (pH = 6.2–7.0). Such soils are populated by pure beech forests. When on pure limestone, the soil scientist Nikola Pavičević named them 'buavice' (Pavićević 1956).

In the profile structure of alpine rendzinas, the accumulation of organic matter can be noted, as well as the formation of semi-raw and in places raw forms of humus (up to 5 cm thick). Clay-forming processes are pronounced, and the clay + silt content is 60–70%. With this type of rendzina, acidification is marked, particularly in the organogenic horizon in mountain pine communities on the upper forest boundary. The content of organic matter is as high as 57%, while the ratio of carbon to N (C/N = 27.8) points to moder-mor (raw) types of humus. The humus-accumulative A horizon is mull in character (C/N = 9.9). The surface horizons are the results of the effects of climate and vegetation, while the humus-accumulative horizon is affected by the calcareous parent substrate (Jović 1969; Antić et al. 1982).

Land use. The ecological production value is low. The main limiting factor to the productivity of steppe rendzinas is their shallow depth and low supply of nutrients, apart from nitrogen, which is released during the decomposition of organic matter. The habitat is dry, and warm and forest vegetation mainly takes the form of low xerothermic woods or thickets. Pure beech forests can usually be found on mountain rendzinas, which have somewhat more ecological value. Alpine rendzinas are relatively dry habitats with low ecological value. Mountain pine (*Pinus mugo*) associations mainly grow on them. However, on this soil, Bosnian pine (*Pinus heildreichii*) can also have an adequate wood volume production (Jović and Jović 1972).

Guidelines for management. Rendzinas on marlstones and moraine have the best production characteristics, whilst those on marly limestone and dolomite are significantly less fertile. They have favourable physical and chemical properties, but often due to their shallow depth, their production capabilities are not great, particularly those that are very skeletal and are characterized by xerothermic conditions. Deep rendzinas provide valuable agricultural land, particularly for use in fruit and wine growing.

7.2.3 Chernozem

The name of the soil type. The name is derived from the Russian language, meaning *black soil*. Genetic soil horizon designations were used for the first time for this soil type, and its study marked the beginning of a period of pedogenetic pedology (Dokuchaev 1883/1948). Chernozems are soils with a thick black surface layer, rich in organic matter. This soil type is also known by different names: Calcareous Black Soils and Kalktschernoseme (Germany), Chernosols (France) and Eluviated Black Soils (Canada). Subtypes of this soil according to the classification system used in Serbia are as follows: chernozem on loess and loess-like sediments; chernozem on calcareous aeolian sand; and chernozem on alluvial deposits (Škorić et al. 1985).

The WRB name. Chernozem.

Genesis and sequence of horizons. Chernozems are soils of semi-arid steppe regions with a mollic A horizon, deeper than 40 cm and a transitional AC horizon (25–30 cm deep), Fig. 7.12. They develop on calcareous loamy and less frequently sandy, loose substrates. The colour of the humus horizon is dark brown. These soils have a clear granular structure, apart from when they are formed on sand. Krotovinas (the burrows of moles, mole rats, and mice) and pseudomicelia are frequent diagnostic criteria. The soil is calcareous from the surface down or in the transitional horizon.

The parent substrate of chernozem is loess material, enriched with varying levels of calcium carbonate. During the deposition of loess and later through wind activity, different forms occurred, which affected the genesis of this soil type to different degrees: loess terraces, loess plateaus, etc. The relief of chernozem is mainly lowland (plains and plateaus) in character, which has caused varying intensities of profile wetting. The main factors in chernozem formation are

Fig. 7.12 Chernozem (*photo* Vasin, J.)

the temperature and moisture of the soil. As summers are warm and dry, microbiological processes are most intensive in the spring and autumn when temperatures are lower and moisture levels higher, which results in the accumulation of a large amount of humus in the profile.

The majority of chernozems form on loess, but they can form on other calcareous substrates too (river and lacustrine sediments). This soil is characterized by the intensive humification of organic matter and retarded mineralization of the humus that is formed, leading to the accumulation of humus in the surface horizon. The humus-accumulative horizon is very deep thanks to the deep penetration of steppe grasses and the activities of soil fauna. The wet period of the year sees the leaching of readily soluble salts and to a lesser extent $CaCO_3$, so the adsorption complex is highly saturated with alkaline earth metals. There is no leaching of colloidal particles. When the soil pH is neutral and during dry summers and cold winters, the chemical decomposition of primary minerals is not intensive. Chernozem forms through the continuous interaction of the humus and mineral colloid component, which is present in the loose loess. In neutral environments with the presence of Ca and Mg ions, stable humus compounds, C and Mg humates, develop. They create a stable clay–humus complex and aggregates with a crumbly structure. The optimal conditions for the genesis of chernozem were in the warm postglacial period from approximately 7000–8000 BC, while today's climatic conditions expose a part of the marginal regions of the chernozem zone not only to processes of degradation, but also lessivation. Hence, in Baranja (Croatia), the presence of not only lessive, but also pseudogleyic chernozem has been noted (Antić and Jović 1965; Antić et al. 1972).

Position in an evolutionary series. Chernozem is most often the mid-point in an evolutionary series on loess. It begins with regosol, which via protoregosol, evolves into chernozem. From this point, evolution can have a bioclimatic character, with the next stages being brownized chernozem, then eutric cambisol, culminating in lessive eutric cambisol (Živković 1955). The other direction evolution can take has a topogenic character (in depressions of lowland pedunculate oak forests with a mixture of Italian and Turkey oaks), where chernozem evolves into lessive chernozem and then pseudogleyic chernozem (Antić and Jović 1965). Classification at the level of varieties and forms is made according to the criteria set out in Table 6.1 (Chap. 6 Soil Classification).

Distribution. In Serbia, the main chernozem zone is situated in Vojvodina, where it marks the most western reaches of the southern European chernozem zone. Smaller areas are found south of the Sava and Danube rivers, between Belgrade and Požarevac, in the valley of the River Morava, Stig (the Stig Valley) and the Danube river valley (Fig. 7.13).

It forms in regions with a continental and steppe climate, characterized by cold, dry winters, wet springs and warm summers. The natural vegetation of chernozems is xerophytic grass-steppe communities dominated by the species *Stipa* sp. and *Festuca sulcata* (Hack.) Beck. In the region of Vojvodina, besides steppe grasses, mesophytes can also be found (*Festuca pseudoovina* Hack. ex Wiesb., *Agropyrum crystatum* (L.) Gaertn., *Adonis vernalis* L., etc.). Today, the majority of the area populated by natural steppe vegetation on chernozem has been converted into fertile agricultural land. Serbia has approximately 10,589.60 km^2 of chernozem, or 12% of the land area.

Physical and chemical properties. The high agricultural productivity of chernozem serves as a standard for comparison with other soils. The physical, chemical and biological characteristics of this soil are close to the theoretical optimum. The profile of chernozem is characterized by a thick humus-accumulative horizon, which diffuses irregularly into the parent substrate—loess. In many cases, this profile structure is disturbed by the presence of an illuvial horizon of $CaCO_3$, which appears in the form of a transitional A/C horizon. The colour of the humus-accumulative layer is dark brown, but for lightly gleyed chernozem it is black. Its thickness ranges from 50 to 80 cm.

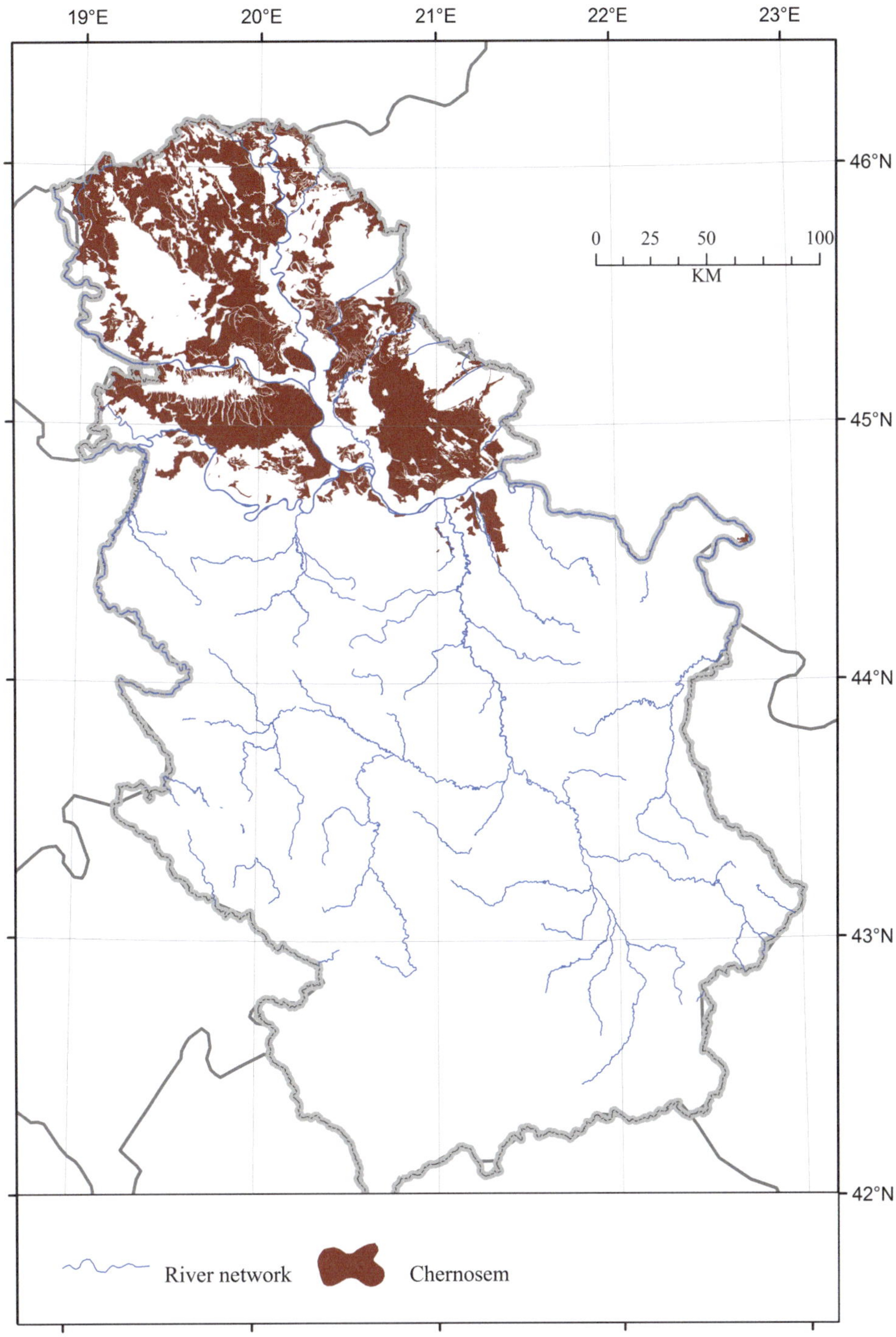

Fig. 7.13 Distribution of chernosem

The presence of krotovinas (the result of the activity of animals which break down the organic remains and mix them with the mineral part of the soil, hence enabling more complete humification and improving the physical characteristics of the soil by creating a stable granular structure) is characteristic for the morphological profile of chernozem and occurs in the form of yellow spots in the humus-accumulative horizon, or black ones in loess. The crumbly structure and high porosity are the basis for the favourable properties of this soil. They have a favourable mechanical composition (Table 7.9).

The chemical properties of chernozem come from its mull humus with a C:N ratio of 8–12 and the domination of Ch over Cf acids with Ch:Cf >1 (Antić et al. 1982). The high nitrogen content is tied to alpha-amino acids, i.e. the form most available to plants. The clay minerals of chernozem are illite or montmorillonite with fewer minerals of the kaolinite group. The humus content ranges from 3 to 5% and, together

Table 7.9 Physical characteristics of chernozem

Horizon	Hydroscopic water, humidity (%)	Granulometric composition (%)					
		2.0–0.2 mm	0.2–0.02 mm	0.02–0.002 mm	<0.002 mm	Total sand	Clay + silt
A							
X	0.83	1.92	73.00	17.59	13.93	64.27	31.52
Min	0.35	0.08	31.20	1.20	1.40	0.00	5.30
Max	1.42	8.70	94.70	45.70	37.80	94.70	73.72
AC/C							
X	0.51	6.13	60.89	19.92	14.94	60.90	34.86
Min	0.35	0.10	29.70	1.20	1.70	29.70	3.70
Max	0.81	11.50	96.30	39.90	41.80	96.30	69.90

Source Košanin and Knežević (2002), Galić et al. (2009) and Marković et al. (2010)

Table 7.10 Chemical characteristics of chernozem

Horizon	pH		Y1 ccm n/10 NaOH	Adsorption complex				Humus (%)	C (%)	N (%)	C/N	$CaCO_3$ (%)	P_2O_5 (mg/100 g)	K_2O (mg/100 g)
	H_2O	$CaCl_2$/KCl		T-S (cmol kg^{-1})	S (%)	T (%)	V (%)							
A														
Average	7.43	7.06	1.09	0.71	50.20	50.91	98.57	2.15	1.53	0.17	11.61	3.72	6.72	11.29
Min	5.45	5.03	0.65	0.43	30.60	31.45	97.30	1.05	0.60	0.09	6.00	0.00	1.00	5.40
Max	8.65	8.15	2.62	1.70	72.80	74.50	99.32	3.54	4.08	0.27	42.19	12.48	26.80	25.40
AC/C														
Average	8.17	8.34						0.68	0.43	0.04	27.06	18.41	1.00	3.70
Min	6.96	7.85						0.35	0.20	0.02	10.43	2.90	0.80	3.00
Max	8.94	8.66						1.32	0.76	0.06	43.69	34.57	1.20	4.20

Source Košanin and Knežević (2002), Galić et al. (2009), Marković et al. (2010) and Kolarić et al. (2014)

with the clay minerals, is the main cause of the high adsorption capacity (up to 50 mg ekv/100 g soil).

Due to the carbonates in loess and the weak leaching of $CaCO_3$, chernozems in Serbia are calcareous from the very surface or from a depth of 20–30 cm. The $CaCO_3$ content in typical chernozem is usually about 3–6%, but is sometimes higher. There is more in the transitional AC horizon (15–20%), but it reaches its maximum in the upper part of the C horizon (Table 7.10). In leached chernozem, the A horizon, and sometimes the whole of the AC horizon, can lose carbonates. With brownized chernozem, $CaCO_3$ occurs only in the C horizon. The soil reaction of typical chernozem is slightly alkaline (pH = 7.6–8.3), and as the $CaCO_3$ content increases with depth, so does the pH.

Chernozems in Serbia have a moderate adsorption capacity due to the humus content and the domination of illitic clay minerals. As the humus content decreases with depth, the total adsorption capacity also falls. Base saturation is high for calcareous chernozems (over 97%), but is lower for leached chernozems (90–95%) and brownized chernozem (80–90%) (Antić et al. 1982). The adsorption complex is dominated by Ca ions (80–85%) and, to a lesser extent, Mg ions. The amount of adsorbed Na ions is minimal (<1%). Chernozem is characterized by a good buffering capacity.

In Serbia, several stages of chernozem can be identified: leached (degraded) chernozem, brownized chernozem, lessive chernozem, pseudogleyic chernozem and salinized variants of chernozem (Antić et al. 1982).

Leached (degraded) chernozem is characterized by the absence of $CaCO_3$ in the humus-accumulative horizon, its accumulation between the A and C horizons and the tendency for loss of structure. The adsorption complex exhibits slight acidification, which is accompanied by a decrease in pH. The colour of the soil is darker, and it has a less favourable mechanical composition that calcareous chernozem. The clay and silt contents are usually 50–60%. In Serbia, they are most widespread in the Bačka region and partly in Banat and the flat parts of Srem. They are connected to loess terraces or deeper depressions where water is retained for longer.

Brownized chernozem has an initial, but not completely developed horizon of brownization A(B) between the A and

C horizons. The depth of the accumulative horizon is reduced, and humus content is also less. This subtype is the result of the effects of a wetter climate and the penetration of forest vegetation onto the steppe. It can be found on the slopes of the Fruška Gora Mt. The process of brownization is characterized by the further acidification of the adsorption complex and a reduction in the pH (pH in H_2O is <7), accompanied by an increase in the clay and colloid fraction (particles < 0.002 mm), which can be as high as 40%, while the total content of clay + silt can amount to 70% (Živković et al. 1972).

Lessive chernozem is connected to forest vegetation, most frequently oak forests. In Serbia, it occurs in the forests of Srem. The surface A horizon is grey-brown in colour and is almost structureless and powdery, with a little humus and a favourable mechanical composition, which is the result of the translocation of clay particles and humus matter. Beneath it, there is a dark brown to black, more compact, alluvial B_t/A horizon, the structure of which is chernozem-like. The transitional section adjoining the parent substrate is dark in colour with grey nuances and a calcium carbonate concretion. Then comes the yellow-grey loess material (Antić and Jović 1965; Antić et al. 1967). The physical and chemical properties show that lessivation is pronounced from the very surface (Tables 7.11 and 7.12). There is a great difference in the colloid fraction between the A and B_t/A horizons. The total content of the smallest fractions (clay + silt) increases with depth, which gives this soil a clayey character. The chemical properties differ to those of calcareous chernozem because carbonates only appear from a depth of 50 cm. This explains the slightly acidic pH of the surface part of the profile and the considerably reduced degree of base saturation.

Pseudogleyic chernozem forms when the B_t/A horizon is made up of heavy clay and when water retention can occur on the surface (Muckenhausen 1962). In Serbia, marked lessivation of chernozem leads to the formation of this subtype, which can be found in the flat parts of Srem. In some localities, in depressions where groundwater is closer to the surface, variants of pseudogley-gley chernozems have been noted.

Salinized variants of chernozem are degraded forms of chernozem in the form of solonchak and solonetz. In solonchaks, free salts appear, which impact on a change in structure, i.e. the physical properties of the soil. Namely, there is an increase in the content of Na ions, which brings about a prismatic structure and lessivation, leading to a clearly defined humus-accumulative horizon, the bottom part of which is more compact and plastic.

Land use. Chernozems are soils with high agricultural productivity on which cereal crops can be grown. They are found in grass and meadow steppes and forest-steppes. In Serbia, chernozems develop in Central European climatic conditions, characterized by high levels of precipitation and high average temperatures. In the zone where chernozem is found, mean annual precipitation is 620 mm and the average annual temperature is about 11 °C. The fact that this soil originated under steppe and forest-steppe vegetation shows that the vegetation introduces large amounts of organic matter, which contributes to the intensive humification of the

Table 7.11 Physical properties of lessive chernozem

Horizon	Depth (cm)	Hydroscopic water, humidity (%)	Granulometric composition (%)					
			2.0–0.2 mm	0.2–0.02	0.02–0.002	<0.002	Total sand	Clay + silt
A_1A_3	0–20	1.75	0.39	40.17	45.44	14.00	40.56	59.44
B_t/A	30–54	2.75	0.27	27.09	36.88	35.76	27.36	72.64
B_tA/C	540–62	1.98	1.36	27.64	38.12	32.88	29.00	71.00
C	62–80	1.45	3.67	29.65	41.44	25.24	33.32	66.68

Source Antić and Jović (1965)

Table 7.12 Chemical properties of lessive chernozem

Horizon	Depth (cm)	pH		Y_1 ccm	Adsorption complex				$CaCO_3$ (%)	Humus (%)	N (%)	C/N	Readily available	
		H_2O	KCl		T-S (M ekv/100 g %)	S (M ekv/100 g %)	T (M ekv/100 g %)	V (M ekv/100 g %)					P_2O_5 (mg/100 g)	K_2O (mg/100 g)
A_1A_3	0–20	6.50	5.45	8.60	5.30	17.6	23.23	75.93	–	3.44	0.23	9	0.35	15.4
B_t/A	30–54	7.25	5.80	–	–	–	–	–	0.36	1.21	–	–	0.25	16.6
B_tA/C	540–62	8.60	7.30	–	–	–	–	–	20.70	1.02	–	–	0.35	9.4
C	62–80	8.75	5.50	–	–	–	–	–	35.29	0.75	–	–	0.3	5.9

Source Antić and Jović (1965)

soil profile. The total amount of plant remains in the steppe ranges from 10 to 30 t/ha (Bazilevič and Rodin 1964), which is several times higher than in forest ecosystems. The abundance of plants enriches the soil with organic matter and mineral elements, particularly Ca, Mg, P and S. High fertility is the result of the existence of a deep and homogeneous physiologically active layer, a favourable mechanical composition, a stable granular structure, which leads to a favourable water, air and temperature regime, favourable chemical properties (high content and good structure of humus, favourable pH levels, good ratio of adsorbed ions), high nutrient content and good microbiological activity. For these reasons, these soils are used in Serbia for intensive agricultural production, the exception being the small areas of leached, lessive and brownized chernozems under forest vegetation.

7.2.4 Vertisol

The name of the soil type. The term vertisol (Latin *vertere*—rotate) comes from the continual mixing of the soil which takes place in the A horizon through the process of self-rotation or pedoturbation. Smonitza (smola—resin, pitch) is the Serbian name for this soil, which is black in colour and sticky like resin. The soil scientist Stebut was the first to officially use the term in the classification of soils (1949, 1950, 1953), and later it became the generally accepted name for this type of soil in south-east Europe. Subtypes of this soil according to the classification system used in Serbia are as follows: calcareous vertisol, which contains carbonate in the humus horizon or only in its lower part; non-calcareous vertisol; and brownized vertisol, which has a cambic horizon shallower than the A horizon (Škorić et al. 1985).

The WRB name. Vertisol.

Genesis and sequence of horizons. There are two viewpoints in Serbia as to the genesis of vertisol. On the one hand, Stebut (1924) wrote about the relict and hydromorphic character of vertisol, drawing on Cvijić's theory on the existence of lakes in valleys, through tectonic changes in more arid climatic conditions, sediment created conditions for the formation of vertisol. As the lakes receded and with later climatic changes, vertisols became terrestrial in character. On the other hand, Neugebauer and Zakosek (1962), and Filipovski and Ćirić (1963) believed that vertisols were of recent origin and had not gone through the stage of hydromorphism. In their opinion, this was proved by analyses of pollen, in which the presence of elements of organic matter of lacustrine or fluvial origin was not detected, i.e. the soils were continually terrestrial in terms of the character of their genesis. Vertisols are estimated to be approximately 2800 years old. The hydromorphic character of vertisols is the result of a cycle of hydromorphic accumulative processes in the formation of soils under grass and grass-woody vegetation, with the indirect involvement of groundwater enriched with Ca and Mg bicarbonates. Concretions of $CaCO_3$ and R_2O_3 are evidence of the existence of hydromorphic periods (Kovda 1973).

Typical vertisols are soils with an A-AC-C profile (Fig. 7.14). They are deep soils, which can be up to 150 cm in depth. The border between the A and C horizons is not usually a horizontal line, but is wavy or zigzag. During dry periods, the volume of the substrate with montmorillonite clay contracts and cracks, leading to the formation of wedge-shaped vertical cracks up to a depth of 1 m and more. In addition to these vertical cracks, narrow horizontal cracks also appear, and hence the prismatic aggregates leach out, which is a typical characteristic of the structure of vertisols. Crumbs of soil from the surface fall through vertical cracks wider than 1 cm to the bottom of the cracks, and hence the humus horizon becomes deeper in the form of wedges. During wet periods, clay swells and the cracks close, but the material which has fallen into the cracks also expands and

Fig. 7.14 Vertisol (*photo* Čakmak, D.)

puts lateral pressure on the aggregates. The infilling of the cracks with the surface layer of soil and processes of pedoturbation contribute to the formation of a deep humus horizon (50–100 cm) and its homogenization. The process of humus accumulation sometimes occurs under anaerobic conditions. In such conditions, plenty of bituminous material and low-molecular weight humic acids form, which together with montmorillonite clay form a stable argillo–humic complex, dark grey to pitch-black in colour. The contraction, expansion and mixing of the soil mass without the influence of fauna prevents the formation of other horizons apart from the A horizon and also clay particle migration. This leads to a continual increase in alkaline earth metals. In the initial stages of development, vertisols usually contain $CaCO_3$ inherited from the substrate, but this is later washed away. Leaching is slow due to the low permeability of these soils. If the substrate is not as rich in $CaCO_3$ and the climate is more humid, the evolution of vertisol progresses more quickly. The content of montmorillonite in the substrate has a major impact on the properties of vertisol and the processes which occur in it. Classification at the level of varieties and forms is made according to the criteria set out in Table 6.1 (Chap. 6 Soil Classification).

Position in an evolutionary series. The evolution of vertisol depends on climatic conditions. One path is similar to the evolution of chernozem, i.e. it evolves in the direction of cambisol eutric, and the other path is in the direction of lessive and pseudogleyic vertisols. Brownized vertisols occur on southern and western exposures, evolving from vertisols with a lighter mechanical composition. This process begins in the lower part of the A horizon, where there is the least humus, and is accompanied by the liberation of colloid iron. In Serbia, this type of vertisol can be found in the valleys of the Sava, and the Western and Southern Morava. More common for vertisol is the process of lessivation and all types are subject to it, apart from those that are calcareous. It takes place from the surface down, but does not stretch deep into the soil profile due to the heavy mechanical composition, which resists both lessivation and also water filtration through the profile. Pseudogleyic vertisols can occur directly even on calcareous vertisols. Due to poor drainage and water retention, pseudogleization starts from the very surface of the soil. The process is accompanied by the destruction of structural aggregates, gradual bleaching, and the formation of Fe and Mn nodules.

Distribution. This soil type occupies large areas of Serbia, mainly in central (Šumadija) and eastern parts of the country (Negotinska krajina) It is also found in the southern province (Kosovo and Metohija), as well as in northern parts of the northern province (Vojvodina), Fig. 7.15. It occurs at altitudes of 200–300 m a.s.l.. The widest expanses are found in Šumadija, Pomoravlje and eastern Serbia. Vertisols form on substrates containing more than 30% clay, mainly of the montmorillonite type, most often on Tertiary lake clays, rich in montmorillonite and limestone (marly clays, calcareous clayey argilloschists), on eluvium of various basic rocks, which during disintegration produce plenty of montmorillonite (basalt, gabbro, amphibolite, diabase and serpentinite), and over volcanic tuff and ash. A second important prerequisite for the formation of vertisols is a climate characterized by alternating wet and dry periods. Plain and plateau or slightly undulating relief is most suitable for their formation as, importantly, it restricts surface drainage. They form at altitudes of 200–900 m a.s.l., most frequently in the zone between 200 and 600 m, i.e. in valleys where fine lacustrine clay builds up. Natural vegetation includes broadleaf forests (most often Italian and Turkey oaks) and grass communities, while semi-hydrophilic vegetation also occurs in places. In Serbia, vertisols cover approximately 7.7% of the land area, i.e. about 6836.65 km^2.

Physical and chemical properties. Vertisols are very widespread soils with high production potential. They are characterized by a heavy, unfavourable structure, montmorillonite clay, the adhesion of structural aggregates, intense expansion and contraction, and the occurrence of a prismatic structure.

The fact that they are hydromorphic in character indicates that they were exposed to intense wetting during their genesis. The parent substrate is Tertiary and sometimes clay, sediments, containing varying levels of $CaCO_3$. They can also occur on lighter sediments, but these are always rich in montmorillonite clay. As they are formed on lacustrine sediments with different mineral compositions, there are great differences in their mechanical composition. Climatic conditions were similar to today's tropical and subtropical climate, which caused the intensive transformation of primary into secondary minerals, mainly of the montmorillonite type, with the presence of $CaCO_3$ in the substrate, i.e. neutral and slightly alkaline chemistry, helping their formation. Such ecological conditions favoured the development of dense wetland vegetation and the accumulation of humus in anaerobic conditions. Vegetation has also played a major role in the genesis of vertisol. It is thought that in the past vertisols were populated with alder and ash forests, while today they are covered with thermophile Italian oak and Turkey oak forests, which has impacted the transformation of these soils.

Vertisols are characterized as soils with an unfavourable chemical composition. In terms of texture, they are clayey or heavy clayey, with the clay + silt fraction ranging from 70 to 80%, sometimes even as high as 90% (in the vicinity of Kraljevo). In addition, the colloid fraction (particles < 0.002), comprised of almost 90% minerals of the montmorillonite type, can be up to 60%, which contributes to the colloidal and plastic character of vertisol. The high content of clay, which swells, means vertisol is structureless, sticky

Fig. 7.15 Distribution of vertisol

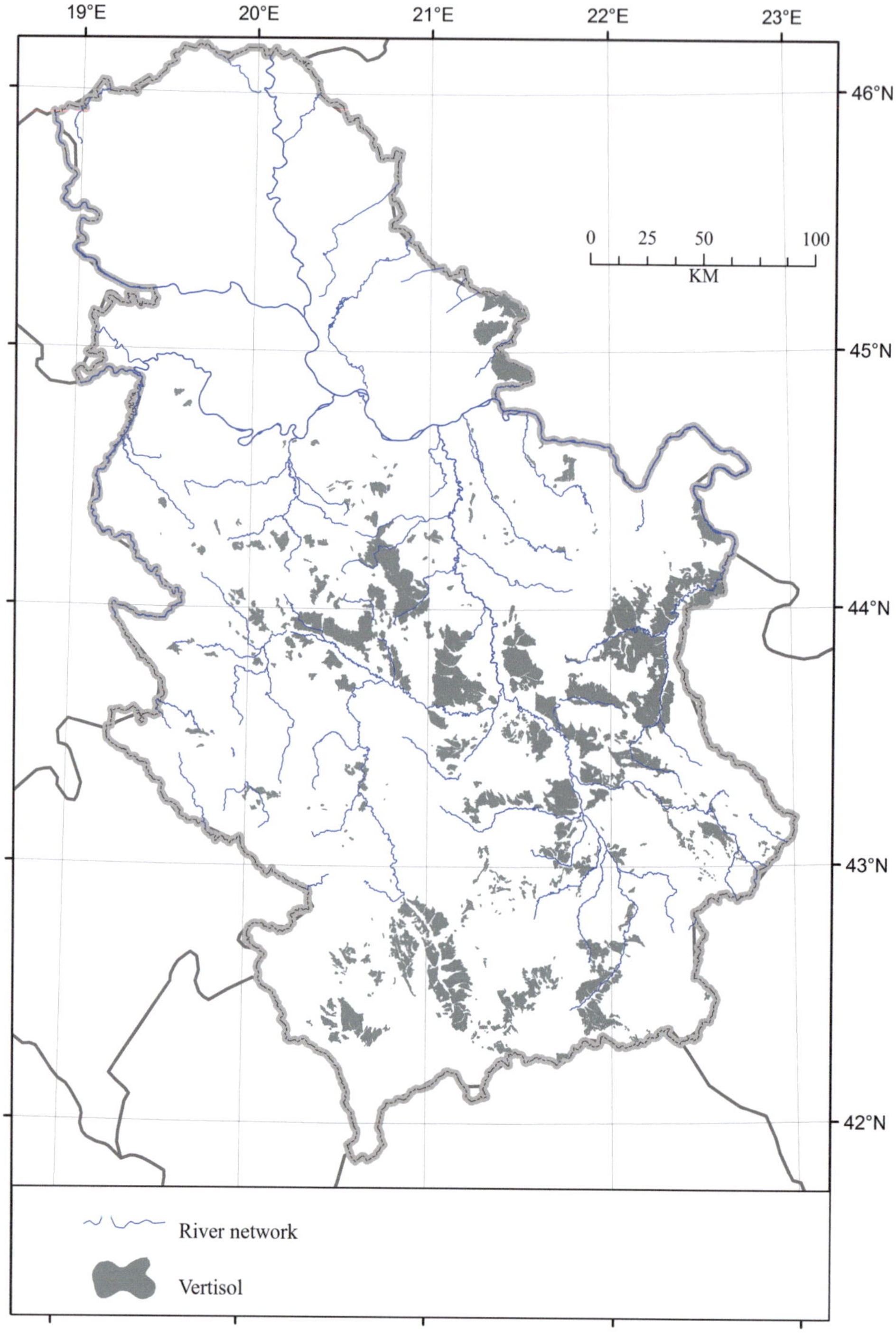

and plastic when wet, but is a compact, cracked and fractured mass when dry (Table 7.13).

Vertisols are soils with favourable chemical characteristics. They are slightly acidic to slightly alkaline (pH 5.8–8.1), depending on the $CaCO_3$ content. Humus content ranges from 2.5 to 4%, and can be 7–8% under natural vegetation. Due to the high content of clay minerals and their properties, and in some cases due to the large amount of humus, vertisols are characterized by a high adsorption capacity (>40 meq), while the adsorption complex is saturated with Ca and Mg ions.

Land use. Vertisols cover large areas of Serbia, most often in lowland, mainly agricultural regions. This type of soil has high potential fertility (a deep homogenous humus-accumulative horizon, a large amount of biogenic elements, a high adsorption capacity, and good chemical properties). However, they have an unfavourable air–water regime, so amelioration is difficult, although the application

Table 7.13 Physical and chemical characteristics of vertisols in Serbia

Soil characteristic	Field			Meadow		
	Mean	Range	Standard deviation	Mean	Range	Standard deviation
pH (H_2O)	7.1	5.8–8.1	0.9	6.9	5.6–8.1	0.9
pH (KCl)	6.0	4.6–6.9	0.9	5.8	4.7–7.0	0.9
Humus content (%)	3.3	2.5–4.0	0.5	3.5	2.0–5.6	1.1
P_2O_5 (mg 100 g^{-1})	7.7	0.6–28.0	8.5	4.2	0.8–17.8	5.0
K_2O (mg 100 g^{-1})	34.4	19.0–59.6	11.8	31.1	20.4–53.5	10.4
CEC (m.e. 100 g^{-1})	25.1	15.5–31.5	5.6	23.8	16.9–34.7	6.6
Sand content (%)	29.6	21.4–36.0	4.8	32.2	22.3–50.5	9.0
Silt content (%)	24.6	18.8–31.2	3.6	22.8	11.9–29.4	5.6
Clay content (%)	45.8	33.5–54.4	7.2	44.9	28.9–64.3	11.1
Silt + Clay content (%)	70.4	64.0–78.6	4.8	67.7	49.5–77.7	9.0

Source Milivojević et al. (2011)

of organic fertilizers and calcification improves their properties. The effective fertility often depends on the distribution of precipitation during the vegetation period (due to their poor physical properties). In areas with long periods of drought, plants on vertisols do not only suffer from a lack of available water, but they also perish due to the splitting of roots because of pedoturbation. Long wet periods are also unfavourable due to the creation of anaerobic conditions and the possibility of erosion intensifying as the infiltration capacity of vertisols is very low when they are very wet. In Serbia, these soils are used for plant production, and more rarely for the cultivation of oak forests with high productivity.

Guidelines for management. Even and moderate wetting gives the best conditions for exploiting the high production potential of vertisols. They require irrigation in order to provide high yields and fertilization with phosphorus.

7.2.5 Kalkomelanosol (Limestone-Dolomite Black Soil)

The name of the soil type. The name originates from the Greek words *kalk*—limestone and *melanos*—dark-coloured. Local names are black soil on limestone, black soil on hard limestone and dolomite, and mountain rendzina. The subtypes of this soil according to the classification system used in Serbia are as follows: organogenic black soil; organo-mineral black soil; brownized black soil; and rubificated black soil (Škorić et al. 1985).

The WRB name. Leptosol.

Genesis and sequence of horizons. These soils remain at the stage with an A-C profile for a long time. They form on hard, compact limestone and dolomite, which undergo pronounced chemical weathering (the dissolution and leaching of carbonates). The mineral part of the soil is formed from a silicate residue following the carbonates being washed away. In pure hard limestone, there can be 99% $CaCO_3$ and 1% silicate residue, from which the solum is formed. During formation, it goes through several stages:

Fig. 7.16 Kalkomelanosol (*photo* Čakmak, D.)

Fig. 7.17 Distribution of kalkomelanosol, regosol and lithosol on limestone

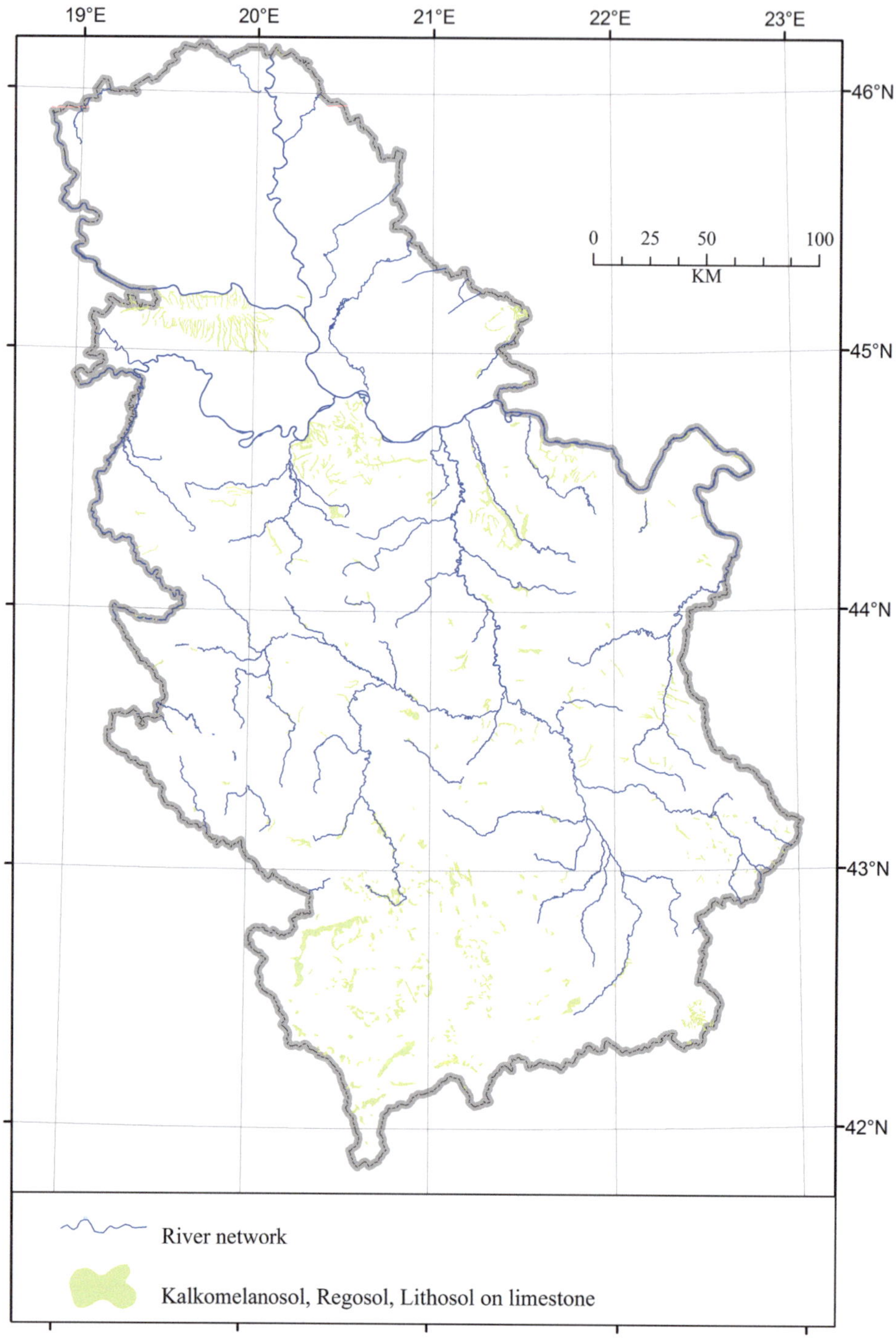

The organogenic stage is the initial stage, characterized by a shallow humus horizon, up to 10 cm in depth, with high humus content (>25%) and the absence of clay–humus complexes. It is most widespread in high mountain zones where the most xerothermic grass species grow.

The organo-mineral stage occurs as the profile deepens (leaching of carbonates) and silicate residue accumulates. This allows the soil to hold more water and for mineralization to intensify, bringing about the mature phase of black soil with mull humus. This phase is characterized by a deeper profile, up to 30 cm, a lower humus content (<25%), and the formation of clay–humus complexes. In the humus horizon, there are coarse fragments of limestone. It is more widespread than the preceding subtype and dominates in regions of high-altitude mountain pastures (Fig. 7.16).

With further evolution, this soil develops to the next stage —brownized black soil or rubificated black soil. In the subalpine zone, with the influence of forest and bushy vegetation, a thick layer of raw humus may form above the humus A horizon (Of). Classification at the level of varieties and forms is made according to the criteria set out in Table 6.1 (Chap. 6 Soil Classification).

Position in an evolutionary series. In Serbia, these soils evolve via brownized black soil into kalkocambisol, and then into lessive soil on limestone and dolomite.

Distribution. It forms at all altitudes, from sea level to over 2000 m a.s.l., but is most widespread in high-altitude mountain regions where the extreme climatic conditions most suit its formation. In Serbia, leptosols are widespread on Devica, the Svrljig Mts, the Balkan Mts, Suva Mt., Vidlič, Vlaška Mt., Rtanj, Tupižnica and Kučajska Mts, on Tara Mt., Junička Mt., Žljeb, Mokra Gora, Paštrik, Koritnik and Ošljak, and on the northern part of Kopaonik in the zone of beech forests (Fig. 7.17). In Serbia, kalkomelanosols and kalkocambisols cover approximately 3.5% of the land mass, i.e. about 3052.4 km^2.

Physical and chemical properties. The leaching of carbonates from compact limestone and dolomite is a long process, with limestone having little silicate residue. As such, in the initial phase, a humus horizon with no organo-mineral complexes forms over the hard limestone. Later, with the accumulation of silicate residue, these complexes form. Humus accumulation is intensive as conditions for its mineralization are unfavourable (drought and extreme temperatures on high mountains). The resultant humic acid is neutralized by $CaCO_3$ and $MgCO_3$, and humates form. The large amount of these humates leads to the formation of stable aggregates in the humus-accumulative horizon. Despite forming on a calcareous substrate, these soils do not contain carbonates. Their pH is neutral to acidic depending on the developmental stage, climatic conditions, and vegetation. As a rule, in the initial stages, they are least acidic. Leptosols which form in higher and wetter mountain regions have a higher degree of acidification. At lower altitudes and in more arid climatic conditions, they are neutral to slightly acidic (less leaching of bases).

They are rich in humus with humus content amounting to 25–50% during the organogenic stage and 10–25% during the organo-mineral stage. As such, they are characterized by a high adsorption capacity (often >50 meq/100 g of soil). In terms of adsorbed ions, Ca and Mg ions predominate. If formed on dolomite, Mg ions may exceed Ca ions. The supply of readily available forms of K is good, but they are very poor in available P. The total N content is high because of the high humus content, but these soils are poor in available forms of N due to the weak mineralization of organic material (Tables 7.14 and 7.15).

Land use. Leptosols are dry, warm soils. As a result, xerophytic vegetation dominates on them. They are porous soils with a high content of non-capillary pores. They are very permeable to water and are well-aerated. The production potential of these soils is greatly affected by their shallowness and high permeability, which means that plants on them perish due to drought. The majority of these soils are populated with xerophytic vegetation of meadows and pastures. Fewer are covered by forest, while the smallest proportion is used for cultivating potatoes, rye, barley and wheat.

Guidelines for management. Due to their high permeability, these soils are not subject to water erosion, but aeolian erosion is an issue, particularly at high altitudes and particularly if vegetation cover is sparse. The erosion of this soil leads to exposed bare karst.

Table 7.14 Physical characteristics of kalkomelanosol

Horizon	Hydroscopic water, humidity (%)	Granulometric composition (%)							
		2.0–0.2 mm	0.2–0.06 mm	0.06–0.02 mm	0.02–0.006 mm	0.006–0.002 mm	<0.002 mm	Total sand	Clay +silt
A									
X	8.88	1.02	14.81	17.58	27.05	12.76	27.03	33.28	66.78
Min	4.60	0.00	2.29	8.80	18.20	4.10	9.60	13.60	32.20
Max	17.04	4.60	44.10	34.20	34.70	21.10	53.10	67.80	86.40
A(B)									
X	7.13	0.95	7.55	8.30	23.52	12.42	47.22	16.80	83.16
Min	5.72	0.06	3.30	4.00	11.10	6.70	36.20	13.30	79.60
Max	9.15	3.83	14.14	11.40	33.70	18.70	64.00	20.40	86.50

Source Knežević and Košanin (2002, 2004, 2009)

Table 7.15 Chemical characteristics of kalkomelanosol

Horizon	pH		Y1 ccm n/10 NaOH	Adsorption complex				Humus (%)	C (%)	N (%)	C/N	P_2O_5 (mg/100 g)	K_2O (mg/100 g)
	H_2O	$CaCl_2$/KCl		T-S (cmol kg^{-1})	S (cmol kg^{-1})	T (%)	V (%)						
A													
Average	6.13	5.31	30.03	19.52	43.88	63.40	67.62	15.82	9.14	0.78	11.39	4.08	17.38
Min	5.10	3.90	10.63	6.91	13.80	23.68	36.68	1.94	1.13	0.12	8.20	0.20	7.30
Max	7.36	6.90	72.50	47.13	97.20	118.70	92.28	43.16	25.03	1.61	15.50	15.60	32.80
A(B)													
Average	7.23	5.71	8.30	5.50	43.52	49.01	87.07	4.98	2.90	0.35	8.10	0.60	15.96
Min	6.45	5.40	6.00	3.90	23.40	31.60	74.05	3.32	1.93	0.28	6.90	0.60	10.50
Max	7.82	6.88	12.00	8.20	60.50	64.40	93.94	7.97	4.62	0.44	10.50	0.60	20.20

Source Knežević and Košanin (2002, 2004, 2009)

7.3 Class III: Cambic Soils with A-(B)-C or A-(B)-R Soil Horizon Sequence

Soils of this class occur as a result of the further development of types from the class of humus-accumulative soils and are characterized by the occurrence of a cambic horizon directly beneath the humus horizon. Beneath the cambic horizon, there can be loose substrate (C) or compact rock (R), so the profile may be A-(B)-C-R (regolithic) or A-(B)-R (lithic). The central diagnostic horizon of the entire class is the cambic (B) horizon, where processes of weathering of primary minerals and argillosynthesis take place. As they do so, an excess of Fe oxide is released and adsorbed on the surface of newly formed clay minerals, giving this horizon its specific brown colour.

7.3.1 Eutric Cambisol

The name of the soil type. The traditional name for these soils in Serbia is gajnjača (Stebut 1923; Živković 1955), referring to those soils on which areas of residual deciduous forests (groves) are found. Due to the great heterogeneity of this soil type, particularly after the inclusion of cambic soils on serpentinite and peridotite under the category of eutric cambisol in the classification system used in Serbia, the term 'gajnjača' has become inadequate, even though it is still frequently used. Subtypes of this soil according to the classification system used in Serbia are as follows: eutric cambisol on loess and loess-like sediments; eutric cambisol on basic and neutral eruptive rocks; eutric cambisol on peridotite and serpentinite; eutric cambisol on lacustrine sediments; eutric cambisol on alluvial, colluvial and aeolian deposits; eutric cambisol on shale; and eutric cambisol on amphibolitic schists (Škorić et al. 1985).

The WRB name. Cambisol, eutric.

Fig. 7.18 Eutric cambisol (*photo* Čakmak, D.)

Genesis and sequence of horizons. Cambic soils are brown, yellowish-brown or rusty-reddish brown, which is characteristic for the (B)v horizon and originates from hydrated iron oxide and clay minerals. Eutric cambic soil has a mollic (Amo) and ochric (Aoh) humus horizon, which lies directly above a cambic (B)v horizon. Varieties with an A-(B)v-R profile usually have a total depth of 50–70 cm, while varieties with an A-(B)v-C-R profile may be deeper than 100 cm) (Fig. 7.18). In the (B)v horizon, the CEC is greater than 50%, while the pH in water is >5.5. It forms in semi-humid regions with mean annual precipitation of 600–700 mm, a marked dry period during the summer, and a mean annual temperature of 10–12 °C. The properties of the parent substrate are of great importance for the formation of eutric cambisol.

In Serbia, they form on a very diverse range of rocks (eruptive, sedimentary and metamorphic), starting from ultrabasic right through to slightly acidic. They have been noted on serpentinite, gabbro, diabase, andesite, certain gneiss and amphibolitic schists, limestone, marl, loess, loess Tertiary lacustrine sediments, and alluvial and diluvial

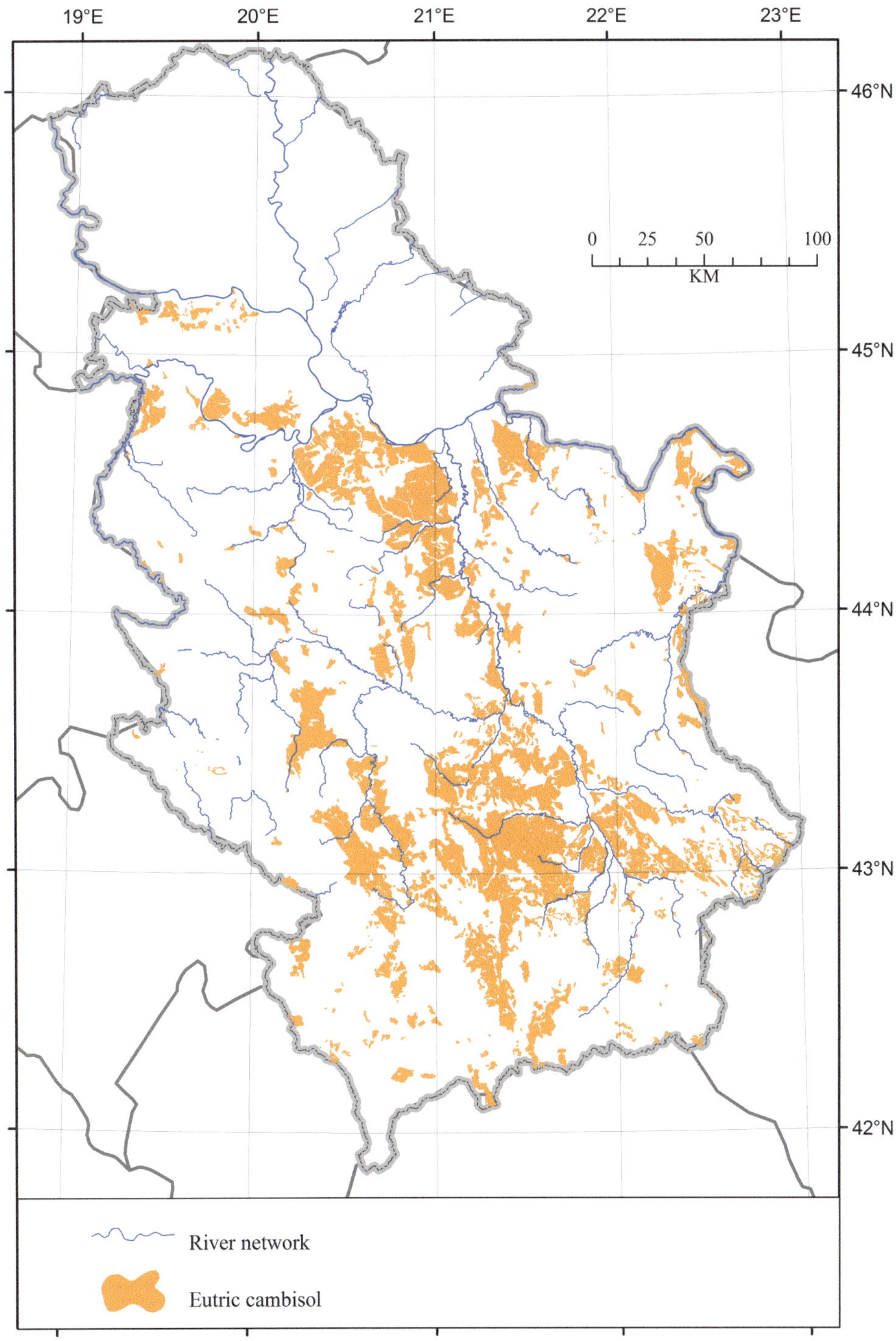

Fig. 7.19 Distribution of eutric cambisol

deposits. If the substrate is calcareous, the leaching of carbonates precedes the process of argillosynthesis. Forest litter accumulates an appreciable amount of base cations which contribute to maintaining the slightly acidic to neutral pH of the environment. In these conditions, supported by CO_2, the transformation of organic matter proceeds unhindered towards the formation of a stable organo-mineral complex in the humus. Iron released during the process of argillosynthesis acts as a binding cation between the clay and humic acid. Carbon dioxide is the main agent in the decomposition of primary minerals, which proceeds more intensely in the deeper parts of the profile. This is why the (B)v horizon is usually somewhat more clayey than the A horizon, while its colour depends on the amount and degree of hydrated iron. The formation of eutric cambisol most often occurs as a secondary process in soils with an A-C profile and proceeds through a series of transitional stages (brownized rendzina, brownized chernozem, etc.).

This process includes dealkalization processes: the occurrence of the opposite, but weaker process of base cation enrichment through biogene accumulation and the upward movement of bases; the formation of an illuvial-calcareous horizon (B)v; the destruction of primary aluminosilicates and simultaneous argillation, which is most pronounced in the middle part of the profile in the (B)v horizon; and the fixation of released excess $Fe_2O_3 \cdot nH_2O$ in the (B)v horizon. Classification at the level of varieties and forms is made according to the criteria set out in Table 6.1 (Chap. 6 Soil Classification).

Position in an evolutionary series. Eutric cambisols occur as a primary soil formation and also through the evolution of brownized chernozem, brownized rendzina or brownized vertisol in lowland regions, or ranker, most frequently mull-moder type, in mountainous regions. Through further evolution, they become lessive soils (luvisol).

Distribution. In Serbia, eutric cambisol forms at altitudes of 50–900 m a.s.l., but most are found beneath 500 m. They occur in Šumadija, Pomoravlje and Mačva (Fig. 7.19). In Serbia, it is a zonal soil covered with climatogenic vegetation of xerothermic Italian and Turkey oaks. Apart from these forests, they are also a habitat for the pubescent oak and oriental hornbeam, the pubescent oak and South European flowering ash, etc. Eutric cambisol in Serbia covers approximately 12.4% of the land mass, i.e. about 10,958.3 km^2.

Physical and chemical properties. The physical properties of eutric cambisol are characterized by their mechanical composition, which lies between chernozem and vertisol. In Serbia, the clayiest variation of eutric cambisol is found on Tertiary sediments. The chemical properties show that this soil has high potential, which is the result of a high content of Ca, Mg and other elements in the parent substrate on the one hand, and the high biological activity of the soil on the other. The exception is eutric cambisol on peridotite and serpentinite substrates, which, despite having a high degree of saturation by bases (mainly Mg), are deficient in K and Ca, as there is none in the parent substrate. From these rocks, the soil can also inherit a high concentration of some oligoelements, even to toxic levels (Ni and Cr). The high content of free iron in eutric cambisols may lead to the demobilization of phosphorus, so reduced levels of active P are not a rare occurrence in these soils. In fields, they most often contain 2–3% humus, and 4–6% in forests.

They are slightly acidic to neutral soils (pH above 5.5, most often around 6.5). They have a high degree of base saturation (70–80%) and a high adsorption capacity for base cations (which reaches 40 mg eqv/100 g). The C/N ratio is good and ranges between 10 and 14 in the humus-accumulative horizon. Physico-chemical characteristics of eutric cambisol are presented in Tables 7.16 and 7.17.

Land use. Eutric cambisols on loess, lacustrine sediments and alluvial-colluvial deposits have the highest productivity. They are the deepest soils, have good physical properties,

Table 7.16 Physical characteristics of eutric cambisol

Horizon	Hydroscopic water, humidity (%)	Granulometric composition (%)					
		2.0–0.2 mm	0.2–0.02 mm	0.02–0.002 mm	<0.002 mm	Total sand	Clay + silt
A							
X	3.12	9.08	40.22	31.92	21.40	46.68	54.43
Min	1.55	0.00	25.00	13.10	11.80	26.10	24.90
Max	4.88	48.20	65.80	43.90	39.10	75.10	73.90
B							
X	2.50	15.46	43.34	29.70	19.99	50.31	49.69
Min	1.47	0.00	25.40	13.60	10.20	25.80	26.40
Max	4.75	46.00	65.70	37.60	40.00	73.60	74.20

Source Košanin and Knežević (2005, 2007), Košanin and Gajić (2008), Pivić et al. (2008, 2014) and Galić (2014)

Table 7.17 Chemical characteristics of eutric cambisol

Horizon	pH		Y1	Adsorption complex				Humus	C	N	C/N	$CaCO_3$	P_2O_5	K_2O
	H_2O	$CaCl_2$/KCl	ccm n/10 NaOH	T-S (cmol kg^{-1})	S (cmol kg^{-1})	T (%)	V (%)	(%)	(%)	(%)		(%)	(mg/100 g)	(mg/100 g)
A														
Average	5.97	5.04	7.96	9.04	21.67	30.71	69.49	4.36	3.50	0.23	11.62	0.14	4.88	12.93
Min	5.20	4.20	6.70	2.27	13.00	19.61	41.67	1.10	1.40	0.08	9.80	0.00	2.05	9.60
Max	6.50	5.69	10.00	18.20	41.00	55.47	78.90	12.64	7.33	0.75	14.30	0.41	8.00	18.00
B														
Average	6.39	5.10	4.75	8.59	21.79	30.37	71.02	2.28	1.51	0.27	9.60	0.14	3.43	7.97
Min	5.45	4.21	0.95	0.91	15.40	21.49	52.74	0.67	0.39	0.09	9.60	0.00	1.35	3.96
Max	8.28	7.20	7.50	14.63	41.40	50.83	87.01	9.23	5.35	0.56	9.60	0.41	5.66	14.20

Source Košanin and Knežević (2005, 2007), Košanin and Gajić (2008), Pivić et al. (2008, 2014) and Galić (2014)

are found on level to partly sloping terrain and all agricultural crops are grown on them, including vineyards and orchards. Eutric cambisols formed on compact neutral and basic eruptive rocks or compact sediments with a similar mineral composition can be one of two varieties: lithic (shallower) and regolithic (deeper). Regolithic varieties fall into the class of the best forest soils. Eutric cambisol on peridotite and serpentinite rock is the least productive soil with low trophicity due to the composition of the rock. In addition, it is a shallow soil with lots of coarse fragments and a dry habitat.

Guidelines for management. In order to achieve a good crop yield, irrigation is necessary during the summer months.

7.3.2 Dystric Cambisol

The name of the soil type. In the Serbian classification, dystric cambisol is called 'brown acid soil'. This soil is characterized by an ochric (Aoh) or umbric (Aum) humus horizon, with the latter characteristic at higher altitudes, which lies directly above the cambic (B)v horizon. The chemistry of the soil is acidic (pH in water is <5.5), with the CEC <50%. It forms on acidic rocks, such as sandstone, phyllite, shale, acid eruptive rocks and other quartz/silicate rocks. V. Avdalović gave a detailed description of the genesis and properties of this soil type in Serbia (Avdalović 1975). The subtypes of this soil according to the classification system used in Serbia are as follows: dystric cambisol, typically with ochric humus; humus-rich dystric cambisol with an umbric humus horizon (mainly at higher altitudes); luvic dystric cambisol with the occurrence of colloidal membranes in the (B) horizon, without a marked E horizon; psuedogleyic dystric cambisol; and podzolic dystric cambisol with the occurrence of initial E and Bh or Bfe horizons (Škorić et al. 1985).

The WRB name: Cambisol, dystric.

Genesis and sequence of horizons. In Serbia, these soils form a high-altitude belt in mountain massifs, at altitudes of 500–1100 m a.s.l., populated by broadleaved and mixed broadleaved/coniferous forests (forests of mountain beech and beech-fir). Climate is moderately humid with precipitation of 700–1000 mm and average annual temperatures of 5–8 °C.

The parent substrate is mainly acidic silicate rocks, which when broken down result in only a small amount of clay, but a high percentage of sand. Most often this is granite, granodiorite, rhyolite, cornite and other igneous rocks, and gneiss, mica-schist, phyllite and other metamorphic rocks. They also form on acidic sedimentary rocks, such as new red sandstone and other acidic sandstones and shales. The profile of dystric cambisol is relatively simple. Depending on the hardness of the rock, it can have lithic or regolithic contact (an A-(B)v-C profile or A-(B)v-R). There is a clearly visible humus-accumulative A horizon, dark brown in colour and usually with a crumbly structure, about 10–20 cm thick. Beneath this horizon, there is a sharp transition to the (B)v horizon, which is ochre or yellow in colour, depending on the iron content and degree of hydratization. The thickness of this horizon ranges from 30 to 80 cm. The transition to the C horizon is gradual, and often a transitional (B)v/C horizon can be seen (Fig. 7.20).

Typical processes in these soils are the liberation of primary minerals from the rocks (disintegration of the parent substrate), which proceeds more quickly than the processes of oxidation, hydratization and hydrolysis. Argillogenesis is retarded due to the acidic environment and the formation of minerals is typical, above all kaolinitic minerals. Classification at the level of varieties and forms is made according

Fig. 7.20 Dystric cambisol (*photo* Čakmak, D.)

to the criteria set out in Table 6.1 (Chap. 6 Soil Classification).

Position in an evolutionary series. Dystric cambisol occurs most often as a primary soil formation and to a lesser extent through the evolution of ranker. In the zonal sense, in Serbia it occurs between the brown podzolic soils occupying a high-altitude belt and lessive soils occupying a lower belt. As such, the evolution of these soils flows in two directions. Due to changes in the bioclimatic complex caused above all by an increase in altitude, one pathway of evolution is via humus dystric cambisol into brown podzolic soil and eventually secondary podzol. Humus-rich dystric cambisol also evolves into dystric cambisol through the succession of vegetation from mountain beech to subalpine beech. Despite there being no visible signs of podzolization in the morphology of the profile, the dynamics of the organic material and of the mobile Fe and Al oxides make this soil closer to podzolic brown soils (Jović 1966; Avdalović 1975). Organic matter decomposes more slowly due to low temperatures and higher humidity. The amount of humus increases, accompanied by an increase in the aggressive fraction of fulvic acid 1a. The R_2O_3:FK ratio narrows in the humus-accumulative horizon, solutions are diluted, while sesquioxides are more mobile, which together with the fraction of fulvic acid 1a, migrate to the lower parts of the profile. However, the mobility of the organo-mineral complex is low, and most often it remains in the lower parts of the A horizon, which can be up to 60 cm thick. Nevertheless, the large amount of humus masks the migration process, so the morphological differentiation of the humus-accumulative horizon is not marked. *Podzolic dystric cambisol* is linked to extremely acidic parent substrates, most often quartzite, which produces sandy material rich in quartz, but poor in sesquioxides and clay when broken down. In this case, the humus-accumulative horizon is composed of acidic moder humus, while the (B)v horizon is a whole, as with other types of acidic brown soils. This soil is characterized by a clear boundary and sudden increase in Fe and Al with the transition from the humus-accumulative to the (B)v horizon. The basic characteristics of this soil is the regular presence of an aggressive fraction of fulvic acid 1a, which is greater than the sesquioxide content in the surface parts of the profile, so the R_2O_3:FK ratio is <1, giving it similar characteristics to podzols. This occurs on sandy substrates with coniferous vegetation (Avdalović 1975).

In the hilly belt in the zone of oak forests, on clayey substrates, the pathway of evolution is via lessive dystric cambisol towards the surface pseudogleization of these soils. *Lessive brown acid soils* occur on siliceous parent substrates, which release more clay when breaking down, such as clayey sandstone and phyllite. The clay index in the (B)/A horizon points to the lessivation process where the content of colloidal clay in the (B)v horizon is twice as high as in the A horizon. The amount of clay is the main factor in lessivation, together with Fe and other cations. This process is characterized by the conditions for redox processes, i.e. the alternation of a distinctly wet with a distinctly dry phase, which results in the transformation of insoluble Fe into the soluble form and its migration through the profile. In this way, microaggregates freed from iron, which is a coagulation agent, disperse more easily and take on a fine colloidal form. At this evolutionary stage, intact hydrophilic colloid particles, mainly montmorillonite, but also illite, may translocate. Fulvic acids dominate humic acids in the humus, as supported by a Ch:Cf ratio <1. In addition to the fraction of fulvic acid (1 + 2), lessive brown acid soils also contain the aggressive fraction of fulvic acid 1a, but in small amounts. The R_2O_3:FK ratio is far wider, particularly in the (B)v horizon. The morphology of the profile is influenced by the relocation of free Fe and Al oxides bound to the mineral component of the soil. The migration of clay, Fe and Al brings about its differentiation, and with it, its different appearance to typical dystric cambisol (Avdalović 1975).

Distribution. In Serbia, these soils are found on mountains (Goč, Željin, Kopaonik, Kosmaj, Bukulja, Cer and Majdanpečka Domena), Fig. 7.21. Dystric cambisol covers approximately 8.2% of the land mass, i.e. about 7289.61 km^2. The vegetation of brown acid soils is beech and beech-fir forests.

Physical and chemical properties. The soil has a favourable mechanical composition with the colloid clay fraction not exceeding 20% and a total clay + silt fraction under 50%, while the coarse sand fraction is approximately 25%, which creates favourable water–air properties (Table 7.18). Likewise, the total porosity of the A horizon reaches 78%, and although decreasing with depth, is still 60% in the (B)v horizon, while the air capacity ranges from 30 to 50%, which points to the effective filtration of precipitation through the profile (Antić et al. 1982). Dystric cambisol is characterized by marked acidity (pH in H_2O is 3.8–5.5) and a low degree of base saturation (<50%). While

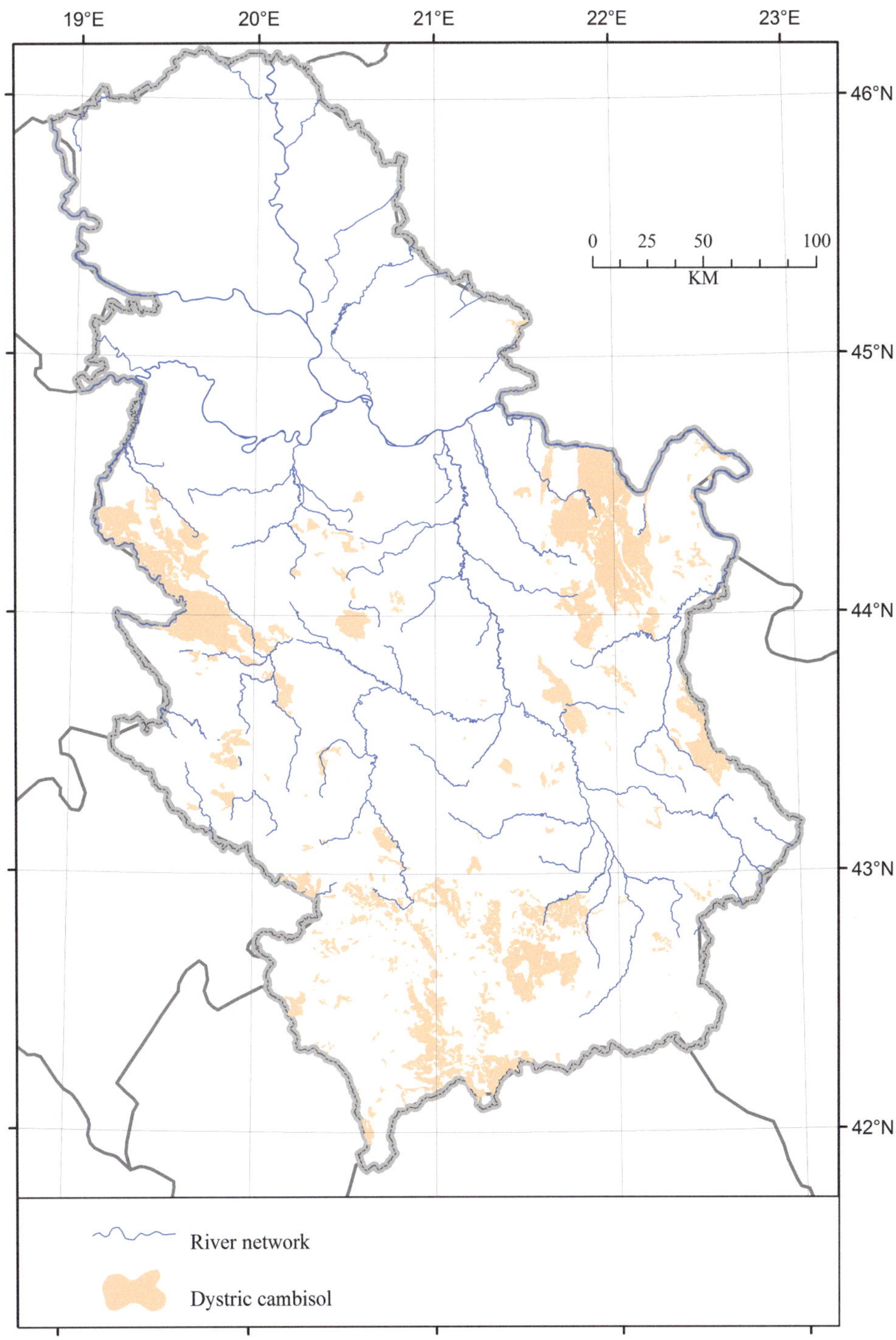

Fig. 7.21 Distribution of dystric cambisol

it is considerably lower in the rest of the profile, the total cation capacity in the A horizon is high (on average up to 40 cmol kg^{-1}) (Table 7.19). This is due to the high humus content, as the proportion of clay here is very low. The humus is mull-moder to moder.

In these soils, the fulvic fraction outweighs humic acid, which is shown by a Ch:Cf ratio <1 across the whole profile and the domination of the aggressive fraction of fulvic acid 1a, which, however, does not exceed the total fulvic acid content (fraction 1 + 2). At the same time, this soil contains large amounts of R_2O_3, enough for the accumulation of the fraction of fulvo acid 1a in the very surface, meaning it does not migrate into the lower layers. As a result, there is a wider R_2O_3:FK ratio in the humus-accumulative horizon, which is always greater than 1 (most often 2–3). In the humus-accumulative horizon, there is not only an elevated sesquioxide content in relation to fulvic acid 1a, but there is also the relative accumulation of Fe. On the other hand, Al migrates from the surface into the lower parts of the profile and accumulates in the (B)v horizon (Tables 7.20 and 7.21).

Land use. The general conditions of this habitat that lead to the quick disintegration of the parent substrate, as well as the favourable transformation of organic matter, and as such the relatively intensive cycling of nutrients, determine the high ecological value of brown acid soils in Serbia as forests. Most often, the forests belong to productivity classes I and II. The beech and beech-fir forests on this soil in Serbia have extremely high biomass production ability. This is true for all parent substrates, except in cases of extremely acidic rocks. Forests populate the majority of brown acid soils in mountain regions, with meadows and pastures covering a smaller proportion, and fields and orchards the smallest area.

Guidelines for management. The physical characteristics of this soil are favourable. Fertilization (N and P) can have

Table 7.18 Physical characteristics of dystric cambisol

Horizon	Hydroscopic water, humidity(%)	Granulometric composition (%)							
		2.0–0.2 mm	0.2–0.06 mm	0.06–0.02 mm	0.02–0.006 mm	0.006–0.002 mm	<0.002 mm	Total sand	Clay + silt
A									
X	3.23	22.42	18.26	15.21	21.94	9.61	13.44	55.17	44.83
Min	0.87	0.10	6.08	5.90	4.30	4.00	7.30	36.74	15.60
Max	5.48	66.30	29.30	31.70	38.10	16.20	25.70	84.40	63.26
Bv									
X	2.16	22.11	17.84	13.63	20.36	11.32	15.24	52.90	47.11
Min	0.73	0.10	7.63	5.10	4.10	1.70	6.70	32.71	21.30
Max	3.64	51.80	29.23	33.70	33.62	24.10	38.30	78.70	67.29

Source Pavlović (1992, 1998), Knežević and Košanin (2002), Knežević and Cvjetićanin (2003), Belanović and Košanin (2005), Košanin and Knežević (2005, 2007), Čokeša et al. (2007), Kostić (2007) and Knežević et al. (2011)

Table 7.19 Chemical characteristics of dystric cambisol

Horizon	pH		Y1 ccm n/10 NaOH	Adsorption complex				Humus (%)	C (%)	N (%)	C/N	P_2O_5 (mg/100 g)	K_2O (mg/100 g)
	H_2O	$CaCl_2$/KCl		T-S (cmol kg^{-1})	S (%)	T (%)	V (%)						
A													
Average	4.83	3.91	47.87	29.29	11.47	40.75	28.17	8.92	5.04	0.45	11.59	3.73	15.82
Min	3.80	3.10	19.00	12.35	1.80	15.78	4.16	2.13	1.24	0.12	6.50	0.50	4.80
Max	5.56	4.67	105.07	68.30	25.60	80.87	50.92	19.89	11.54	1.08	15.70	12.80	35.60
Bv													
Average	5.23	4.08	22.54	15.10	7.21	22.54	31.56	1.29	0.73	0.15	7.78	1.19	7.21
Min	4.40	3.30	9.79	6.36	1.00	12.13	7.45	0.20	0.12	0.04	5.00	0.00	2.00
Max	5.80	5.00	53.72	34.92	18.39	44.80	65.62	6.10	3.54	0.55	13.00	9.60	18.20

Source Pavlović (1992, 1998), Knežević and Košanin (2002), Knežević and Cvjetićanin (2003), Košanin and Knežević (2005, 2007), Čokeša et al. (2007), Kostić (2007) and Kostić et al. (2012b)

Table 7.20 Group and fractional composition of humus of dystric cambisols in an *Abieto-fagetum* forest on Mt. Maljen

Horizon	Depth (cm)	Total C	Humic acids				Fulvic acids						Sum of all acids	Insoluble residue	Ch/Cf
			1	2	3	Σ	1a	1	2	3	4	Σ			
Aoh	0–4	32.82	4.23	0.72	0.30	5.25	0.62	4.39	0.86	0.38	0.42	6.67	11.92	20.90	0.78
		100.00	12.89	2.19	0.91	15.99	1.89	13.37	2.62	1.15	1.27	20.32	36.31	63.68	
A	4–16	9.06	1.89	0.00	0.11	2.00	0.35	1.92	0.40	0.15	0.20	3.02	5.02	4.04	0.66
		100.00	20.86	0.00	1.21	22.07	3.86	21.19	4.41	1.65	2.20	33.33	55.40	44.60	
(B)v	16–31	3.14	0.52	0.00	0.08	0.60	0.13	0.59	0.20	0.10	0.12	1.14	1.74	1.40	0.52
		100.00	16.56	0.00	2.54	19.10	4.14	18.78	6.36	3.18	3.82	36.30	55.41	44.58	
(B)v	31–65	0.78	0.11	0.00	0.02	0.13	0.05	0.18	0.04	0.01	0.03	0.31	0.44	0.34	0.41
		100.00	14.10	0.00	2.56	16.66	6.41	23.07	5.12	1.28	3.84	39.74	56.42	43.58	

Source Pavlović (1998)

Table 7.21 Composition of 0.05 M H_2SO_4 extract of dystric cambisols in an *Abieto-Fagetum* forest on Mt. Maljen

Horizon	Depth (cm)	pH H_2O	Fulvic acid 1a (C × 2)	R_2O_3 (%)	Fe_2O_3 (%)	Al_2O_3 (%)	R_2O_3/FK	Al_2O_3/Fe_2O_3	Fe (B)v/Fe A
Aoh	0–4	5.02	1.24	–	–	–	–	–	–
A	4–16	4.73	0.70	0.91	0.50	0.41	1.30	0.82	–
(B)v	16–31	5.15	0.26	0.87	0.11	0.76	3.35	6.90	0.22
(B)v	31–65	5.16	0.10	0.74	0.07	0.67	7.40	9.57	0.14

Source Pavlović (1998)

significant ameliorative effects, meaning potatoes, oats, barley and rye can be cultivated. However, these are typical forest soils, yielding a high wood volume production.

7.3.3 Kalkocambisol (Brown Soil on Limestone and Dolomite)

The name of the soil type. The name comes from the Latin word *calxis* (lime) and *cambisol* (*cambio*—change). This soil has a mollic (Amo) humus horizon, which lies directly above a brown cambic (B)rz horizon. It forms on pure limestones and dolomites, which are most often karstified. The whole solum is non-calcareous. It has a loamy or heavier granulometric composition and a very strong polyhedral structure. Subtypes of this soil according to the classification system used in Serbia are as follows: typical (there are no signs of processes other than those characteristic for this soil type) and luvic (the beginnings of an E horizon) (Škorić et al. 1985).

The WRB name. No adequate name.

Genesis and sequence of horizons. In Serbia, these soils are mainly linked to the more mesophilic forests of mountain beach, beach and fir, and less commonly to mesophilic oak forests. This is why this soil is characterized by a relatively humid summer pedoclimate as opposed to *terra rossa*. It forms on hard pure limestones and dolomites, which have less than 1% insoluble residue. These are mainly relict soils, linked to the palaeoclimatic conditions of the Tertiary. In Serbia, this soil is often called *terra fusca*—a term first used by Pavićević (1953), while studying the soils of Suva planina. The name comes from 'kalksteinbraunlehm' (*terra fusca*), which Kubiëna introduced (1948), wanting to highlight the link to *terra rossa*, while stressing that this is a reddish-brown (*fusca*—brown, dark) soil. It has a simple profile structure—A-(B)rz-R or C/R (Fig. 7.22). In the morphology of the profile, there is a prominent cambic (B)rz horizon, but not a (B)v horizon. Specifically, the insoluble residue is not subject to the process of argillosynthesis; instead the clay accumulates after the leaching of $CaCO_3$, and a (B)rz horizon forms, characterized by the residual accumulation of clay as the limestone and dolomite break down. The basic characteristics of this horizon are its heavy mechanical composition with a large colloid fraction, plasticity and poor drainage. The humus-accumulative horizon is up to 15 cm thick, while the (B)rz horizon is between 30 and 50 cm in depth. There may also be a transitional A/(B)rz horizon. The total depth of this soil rarely exceeds 60 cm. Classification at the level of varieties and forms is made according to the criteria set out in Table 6.1 (Chap. 6 Soil Classification).

Position in an evolutionary series. This soil type arises from the evolution of kalkomelanosol, with the transition accompanied by the enrichment of the profile with clay (the

Fig. 7.22 Kalkocambisol (*photo* Čakmak, D.)

insoluble residue contains up to 90% clay), meaning that this soil has low permeability for water and less aeration, as well as less biological activity compared to the previous stage. Kalkocambisols are relatively stable soil formations and with a change in bioclimatic conditions, they become subject to lessivation. In Serbia, on the limestone part of Mt. Tara, all the genetic evolutionary stages in the development of this soil on limestone and dolomite have been noted: kalkomelanosol → brownized kalkomelanosol → *terra fusca* → lessive *terra fusca* → *terra fusca*—pseudogley (Antić et al. 1963).

Distribution. In Serbia, these soils are found on mountains (Tara and Suva planina). On the soil map of Serbia, the same colour is used to denote these soils as kalkomelanosols, clearly highlighting that these form an evolutionary series on limestone and dolomite (Fig. 7.23). Together, kalkomelanosol and kalkocambisol cover approximately 3.5% of the land mass in Serbia, i.e. around 3052.40 km^2. The vegetation of these soils comprises beech and beech-fir forests on limestone and dolomite.

Physical and chemical properties. These soils are characterized by a high clay + silt content and, in terms of their mechanical composition, can be classed as clayey loam, clay and heavy clay soils (Table 7.22). Kalkocambisols are non-calcareous soils. They have less humus than the previous evolutionary stage (kalkomelanosol) and brownized kalkomelanosol. The amount of humus is usually up to 10% in forests. The humus content is also high in the (B) horizon, up to 3.75%. These soils have a high adsorption capacity, while the degree of base saturation is over 50%, and can be over 80%. They are poorly supplied with phosphorus, but have a much better supply of K_2O (Table 7.23).

Land use. The ecological production value of this soil for forestry is relatively good. If there is enough soil depth and if conditions at the habitat are quite mesophilic, wood volume production can be at a satisfactory level. In Serbia, the majority of these soils are covered by forests, with a smaller area used as pasture, and fields covering the smallest area.

Guidelines for management. This soil is subject to erosion and it is necessary to afforest these areas so as to prevent this. Fertilization with phosphoric fertilizers is also important.

7.3.4 Red Soil (*Terra rossa*)

The name of the soil type. This soil gets its name from its colour (*terra*—soil, *rossa*—red). Red soil is soil from the Mediterranean and sub-Mediterranean regions with an ochric humus horizon (A_{oh}) lying directly above a cambic $(B)_{rz}$ horizon and is red in colour. It forms on pure limestones and dolomites which are karstified, and the solum is non-calcareous. Its texture is heavier than loamy, while its structure is stable polyhedral. Subtypes of this soil according to the classification system used in Serbia can be typical (there are no signs of other processes other than those characteristic for this type), luvic (the beginnings of an E horizon) and ferralic: up to a depth of 1 m pisolithic concretion occurs and/or hardening on exposure to air (Škorić et al. 1985).

The WRB name. No adequate name.

Genesis and sequence of horizons. In Serbia, these soils are not at all widespread. In the past, the influence of the Mediterranean climate extended along the valley of the Velika Morava to the foothills of the Kučaj and Rtanj mountains. As the Mediterranean climate retreated to today's boundaries, a part of this soil found itself subject to different climatic conditions and, according to Stebut (1953), red soil in Serbia has a relict and cryptorelict character. Of exceptional significance for the genesis of red soil is the parent substrate. Namely, typical red soils form on hard limestone, particularly from the Cretaceous period. This substrate is characterized by a low level of insoluble residue (0.2–0.4%), accompanied by low SiO_2 content, which is also characteristic for other soil types from more humid regions. Ćirić and Aleksandrović (1959) believed that of greater importance was the SiO_2:R_2O_3 ratio in the insoluble residue of limestone, which is 2 or <2 for typical red soil. The basic pedogenetic process in red soil is rubification. The essence of this process is the dehydration and crystallization of Fe oxide, which is absorbed to the surface of clay minerals as they are freed from the limestone as insoluble residue. At this point, the mineral hematite crystallizes, resulting in the

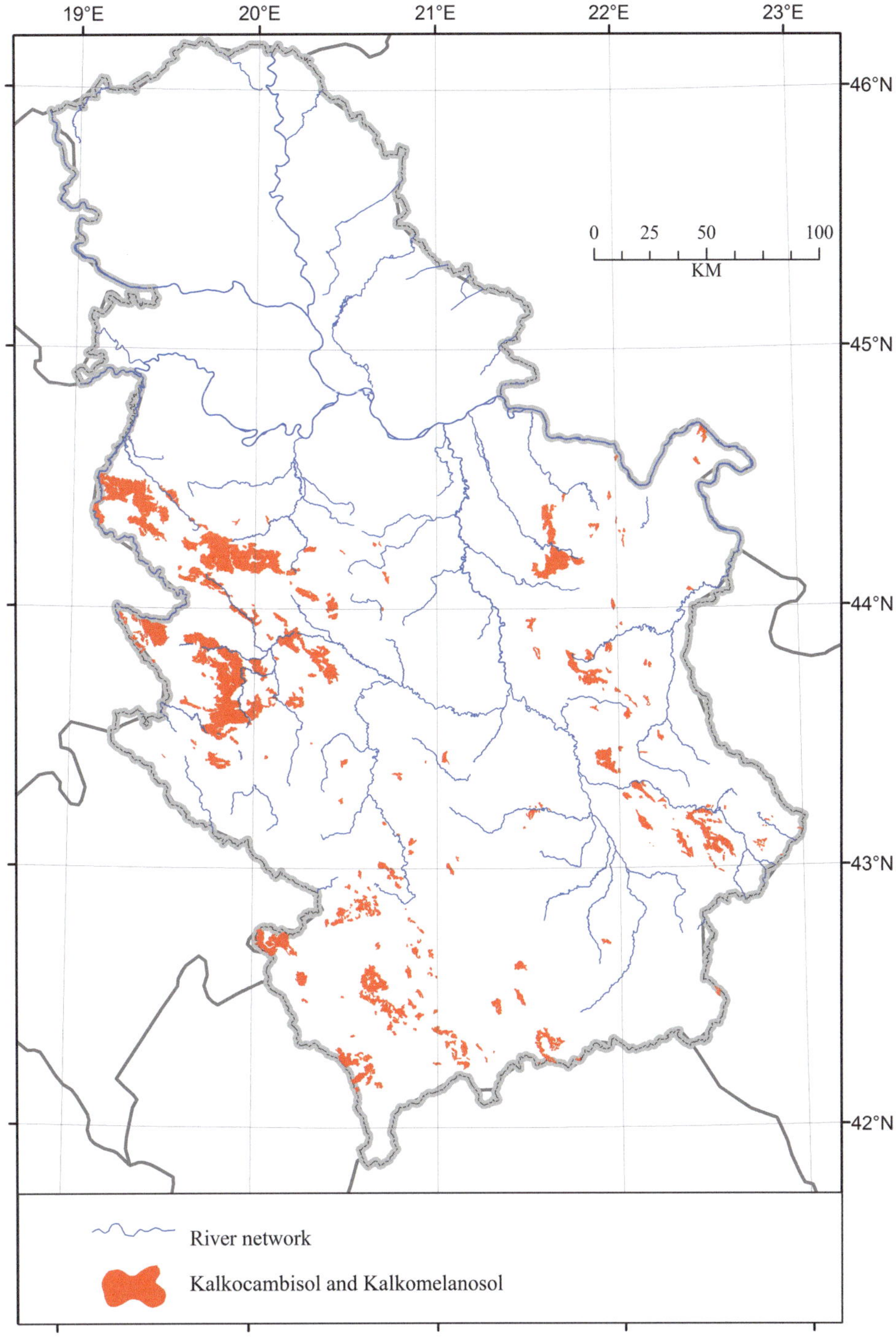

Fig. 7.23 Distribution of kalkocambisol and kalkomelanosol

red colour. This process requires a period that is wet enough to allow for the quick leaching of carbonates and the accumulation of residue, but also a dry period during which the process of dehydration can take place. An important factor is the content of organic matter, which in soils on limestone and dolomite determines whether rubification or brownization will occur. When there is more organic matter, an organo-mineral complex forms, from which iron is freed through slow biodegradation and goethite crystallizes, resulting in a brown colour. With less organic matter, cryptocrystalline ferrihydrite forms, which quickly transforms into red hematite. Red soils are soils with an A-(B)-C profile. Their morphology is characterized by a relatively simple profile whose genetic horizons are difficult to differentiate. Even so, the surface part, the A horizon, differs from the rest of the profile. It is dark brown in colour, with a clear granular or even crumbly structure. This kind of humus-accumulative horizon is found mainly under

Table 7.22 Physical characteristics of kalkocambisol

Horizon	Hydroscopic water, humidity (%)	Granulometric composition (%)							
		2.0–0.2 mm	0.2–0.06 mm	0.06–0.02 mm	0.02–0.006 mm	0.006–0.002 mm	<0.002 mm	Total sand	Clay + silt
A									
X	4.79	1.62	12.58	19.92	26.57	12.28	26.99	34.12	65.88
Min	2.31	0.20	4.81	9.90	14.50	8.90	9.10	20.80	49.80
Max	6.95	4.24	20.30	30.70	35.50	21.60	46.20	50.20	79.20
B									
X	4.45	1.13	10.74	17.46	27.57	11.58	31.60	29.35	70.65
Min	1.63	0.10	1.47	5.00	9.30	3.70	12.80	12.50	53.50
Max	7.21	4.23	16.80	32.20	47.40	17.20	54.40	46.50	87.50

Source Cvjetićanin and Knežević (2000), Knežević and Košanin (2002, 2004), Knežević and Cvjetićanin (2003) and Košanin and Knežević (2007)

Table 7.23 Chemical characteristics of kalkocambisol

Horizon	pH		Y1 ccm n/10 NaOH	Adsorption complex				Humus (%)	C (%)	N (%)	C/N	P_2O_5 (mg/100 g)	K_2O (mg/100 g)
	H_2O	$CaCl_2$/KCl		T-S (cmol kg^{-1})	S (cmol kg^{-1})	T (%)	V (%)						
A													
Average	5.74	4.67	24.22	14.98	30.14	45.89	63.67	5.68	3.30	0.33	10.67	0.92	18.29
Min	4.36	2.63	4.76	3.10	11.60	22.49	38.72	1.83	1.06	0.12	8.80	0.10	6.10
Max	7.15	6.05	87.66	56.98	63.60	92.98	94.94	9.61	5.57	0.77	13.30	1.80	37.00
B													
Average	5.74	4.42	16.59	10.75	26.94	37.68	65.55	1.76	1.02	0.21	7.25	0.43	9.87
Min	5.12	3.14	3.27	2.13	9.80	16.18	44.33	0.77	0.45	0.14	5.20	0.00	3.50
Max	7.17	6.10	32.40	20.41	67.00	69.13	96.92	3.74	2.17	0.42	8.30	0.80	19.00

Source Cvjetićanin and Knežević (2000), Knežević and Košanin (2002, 2004), Knežević and Cvjetićanin (2003) and Košanin and Knežević (2007)

preserved forest vegetation. The transition to a (B) horizon is gradual, but the latter differs clearly—it is red, of a somewhat heavier mechanical composition, and has a less marked structure. The boundary with the C horizon is quite clear, with the (B) horizon lying almost directly on the limestone parent substrate. Classification at the level of varieties and forms is made according to the criteria set out in Table 6.1 (Chap. 6 Soil Classification).

Position in an evolutionary series. The evolution of red soil on many substrates takes different directions, but one of the most characteristic is towards lessive red soil: kalkomelanosol → terra rossa → luvisol.

Distribution. These soils are found to a limited extent on Raška, Rtanj, Suva planina and in the foothills of the Kučaj mountains. In Serbia, they are extra-zonal soils (Fig. 7.24). Xerothermic vegetation grows on them.

Physical and chemical properties. Typical red soil are relatively shallow, with a depth of 50–60 cm. They have a high content of clay + silt, and particularly of the fraction <0.001, which can be as high as 63%. Red soil, even when their mechanical composition is heavy, is permeable and well-aerated due to their structure. The water holding capacity is not especially high (30–40%), while the air capacity is relatively favourable, being highest in the A horizon (15–17%), which is the result of its porosity. Of the total pore count (45–50%), almost a third are air-filled.

The chemical properties of red soil are characterized by a significantly lower SiO_2 content in relation to brown soil. It does not exceed 50%, and may be only 40%. On the other hand, the sesquioxide content is considerably elevated. This most often relates to Fe oxide, which is found at levels two to three times higher than in brown soils. It is a similar case for Al oxide. Even so, red soil has retained its silicic character and more broadly can be categorized as a ferrosilicic soil. Only when the sesquioxide content is extremely high does it have a ferralic or allic character (bauxite red soil). Typical red soil has a relatively high CEC (T), which sometimes reaches 50 mg ekv/100 g of soil. This is related to the silicic character of this soil. At the same time, they have a high sum of base cations (S), i.e. low levels of

Fig. 7.24 Red Soil (*photo* http://tihomirstojanovic.com/2008/12/zemlja-crvenica.html)

non-saturation of the adsorption complex. From this derives a high degree of base saturation (V = 80–90%, and in some cases >90%). This results in a neutral or very slightly acidic chemistry. As a rule, they do not contain $CaCO_3$ in the profile. They have a very low humus content and are also poor in nitrogen and particularly poor in available phosphorous as phosphoric acid binds to iron, rather than their being a lack of it in the soil.

Land use. These are warm habitats populated by xerothermic vegetation with a small amount of litter and are subject to intensive mineralization. On the other hand, the lack of depth presents a problem for the radication of vegetation, made worse by the process of 'deep erosion', which involves whole areas of soil falling into the cracks in the karst bedrock, from which sinkholes later form.

Guidelines for management. Red soil are characterized by average production ability. Due to the high content of positively charged colloid particles, there is the marked fixation of phosphoric acid and the occurrence of chemosorption (the formation of Al and Fe phosphates), and as such, these soils are considered to be poor in phosphorus. The agricultural value of red soil depends on its depth. They are excellent for cultivating vineyards and fruit trees, and vegetables, too, with irrigation.

7.4 Class IV: Eluvial-Illuvial Soils of A-E-Bt-C or A-E-Bt-R Soil Horizon Sequence

The characteristic the soils in this class share is the clearly marked process of eluvial-illuvial migration, which, besides leaching base cations, affects clay minerals, humus or the constituent elements of the aluminosilicate core (Si, Al and Fe). This is why there are two mutually connected diagnostic horizons in the profile of the soils of this class: the E horizon, from where the most intensive leaching takes place; and the B horizon, where the migrating material is retained and accumulates.

7.4.1 Luvisol (Luvic Soil)

The name of the soil type. The name comes from the Latin *ilimare*, which means *to wash out clay*. In pedological literature from ex-Yugoslavia, these soils have been described using various names (luvic brown soil, podzolic brown soil, podzolic soil and weak podzols). The subtypes of this soil according to the classification system used in Serbia are as follows: luvisol on siliceous and calcaro-siliceous substrates; and luvisols on pure limestone and dolomite (Škorić et al. 1985).

The WRB name. Luvisol.

Genesis and sequence of horizon. Luvisols are soils which form on loess and old alluvial and glacial deposits. Mainly occurring on level to partly sloping ground and in lower belts, they form in humid climates. In the conditions in Serbia, these are regions with over 700 mm annual rainfall and a mean annual temperature of 8–11 °C. The morphology of this profile in forests is Olf-A-E-Bt-C (Fig. 7.25).

A clear morphogenetic sign is the bleached E horizon. It might start at the very surface when there is no humus-accumulative horizon. The illuvial Bt horizon, enriched with colloid clay, is the complete morphological opposite, being brownish coloured. In broad terms, luvisol forms through the decomposition of structural aggregates, peptization of the colloidal clay particles, their migration and the illuviation of clay. Peptization is linked to soil salinity, the stability of structural aggregates, the type of clay minerals and the protective action of the colloid. A decrease in the electrolyte concentration in the soil solution can aid and accelerate peptization. This occurs after heavy rain and melting snow when the thickness of the aqueous film around the colloid particles increases severalfold.

Fig. 7.25 Luvisol (*photo* Čakmak, D.)

Montmorillonite clay minerals swell more and peptize more easily than kaolinite and illite. Peptization is also accelerated by acidification of the adsorption complex as the replacement of base cations with H-ions reduces the stability of the structural aggregates. This process only occurs with pH values up to 4.5. Below this level, Al^{+++} and Fe^{+++} ions are released, which are powerful coagulators and maintain the stability of the structural aggregates. Therefore, the formation of luvisol develops in slightly acidic environments (pH 6.5–4.5).

The clay particles migrate in downward direction through macropores, or cracks that appear during long dry periods. When conditions for the migration of the clay particles cease, illuviation begins. This may be with the migration of the clay minerals suspended in soil solution from the zone of large pores into the zone of medium-sized and small pores when the clay particles are deposited on the pore walls. Salinity usually increases as the colloidal solutions migrate into the deeper parts of the profile, with the downward migration of colloid particles inversely proportional to the electrolyte concentration. When illuviation of the colloid clay is strong, reduced drainage capacity can be noted as well as characteristic visible dark brown spots of Mn/Fe which increase with depth. Classification at the level of varieties and forms is made according to the criteria set out in Table 6.1 (Chap. 6 Soil Classification).

Position in an evolutionary series. On the whole, luvisols are not primary, but secondary soil formations. In Serbia, they occur through the evolution of soils with an A-(B)-C profile (eutric cambisol, kalkocambisol and dystric cambisol). The further evolution of luvisol takes place in two directions. One pathway is on substrates with a lighter granulometric composition in colder regions and at higher altitudes, where leaching and acidification take place in the surface layers to such a degree that conditions for podzolization are created. The other pathway is when, in more clayey luvisols, the accumulation of clay in the Bt horizon leads to a reduction in permeability to water and the stagnation of surface water in the profile, and when these luvisols evolve into pseudogley and pass into the class of hydromorphic soils.

Distribution. In Serbia, luvisols occur in a belt from 200 to 1000 m a.s.l. They occupy areas of western, central and eastern Serbia, as well as parts of Kosovo and Metohija (Fig. 7.26).

They cover approximately 5.1% of Serbia's land mass, i.e. about 4491.42 km^2. Vegetation associated with these soils is forests of Italian and Turkey oak, oak-hornbeam,

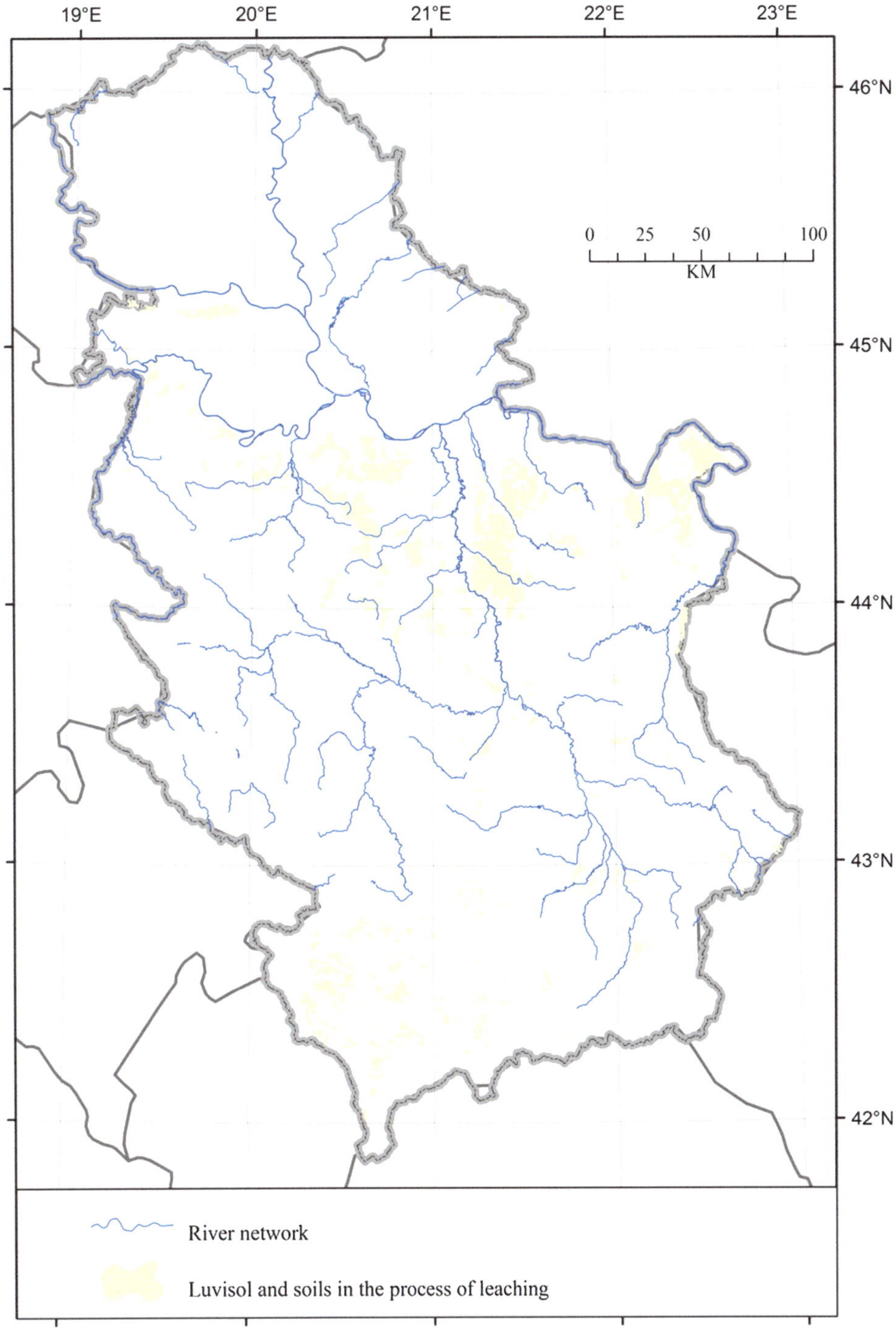

Fig. 7.26 Distribution of Luvisol and soils in the process of leaching

oak-beech and beech-fir, and more rarely polydominant beech-fir-spruce forests.

Physical and chemical properties. Luvisols can have lithic or regolithic contact with the parent substrate and a total depth of over 70 cm. Beneath the O horizon, there is a humus-accumulative horizon with a depth of 5–15 cm. The depth of the E horizon ranges from 10 to 20 cm. This horizon is yellowish grey in colour. The depth of the Bt horizon is between 30 and 80 cm and is brown in colour with a nutty or prismatic structure. The surface horizons are usually sandy or silty loam. The clay content in the E horizon is somewhat higher than in the A horizon. The granulometric composition of the Bt horizon is loam or clay loam, and there is 1.2–2.5 times more clay than in the E horizon. The humus content ranges from 4% in forests at lower altitudes, up to 9% in forests in higher altitude regions. In rare cases, it can even be as high as 20%. When cultivated, the humus content is from 1 to 2%. The chemistry is slightly to moderately acidic (pH 5–6). It has a moderate supply of nitrogen and potassium, but the level of available phosphorus is low. Physico-chemical characteristics of luvisols are presented in Tables 7.24 and 7.25.

Land use. On level or gently sloping land, luvisols are used as arable land or meadows, while on slopes, they are used for vineyards and orchards. These soils are generally less productive than those they evolve from.

Guidelines for management. Through increasing the depth of the plough layer and intensive fertilization, the water–air properties and chemistry of luvisol can be improved, which then means it is possible to produce satisfactory yields.

7.4.2 Podzol

The name of the soil type. The name comes from the Russian word *zola*—ash, relating to podzol's E horizon, which is ashy in colour. These are soils with a marked destructive pedogenetic process, which progresses primarily under the influence of raw humus. In ex-Yugoslavia, this name was used to describe lessive soils, pseudogley and podzol. Subtypes of this soil according to the classification system used in Serbia are as follows: ferrous podzol, which has an illuvial B_{fe} horizon; and humus-ferrous podzol, which has an illuvial horizon with separate B_h-B_{fe} subhorizons (Škorić et al. 1985).

Table 7.24 Physical characteristics of luvisol

Horizon	Granulometric composition (%)					
	2.0–0.2 mm	0.2–0.02 mm	0.02–0.002 mm	<0.002 mm	Total sand	Clay + silt
A						
X	11.24	27.94	38.28	21.91	39.84	60.16
Min	0.20	8.00	16.00	14.10	24.90	34.10
Max	24.80	63.30	55.00	30.00	65.90	75.10
E						
X	2.73	26.92	42.70	25.15	29.65	70.35
Min	0.24	16.74	28.50	18.70	17.00	55.40
Max	11.60	42.70	54.90	31.00	44.60	83.00
Bt						
X	2.00	24.84	33.05	38.45	26.83	73.17
Min	0.40	15.65	27.60	28.60	16.20	58.00
Max	6.90	35.10	42.00	44.60	42.00	83.80

Source Knežević and Košanin (2006), Čokeša et al. (2007), Knežević et al. (2011) and Gajić et al. (2013a, b)

Table 7.25 Chemical characteristics of luvisol

Horizon	pH		Y1 ccm n/10 NaOH	Adsorption complex				Humus (%)	C (%)	N (%)	C/N	P_2O_5 (mg/100 g)	K_2O (mg/100 g)
	H_2O	$CaCl_2$/KCl		T-S (cmol kg^{-1})	S (cmol kg^{-1}	T (%)	V (%)						
A													
Average	4.99	3.52	53.88	29.18	4.37	33.56	15.82	6.80	4.33	0.29	22.96	1.48	8.70
Min	4.20	3.20	39.75	12.84	0.00	19.20	0.00	4.15	2.87	0.08	11.50	1.20	3.50
Max	5.74	4.00	67.50	43.88	9.20	44.18	38.40	9.04	5.23	0.45	41.10	1.80	17.55
E													
Average	4.90	3.75	34.26	23.74	1.61	25.35	6.48	1.63	0.95	0.08	15.10	1.00	3.67
Min	4.44	3.40	27.50	17.88	0.00	20.29	0.00	0.83	0.48	0.03	11.80	0.90	2.08
Max	5.35	3.95	41.50	29.05	4.55	30.26	15.02	2.48	1.44	0.21	18.90	1.13	5.75
Bt													
Average	5.32	4.32	19.62	16.11	8.33	24.44	28.56	0.89	0.51	0.11	11.87	0.80	7.81
Min	4.90	3.70	13.00	9.05	0.00	19.48	0.00	0.34	0.20	0.02	8.40	0.65	2.22
Max	5.95	5.05	25.96	23.54	27.50	36.55	75.24	1.59	0.92	0.33	16.90	0.88	16.00

Source Knežević and Košanin (2006), Čokeša et al. (2007), Knežević et al. (2011) and Gajić et al. (2013a, b)

The WRB name. Podzol.

Genesis and sequence of horizons. Podzols are soils which occur at altitudes of over 900 m with a mean annual temperature of below 7 °C. They form on the following substrates: quartz sandstone, quartzite, chert, phyllite and mica schist. The formation of podzols is linked to the presence of raw humus and its effect on clay minerals. As opposed to luvisols, they are subject to destruction and liberate SiO_2 and R_2O_3. Two groups of processes characterize these soils. The first are processes which destroy the aluminosilicate core under the influence of raw humus and the migration of products from the breakdown of clay minerals together with organic acids. The second group are processes involving the accumulation of migration products in certain places in the profile. The first group of processes are impacted by readily soluble low-molecular organic acids (oxalic and citric), which causes acid hydrolysis of primary and secondary minerals. These acids with Al and Fe from the aluminosilicate core form chelate complexes, which easily migrate downwards. This takes place immediately below the layer of raw humus, from where the Al and Fe migrate, and what remains are bare particles of quartz which are ash coloured (the E horizon). The mobility of Al and Fe oxides depends on the R_2O_3:FK ratio. Hence, Al migrates when the Al_2O_3:FK ratio is less than 3:1, while Fe migrates when the Fe_2O_3:FK ratio is less than 15:1. When these ratios are met, accumulation begins, with Fe first accumulating in the Bfe horizon, and then Al beneath it. The typical morphology of the soil profile for podzols is a layer of raw humus with an E horizon beneath it and an illuvial B horizon beneath that one (O-E-B-C-R). Classification at the level of varieties and forms is made according to the criteria set out in Table 6.1 (Chap. 6 Soil Classification).

Position in an evolutionary series. This soil can be the next evolutionary stage after ranker with primary podzol being formed, or it can be formed through the podzolization of brown acid soil, continuing on to humus-rich brown acid soil and brunipodzol, eventually becoming secondary podzol.

Distribution. Podzols are zonal soils, characteristic for the boreal belt, although they very occasionally occur intrazonally, too. In Serbia, they are found sporadically in very small areas on the mountains of Goč, Golija and Kopaonik. Vegetation is acidophilic, oligotrophic and mesophilic (forests of spruce and Scots pine).

Physical and chemical properties. Podzols are extremely skeletal and sandy, permeable to water and well-aerated. They are very acidic soils (with a pH often below 4), and the degree of base saturation rarely exceeds 20%. The total adsorption capacity is very low and lowest in the E horizon. The C:N ratio in the raw humus is wide, more than 30, which prevents the biotransformation of organic material. The soil is poor in all nutrients, including micronutrients (Mn, Co, Cu and Zn).

Land use. In Serbia, this is a forest soil.

Guidelines for management. High acidity and the scarcity of nitrogen and other nutrients mean that podzols require amelioration in the form of biochemical and agromeliorative measures.

7.4.3 Brown Podzolic Soil (Brunipodzol)

The name of the soil type. Brown podzolic soil is very similar to podzol, but as opposed to podzol, where the morphology of the profile is dominated by an ashy colour resulting from the presence of an E horizon, brown podzolic soil is dominated by the colour brown due to the absence of an eluviated E horizon. For this reason, these soils are called brown podzolic (from the German *Braunpodsol*—brown podzolic soil). The genesis and properties of this soil type in Serbia were described in detail by Jović (1973). Subtypes of this soil according to the classification system used in Serbia are as follows: on quartz sandstone; on quartzite; on chert; on acidic eruptive rocks; on phyllite; and on sandstone.

The WRB name. Podzol, humic.

Genesis and sequence of horizons. The parent substrate of these soils is acidic rocks (granite, quartz monozonite, granodiorite, quartz diorite, diorite, phyllite, dacite, schists, mica schist, and quartz and other sandstones). Most commonly, they form in perhumid conditions, and more rarely in humid climatic conditions. Mean annual rainfall ranges from 900 to 1400 mm, while the average annual temperature is between 3 and 6 °C. In Serbia, they are found at altitudes of 1400–1900 m a.s.l.. Occasionally populated by broadleaf vegetation, they are more frequently covered by broadleaf-coniferous and coniferous vegetation. They are typical forest soils with an O-A/E-Bh-Bfe-C profile structure. The decomposition process of organic residue is somewhat more intensive in brown podzolic soils than in podzols, so the O horizon is characterized by partially decomposed (moder) humus. The podzolization process itself is less intensive than for podzol, so an E horizon does not form, with eluviation occurring from the humus horizon, which then becomes a mixed A/E (humus/podzolic) horizon. This horizon is what makes this soil different to dystric cambisols on the one hand and podzols on the other. The illuvial B horizon is the thickest horizon of brown podzolic soils and is chocolate coloured in its upper reaches due to the accumulation of humus-organo-mineral compounds (Bh). In terms of its texture, it is heavier than the A/E horizon. The lower part of the B horizon is brown to ochre in colour, and here the content of organo-mineral compounds with iron increases with depth (Bfe). The parent substrate is acid siliceous rocks. Classification at the level of varieties and forms is made according to the criteria set out in Table 6.1 (Chap. 6 Soil Classification).

Position in an evolutionary series One pathway for the evolution of brown podzolic soils is their direct development from dystric ranker at altitudes of over 1400 m in the belt of spruce, Macedonian pine/spruce and Macedonian pine forests. In the conditions of the wet, high-altitude perhumid climate on Mt. Golija, beneath the boreal spruce forests, when the process of eluviation is more dominant than accumulation, brown acid soil evolves into podzol. The second direction evolution can take is via dystric cambisol, with first the humus-rich subtype of dystric cambisol forming as a mid-stage in its development. This occurs when substrates are extremely acidic (quartz sandstones, quartzite), with an increase in altitude, a decrease in temperature, and an increase in atmospheric deposition and with the transition from the belt of hill beech forest to that of subalpine beech forest and beech and mountain maple forest. Organic material transforms more slowly, the amount of humus in the soil increases, and the character and pattern of the profile changes. The amount of the mobile and aggressive fraction of fulvic acid 1a increases in particular. The ratio of this fraction of fulvic acid to R_2O_3 widens, solutions become more diluted and sesquioxides become mobile, so they start to be leached from the surface horizons and accumulate in the deeper horizons of the soil. The further development of brown podzolic soil can result in the formation of podzol.

Distribution. In Serbia, these soils are found above the zone of dystric cambisol, at altitudes over 1400 m (Kopaonik, Golija and Prokletije). These are zones of mixed forests of beech/spruce, spruce, subalpine zones of mixed forests of winter flowering heather, juniper and spruce, beech and mountain maple, Macedonian pine/spruce and Macedonian pine forests. In Serbia, they are found sporadically in very small areas.

Physical and chemical properties. The texture of these soils is sandy to loamy sandy, and they are often skeletoid (Table 7.26). The humus-podzolic horizon (A/E) differs clearly in its lighter texture than the illuvial humus/iron horizon. In this horizon, there is the highest content of skelet and coarse sand. This soil is well-aerated and permeable to water, and has a relatively low degree of water retention. It is completely non-calcareous in all horizons. The O horizon is characterized by extremely high hydrolytic acidity. The humus-podzolic horizon is the most acidic with pH varying between 3.75 and 4.75. As a rule, acidity in the illuvial B horizon is lower. It is also the least saturated with bases. This type of soil is characteristic for being humus-rich throughout the whole profile, and hence with the transition from the B to the C horizon, this soil has approximately 2% humus. The soil has a moderate supply of readily available forms of P_2O_5 and K_2O (Table 7.27). In all the horizons, fulvic acid predominates over humic acid in the humus. The humus is mostly linked to sesquioxides and fraction 1 of humic acids and fraction 1 of fulvic acids dominate. In the illuvial horizon, the content of the fraction of 1a fulvic acid is greater than that of all the remaining fulvic acids. The total nitrogen

Table 7.26 Physical properties of brunipodzol under forests of *Picetum excelsae serbicum* Rud. subass. *oxalidetosum* on Mt. Kopaonik

Horizon	Depth (cm)	Hydros. humidity (%)	Granulometric composition in %								
			>2.0 mm	2.0–0.2 mm	0.2–0.06 mm	0.06–0.02 mm	0.02–0.006 mm	0.006–0.002 mm	<0.002 mm	Σ Clay + silt	Σ Total sand
O	0–4	5.01	–	–	–	–	–	–	–	–	–
A/E	4–14	2.37	20.53	39.29	33.41	2.50	9.80	3.50	11.50	24.80	75.20
Bh	14–30	2.06	23.65	33.75	34.85	6.50	4.00	7.10	13.80	24.90	75.10
Bse	30–55	2.23	17.97	32.13	30.67	6.20	7.00	7.20	16.80	31.00	69.00
B/C	55–85	2.21	22.54	37.43	34.33	7.30	5.70	4.00	11.20	20.90	79.10

Source Jović (1973)

Table 7.27 Chemical properties of brunipodzol under forests of *Picetum excelsae serbicum* Rud. subass. *oxalidetosum* on Mt. Kopaonik

Horizon	Depth (cm)	pH		Y_1 (ccm)	Adsorption complex				Humus (%)	C (%)	N (%)	C:N	P_2O_5 (mg/100 g)	K_2O (mg/100 g)
		H_2O	KCl		(T-S) (mg ekv/100 g)	S (mg ekv/100 g)	T (mg ekv/100 g)	V (%)						
O	0–4	4.80	3.98	87.26	56.72	20.49	77.21	26.53	36.07	20.97	0.53	39.50	10.00	18.50
A/E	4–14	4.62	3.84	46.07	29.94	1.32	31.26	4.22	8.72	5.07	0.34	14.90	7.50	7.40
Bh	14–30	4.85	4.10	31.98	20.79	1.21	22.00	5.50	5.60	3.25	0.26	12.50	7.50	5.00
Bse	30–55	5.00	4.28	24.66	16.03	0.61	16.64	3.66	3.30	1.92	–	–	10.00	4.50
B/C	55–85	5.10	4.38	–	–	–	–	–	2.59	1.50	–	–	–	–

Source Jović (1973)

content in these soils is high, but due to their acid chemistry, nitrification is weak and the amount of available nitrogen compounds is low.

Land use. The ecological production value of brown podzolic soils is not high when it comes to agriculture. They are typical forest soils and their productivity for spruce, Macedonian pine and Scots pine is satisfactory. With the destruction of coniferous forests, some areas of these soils have become high-altitude pastures, populated by acidophilic grass communities.

Guidelines for management. These are typical forest soils and hence can be used only for forest cultivation and for pasture following their destruction.

7.5 Class V: Anthropogenic Soils with P-C Soil Horizon Sequence

7.5.1 Rigosol

The WRB name. Anthrosol.

These soils occur as a result of deep ploughing, usually to a depth of 60–80 cm, but up to 100–120 cm in some cases. This mixes two or more horizons of autochthonous soil, resulting in the formation of an anthropogenic horizon (P). Subtypes of this soil according to the classification system used in Serbia are as follows: vineyard soil (vitisol), intensive orchard soil and crop soil.

7.5.2 Hortisol

The WRB name. Anthrosol, hortic.

These garden soils originate during the long-term growing of vegetables and flowers. They have an anthropogenic P horizon with a much higher content of humus than the soil type they originated from. They are approximately 40 cm deep and are characterized by high levels of plant nutrients and high biological activity.

7.6 Class VI: Technogenic Soils (Technosols)

7.6.1 Soil of Disposal Sites (Deposol)

The WRB name. Anthrosol.

These soils form from the deposition of materials with various textures and chemical compositions. They are divided into subtypes according to the classification system used in Serbia based on the type of material deposited: ore, slag and ash.

Anthropogenically formed ash and slag deposit sites arising as a result of coal combustion at thermal power plants, next to vast expanses occupied by opencast lignite mines lead to the permanent loss of fertile land, the pollution of soil, groundwater, air and vegetation, and erosion. Serbia has a total of 8 thermal power plants ('Nikola Tesla A', 'Nikola Tesla B', 'Kolubara', 'Kostolac A', 'Kostolac B', 'Morava', 'Kosovo A' and 'Kosovo B') with a total capacity of 5171 MW, which burn low calorie lignite and produce approximately 5.5 Mt of fly ash and slag annually. As thermal power plants are considered to be one of the biggest sources of environmental pollution, particular attention is given to the deposit sites of fly ash and slag from thermal power plants and possibilities for their stabilization and rehabilitation. For over a decade and a half, complex research has been taking place at the 'Nikola Tesla A' thermal power plant, which includes analysis of the chemical characteristics of lignite, native ash and ash weathered for different amounts of time (Tables 7.28 and 7.29; Fig. 7.27), as well as an analysis of vegetation that spontaneously colonizes these habitats or is sown/planted for the purposes of biological recultivation (Pavlović et al. 2004; Mitrović et al. 2008; Kostić et al. 2012a; Gajić et al. 2013a; Mitrović et al. 2012).

Research is being undertaken with the aim of investigating the adaptive characteristics of this spontaneous vegetation and could be used for the adequate management of these vast areas in order to prevent the pollution of surrounding agricultural land. Research is based on measurements (in vivo and in situ) and analyses of the correlation between the physiological vitality of plants (Fig. 7.28), physiological and morphological damage symptoms in plants and the accumulation of chemical elements in their tissue.

It has been established that the most important limiting factors for plant survival at fly ash deposit sites are as follows: a lack of essential nutrients (a deficit of nitrogen and low availability of phosphorus and potassium); reduced availability of Cu, Mn and Zn; toxicity induced by high pH and/or high concentrations of soluble salts; toxicity induced by high concentrations of B, As, Mo, Se and other potentially toxic trace elements; the formation of compact and/or cemented layers; a reduction in the number of free-living and symbiotic N_2-fixing soil microorganisms; and extreme microclimatic conditions (drought and high temperatures) (Pavlović et al. 2004, 2007; Djurdjević et al. 2006; Mitrović et al. 2008, 2012; Kostić et al. 2012a; Gajić et al. 2013a; Pavlović and Mitrović 2013).

Recent research into this problem has clearly shown that strategies for the recultivation of these types of deposols must be based on sound knowledge of the physical and chemical properties of fly ash and slag, the deposition processes used (in the case of power plants in Serbia, this is wet deposition), and the technical characteristics of the deposit sites. Furthermore, it is important to know that the chemical properties

Table 7.28 Chemical characteristics of lignite and ash at the 'Nikola Tesla A' thermal power plant in Obrenovac (Serbia)

Silicate analysis of lignite (%)		Silicate analysis of ash (%)		Chemical analysis of lignite (mg kg^{-1})		Chemical analysis of ash (mg kg^{-1})	
SiO_2	62.5	SiO_2	60–70	As	1705	As	172
Al_2O_3	20	Al_2O_3	11–27	B	1244	B	410
Fe_2O_3	6	Fe_2O_3	8.87	Cu	70	Cu	225
CaO	4.4	CaO	4–8	Mn	201	Mn	812
MgO	2	MgO	1.6–2.8	Mo	74	Mo	16
SO_3	2.7	SO_3	0.7–1.8	Se	<2	Se	<2
P_2O_5	0.3	P_2O_5	0.05–0.2	Zn	186	Zn	125
Na_2O	0.6	Na_2O	0.2–0.7				
K_2O	0.5	K_2O	0.5–0.7				
TiO_2	0.7	TiO_2	0.3–1.0				
Cl	0.01	Cl	0.02–0.1				
	62.5		60–70				

Table 7.29 Electrical conductivity (EC) and pH of the control soil (fluvisol) and ash (L0—unweathered ash, L1—ash weathered for 3 years and L2—ash weathered for 11 years)

Localities	EC (dS m^{-1})	pH
Control soil (fluvisol)	0.184	7.54
L0—unweathered ash	0.720	8.03
L1—ash weathered 3 years	0.203	7.78
L2—ash weathered 11 years	0.153	7.72

Source Pavlović and Mitrović (2013)

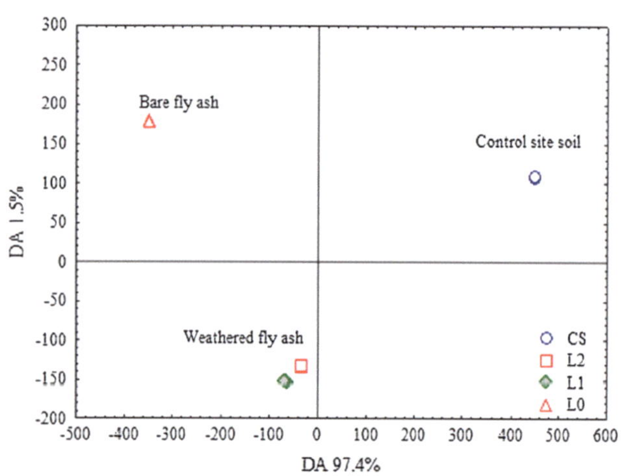

Fig. 7.27 Canonical discriminant analysis based on variations in the content of chemical elements (B, As, Mo, Cu, Mn and Zn) in lignite, fly ash (L0 unweathered ash, L1 ash weathered for 4 years, L2 ash weathered for 12 years) and fluvisol from the control site CS (Pavlović and Mitrović 2013)

of the ash change over time due to weathering and as such, conditions become more favourable for plant growth. In brief, the ecological recultivation of this type of deposol should be based on the following principles: plants should only be grown on the ash so as to form a plant cover, and never for the production of food or medicinal plants; plant succession on recultivated areas and natural recolonization is of key applicative importance as a plant cover formed through natural succession is certain to offer these deposols protection from aeolian erosion; the introduction of new adaptive species (grasses, legumes, shrubs and trees) leads to plant diversity; the choice of species must be based on the ecological potential of each species and must be aligned with the ecological potential of deposol; such species should be sought among those which spontaneously colonize the ash; and the adaptability of each species should not depend too much on agronomic measures (Figs. 7.29 and 7.30).

Hence, research was aimed at devising an optimal ecological and economic model for the biological recultivation of fly ash deposit sites during the working life of thermal power plants (connected to the periodic sluicing of ash in cycles of 3–11 years), as well as an end model for the recultivation of ash deposit sites when the power plant ceases to operate. It was concluded that the optimal model rests on the ecological, vegetative and floristic potential of the area around the deposol, which abounds in a wealth of life forms and ecological types of plants. In addition to the choice of species, attention should be devoted to the dates for planting and sowing so as to have plant cover that is functional all year round and hence to prevent erosion.

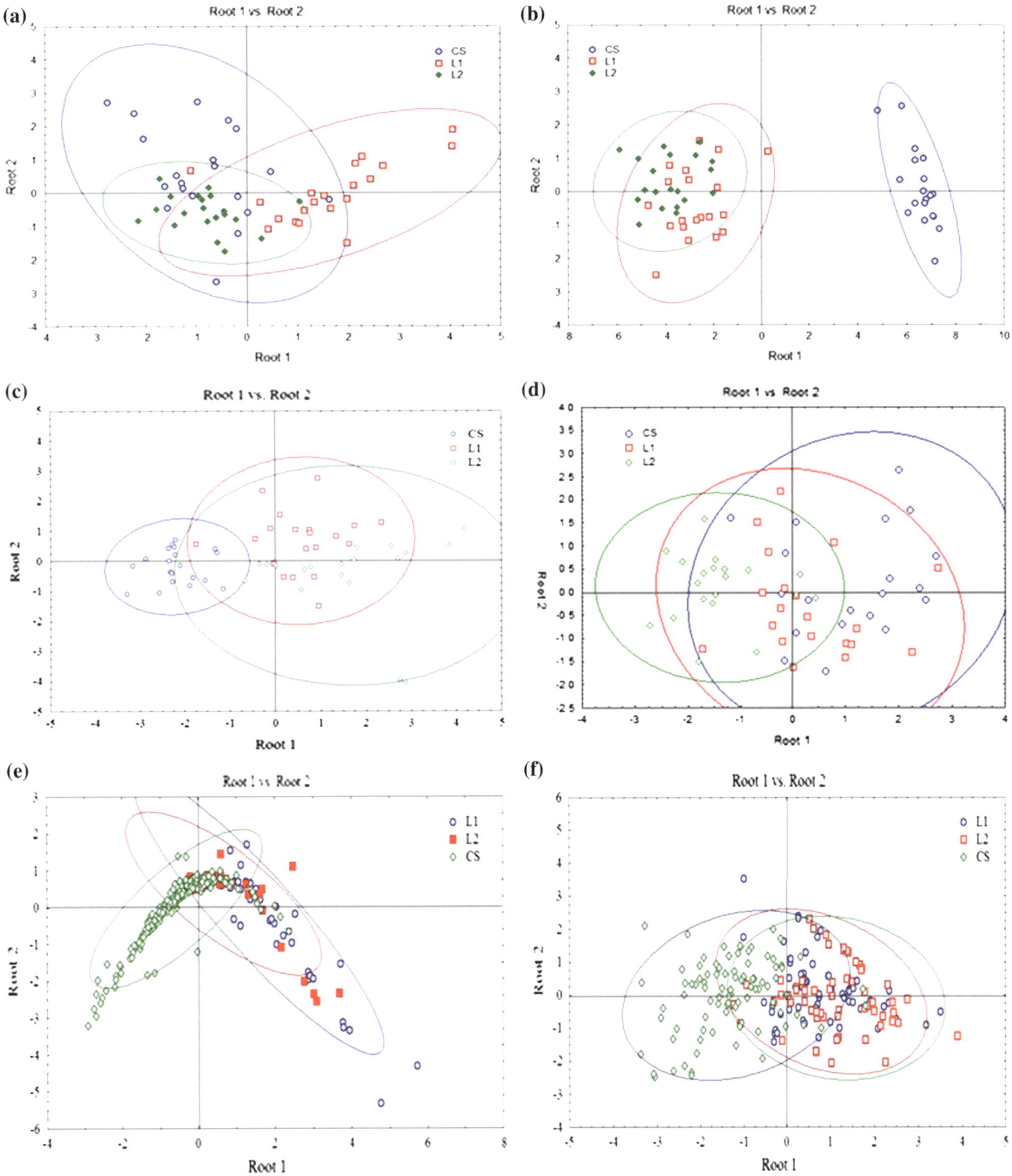

Fig. 7.28 Discriminant analysis based on variations in photosynthetic efficiency parameters (Fv/Fm, F_0, Fm and $t_{1/2}$) for *Tamarix tentandra Pall.* (**a**), *Amorpha fruticosa* L. (**b**), *Populus alba* L. (**c**), *Robinia pseudoacacia* L. (**d**), *Festuca rubra* L. (**e**) and *Calamogrostis epigejos* (L.) Roth (**f**), growing at fly ash deposit sites weathered for 4 years (L1), 12 years (L2) and in fluvisol from the control site (CS) (Mitrović et al. 2008, Kostić et al. 2012a; Pavlović and Mitrović 2013)

Moreover, attention should be paid to the fact that varying sizes of ash particles can be found at ash deposit sites (the coarsest fractions on the side dykes, and the finest in the central parts), and this needs to be borne in mind when selecting species to be sown/planted in the dykes or central parts of the deposit sites. In this way, the vegetation cover formed will take up the nutrients from the substrate and hence prevent their being leached into the deeper layers of the ash deposits, instead returning them through the decomposition of organic matter, creating conditions for the formation of the precursor to soil (humus), and with it, the preconditions for the establishment of a stable vegetation cover.

Only 2–3% of the huge amount of ash is used for other purposes, e.g. for cement production. Recently, its use in motorway construction has been explored. One important potential use for fly ash is the amelioration of acid soils given the intensification of soil acidification processes in Serbia. This would contribute to increasing the area of arable soils, which could be used for agricultural production in accordance with practices going back several decades in countries in the West, but also in the East (India).

7.6.1.1 Flotation Material, Deposits from Waste Water (Flotisol)

These soils form through the deposition of material through waste water and subtypes of this soil according to the classification system used in Serbia are divided according to depth and type of deposit.

7.6.1.2 Atmospheric Deposition—Aeroprecipitates

These soils form through technogenic deposition and deposition from the air.

Andosol

In addition to the soil types already described that are included in the Serbian classification system, Andosols developed on volcanic pyroclastics are also found in Serbia. They were discovered in Serbia for the first time during investigations on Mt. Radan, developed on pyroclastics of the large Lece andesite massif (Pavlović and Topalović 2002; Topalović et al. 2001, 2004). At the same time, the mineral and chemical composition of vitric andosols developed on volcanic tephra from Radočelo, Takovo, Igroš and Fruška Gora was investigated (Kostić et al. 2005; Tomić et al. 2005, 2008).

The name of the soil type. The name *andosol* is derived from the Japanese words *an*—black, dark and *do*—soil. They are developed on pyroclastic materials of various chemical compositions on undulating and mountainous landscapes, under climatic conditions ranging from boreal to the humid tropics, but are most common in mountain

Fig. 7.29 Fly and bottom ash layers in 'Nikola Tesla A' recultivated ash pond in Obrenovac (*photo* Kostić, O.)

Fig. 7.30 Fly and bottom ash layers in 'Kolubara' recultivated ash pond in Lazarevac (*photo* Dželetović, Ž.)

climates, with a wide range of vegetation types. Parent material is mainly volcanic tephra, which comprises volcanic glass, ash, lapilli, tuff, cinders, pumice and other volcanic ejecta.

The WRB name. Andosols, andic and vitric.

Genesis and sequence of horizons. Andosols are developed on volcanic tephra when the short-range order minerals (allophane and imogolite) and/or metal-humus complexes are precipitated during the course of rapid weathering. Halloysite is a common mineral in andosols, particularly those rich in silica, and is associated with environments with distinct dry seasons. Well-developed andosols have high organic matter content (>6% C_{org}) in both the A and B horizons, where Al^{3+} and Fe^{3+} form stable metal-humus complexes.

The fate of liberated silica depends on the extent to which alumina is tied up in organo-metallic complexes. If alumina is combined with organics, then silica is either washed out or precipitated as opaline silica. If alumina is not tied up in complexes, the remainder could precipitate with silica and form allophane often in association with imogolite.

Andosols are characterized by the presence of either an andic or a vitric horizon. An andic horizon is rich in poorly ordered silicate minerals such as allophane and imogolite or Al-organo-mineral complexes, whereas a vitric horizon contains abundant volcanic glass. Identification of andosols is based on the identification of an 'andic horizon' (WRB–FAO 1998). The extraction of poorly ordered colloid materials from andosol by acid ammonium oxalate is used to estimate amounts of allophone, imogolite and ferrihydrite (Blackmore et al. 1987).

The diagnostic criteria for an andic horizon is that it has (Al_{ox} + ½ Fe_{ox}) $\geq$ 2%, a bulk density at field capacity (no prior drying) of <0.9 Mg/m^3, PO_4-retention >70%, and a horizon thickness of >30 cm. The 'vitric horizon' should be at least 30 cm thick, containing 10% or more volcanic glass in the fine-earth fraction (Al_{ox} + ½ Fe_{ox}) $\geq$ 0.4%, bulk density = >0.9 Mg/m^3, and PO_4-retention >25%.

The WDR divides andic horizons into either allophanic (sil-andic), dominated by organo-metal complexes (alu-andic), or vitric. In addition, the tephric material is also defined as unweathered pyroclastics.

The soil is characterized by andic or vitric horizons starting within 25 cm of the soil surface, but on occasions when they are absent, there could be mollic, umbric, melanic, histic or cambic horizons. Rapid weathering of porous tephric material results in the accumulation of organo-mineral complexes or allophane, imogolite and ferrihydrite in the profile with the following sequence A-C or A-B-C, with a dark melanic A horizon up to 50 cm thick on top of a brown B or C horizon. The melanic horizon can contain between 5 and 6% C_{org}.

The thickness of the andosol investigated on Mt. Radan was 60–70 cm at sites near the summit, gradually increasing to 75–80 cm further downhill. The profile has three distinct horizons. The Olfh—leaf litter horizon is about 2 cm thick. The A horizon is relatively thin, 10–12 cm, with a weak granular structure and loam texture. The B horizon is 60–70 cm thick and has a predominantly subangular blocky structure with a clayey loam and silty–clayey loam texture. The skeleton content gradually increases with depth with the profile ending up as brittle to hard andesitic tuff, C or R horizon.

Mt Radan is part of the Lece andesite complex with several volcanic craters and calderas, covering about 780 km^2. It is composed of pyroclastics: ash, tuff, breccias, agglomerate, andesite flows and dikes. The climate of the area is continental temperate, but changes with altitude to a colder mountain regime and is strongly influenced by orographic conditions. The mean annual temperature is about 9 °C, and the mean annual rainfall is about 630 mm. Climate diversity is caused by the change in elevation, which causes moisture division and induces precipitation and temperature gradients along steep slopes. Mountain winters are long and very cold with maximum precipitation, frosts are frequent in early autumn, and late spring and summers are dry and cool with occasional storms. The native vegetation is a dense deciduous formation dominated to a great extent by beech forest, which extends from 600 m up to the summit at 1400 m (Dinić et al. 2002).

At the site, *Fagus sylvatica* L. dominates in the upper tree layer and is 15–35 m in height. In the shrub layer, *Rubus hirtus* W. et K. and *Acer pseudoplatanus* L. are found. In the herbaceous layer, the following species dominate: *Rubus hirtus* W. et K., *Paris quadrifilia* L., *Fagus sylvatica* L., *Acer pseudoplatanus* L., *Nephrodium filix mas* (L.) Rich, *Cardamine bulbifera* (L.) Crantz, *Athyrium filix femina* (L.) Roth, *Stellaria nemorum* L., *Rubus idaeus* L., *Asperula odorata* L., *Circaea lutetiana* L., and *Pulmonaria officinalis* L.

Position in an evolutionary series. Andosols are azonal soils found in all climates and at all altitudes. Consequently, they occur together with almost any other soil group. Allophane stays stable in weakly acid to neutral conditions, whereas Al-organic complexes are dominant in more acid conditions. The clay content in andosols changes over time, especially in subsoil as allophane and imogolite are transformed into halloysite, rare disordered kaolinite, and, in extremely acid conditions, into gibbsite. An andosol may eventually grade into cambisol, luvisol or even podzol depending on precipitation. A typical configuration on mountain slopes would have andosols at the summit and the higher end of slopes, cambisols and luvisols at mid-slope positions, and vertisols (basaltic and andesitic tephra) or acrisols (riolitic and dacitic tephra) near the foot of the slope.

Distribution. Andosols are soils that form in volcanic regions of the world (WRB 1998). In Europe, andosols occur in Italy, France, Germany and Iceland, as well as in Portugal (Azores), Spain (Tenerife) and Greece (Santorini), COST (2002). In Serbia, the complex of volcanic, dominantly pyroclastic, dacite–andesite rocks extends over 2880 km^2, or about 4% of the territory. It extends intermittently from Fruška Gora in the north, then across Šumadija around the Rudnik–Borač–Kotlenik range, to Stari Ras and the Kopaonik–Rogozna area, and further to Radan Mt., and to Vranje, Pčinja and Binička Morava in the south. In eastern Serbia, there is a large Timok Eruptive Area, which extends from Majdanpek in the north to Pirot in the south and further to Bulgaria, as well as a few smaller masses elsewhere. From the petrological point of view, more basic types are found in the east, in the Timok Eruptive Area (860 km^2), intermediate types are the vulcanites of Radan (720 km ^{2}), Kopaonik (740 km^2), Rogozna, Rudnik and Kotlenik Mts., and the most acid are those in the Surdulica, Golija and Borač Mts. In Serbia, andosols can be found at altitudes between 250 and 1650 m, but generally prevail between 450 and 1100 m.

Physical and chemical properties. Andosols are usually dark soils, rich in organic matter. The colour of vitric andosols is determined by the nature of the tephra materials, which range from black basaltic to light coloured rhyolitic materials. The andesitic rock and tuffs are usually grey to dark grey in colour. Surface horizons are dark reddish-brown to dark yellowish brown and subsurface material commonly has chromas 1–2 units higher than the overlying A horizons, ranging from moderate brown and pale yellowish brown to light brown, greyish orange and light yellowish brown (Table 7.30).

Most andosols have good internal drainage, high porosity, high water infiltration and good aggregate stability. All this results in low bulk density, which is one of the diagnostic criteria for andosols, with bulk density as low as <0.9 Mg m^{-3} required.

The soils exhibit dystric, moderately and slightly acid characteristics, and have a lower pH in water in subsurface than surface horizons. Potassium chloride pH is generally 0.7–0.8 pH units lower than the pH in water at the summit, whilst the difference varies from 0.2 to 0.8 in the profile situated downhill from the summit. Sodium fluoride pH was greater than 10 in the subsurface horizons, which is indicative of andic material (Fieldes and Perrott 1966; Shoji et al. 1996). Surface horizons exhibit slightly lower pH values from 9.2 to 9.6, indicating a lower content of poorly ordered minerals due to Al migration down the profile (Table 7.31). The A horizons contain between 7.19 and 9.93% humus (4.16 and 5.77% C organic) and for the most part humus/organic C decreases with depth (Table 7.31). Humus/organic C is probably associated with poorly ordered minerals, such as allophane, which is more stable than Al- and Fe-organic complexes. Andosols have a range of soil pH (H_2O) reactions. Organo-metal (Al- and Fe-) complexes tend to be acid with pH <5, with low base saturation and signs of Al-toxicity, whilst andosols with allophane have pH 5.5–6.5, particularly if fresh parent materials are still present.

The sum of exchangeable bases (arranged in decreasing order: Ca, Mg, K, Na) for the soils investigated ranges from 2.0 to 8.8 cmol(+)/kg. Cation-exchange capacities (CEC) vary between 17.3 and 39.1 cmol(+)/kg (Table 7.31). In well-developed andosol near the volcano summit (profile 1), lower CEC values are found in the surface horizon with higher values in deeper horizons, which correlates well with data for PO_4 retention. Down the slope, profile 4 does not exhibit such a close correlation between CEC and PO_4 retention values due to variations in texture and skeletal content.

Table 7.30 Selected morphological and physical properties of the soils investigated

Profile, Horizon	Depth (cm)	Moisture colour	Coarse Sand (%)	Fine Sand (%)	Silt (%)	Clay (%)
At the summit—1408 m						
Profile 1 Olfh	0–2	–	–	–	–	–
Profile 1 A1	2–14	7.5 YR 5/2	25.50	29.20	23.50	21.80
Profile 1 Bw1	14–25	7.5 YR 4/4	17.60	19.00	27.50	35.90
Profile 1 Bw2	25–70	7.5 YR 5/4	19.00	21.60	26.50	32.90
At the sawmill—1160 m						
Profile 4 Olfh	0–2	–	–	–	–	–
Profile 4 A1	2–12	10 YR 3/2	19.30	31.50	25.40	23.80
Profile 4 Bw1	12–20	10 YR 6/4	21.00	28.20	27.50	23.30
Profile 4 Bw2	20–40	10 YR 6/4	22.20	26.10	26.40	25.30
Profile 4 Bw3	40–75	10 YR 7/4	22.10	28.80	24.30	24.80

Source Topalović et al. (2004)

Table 7.31 Selected chemical properties of the soils investigated

Profile, horizon	pH (H_2O)	pH (KCl)	pH (NaF)	Humus (%)	Exch. Bases	CEC	PO_4- retent. (%)
At the summit—1408 m							
Profile 1 Olfh	5.7	5.0	–	43.45	–	–	–
Profile 1 A1	5.2	4.5	9.2	7.19	8.8	19.51	23.5
Profile 1 Bw1	4.8	4.4	10.7	6.25	2.8	37.34	91.5
Profile 1 Bw2	4.7	4.4	10.7	4.07	2.0	36.61	92.0
At the sawmill—1160 m							
Profile 4 Olfh	5.9	5.4	–	33.12	–	–	–
Profile 4 A1	4.6	4.2	9.6	9.93	6.9	39.11	70.0
Profile 4 Bw1	5.0	4.7	10.1	5.00	8.3	25.64	58.0
Profile 4 Bw2	5.2	4.7	10.4	2.73	6.8	28.30	64.5
Profile 4 Bw3	5.1	4.6	10.4	1.44	5.4	17.34	56.5

Source Topalović et al. (2004)

Phosphate retention >70% is an important diagnostic criterion for andosols (WRB 1998). The phosphate retention of the soils investigated exhibits variations between 23.5 and 92%. Those soils developed at the summit display higher phosphate retention of about 92%, whereas the soils developed downhill from the summit have phosphate retention of between 56.5 and 70% due to their more sandy texture and higher skeletal content (Table 7.31).

The levels of Al, Fe and Si extracted from the soils by ammonium oxalate–oxalic acid (Al_{ox}, Fe_{ox} and Si_{ox}) are not so high, suggesting low-to-moderate amounts of allophane and non-crystalline or poorly crystalline Fe oxides. The Al_{ox} was between 0.43 and 1.69% for the summit pedon and between 0.41 and 0.68% for the downhill pedon. The Fe_{ox} ranged from 0.08 to 0.29%, with higher values found at the summit, and lower values in the profile on downhill slopes (Table 7.32). Low Fe_{ox} contents may be associated with more andesitic tephra compared to more basaltic-mafic tephra, but variations in precipitation and leaching potential may also be the cause.

The levels of Al_{py} and Fe_{py} are generally low, between 0.06 and 0.15, indicating a low Al- and Fe-organic complex (Table 7.32). The soil has low Al_{py}:Al_{ox} ratios, indicating the low-to-moderate weathering rates of the investigated andosols (Fig. 7.31a). The Al_{py}:Al_{ox} ratio also gives some indication as to the occurrence of allophane and/or imogolite. A lower ratio (close to zero) indicates the presence of allophane and/or imogolite (Mizota and van Reeuwijk 1989). The allophane content is lower in the topsoil than in the subsoil (Fig. 7.31a). The lower content of allophane in the topsoil is probably due to the higher content of organic matter in the A horizon than the underlying B and C horizons. The low amount of annual precipitation that reduces the loss of silica from the soils favours allophane formation rather than ferrihydrite and/or gibbsite (Parfitt et al. 1988).

The content of Si_{ox} ranges from 0.32 to 1.35%. The difference in oxalate extractable values is likely to be inherited from differences in tephra composition and slope conditions, as well as differences in weathering processes. The formula for estimating allophane suggested by Parfitt et al. (1988) indicates 4.8–10.2% of allophane at the summit and 2.4–3.0% in the downhill pedon (Table 7.32).

The mineralogical examination of soils under a polarizing microscope revealed appreciable amounts of volcanic glass. The mineral composition of the soils is entirely similar to the composition of the parent rock on which the soil is developed. Feldspars—plagioclase, hornblende, chlorite and augite were identified in the silt fraction of all soils. These minerals are also identified in the fine sand, in which variously coloured volcanic glass was predominant. The mineralogical composition of the clay fraction of the soil does not change much with depth. The soils are characterized by the presence of cristobalite, tridymite, feldspars and halloysite, as well as opaline silica, which give a broad diffuse XRD spectrum centred at about 0.41 nm. Opaline silica is a product of the early stages of the weathering of volcanic tephra. The presence of cristobalite and tridymite with opaline silica is indicative of CT-opal, a disordered low cristobalite and low tridymite occurring in younger volcanic tephra (Jones and Segnit 1971).

Table 7.32 Extractable Al, Fe and Si in oxalate and pyrophosphate (%), volcanic glass content in the <2 mm fraction (%) and allophane (%) estimates for soils investigated

Profile, horizon	Al_{ox}	Fe_{ox}	Si_{ox}	Al_{py}	Fe_{py}	Al_{py}/Al_{ox}	Al_{ox} + ½ Fe_{ox}	Volcanic glass (%)	Allophane (%)[a]
At the summit—1408 m									
P—1 A1	0.43	0.16	0.64	0.10	0.15	0.23	0.51	23.84	4.80
P—1 Bw1	1.57	0.29	1.16	0.15	0.10	0.09	1.72	23.44	8.70
P—1 Bw2	1.69	0.17	1.35	0.10	0.10	0.95	1.78	23.34	10.20
Saw–mill—1160 m									
P—4 A1	0.41	0.18	0.40	0.15	0.12	0.36	0.50	26.00	3.00
P—4 Bw1	0.54	0.10	0.32	0.12	0.10	0.22	0.59	31.36	2.40
P—4 Bw2	0.68	0.10	0.38	0.08	0.06	0.11	0.73	29.82	2.90
P—4 Bw3	0.55	0.08	0.35	–	–	–	0.59	32.26	2.60

Source Topalović et al. (2004)
[a]N.B. Allophane = 7.52 × Si_{ox} Parfitt et al. (1988)

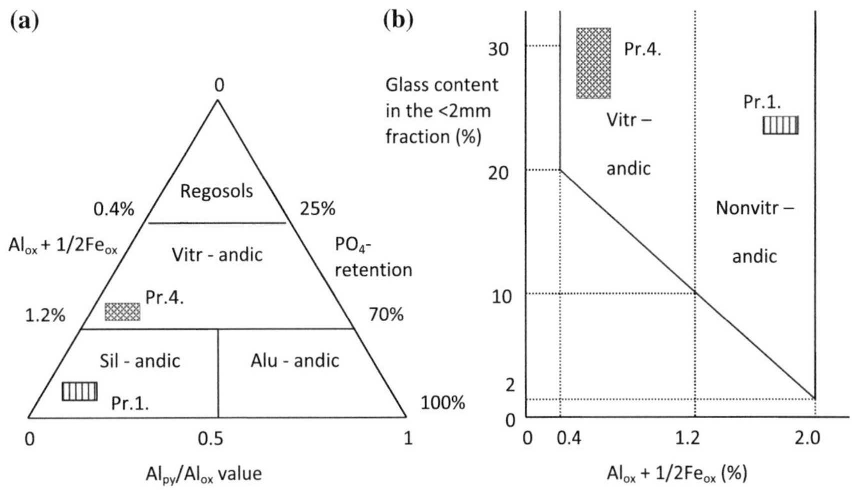

Fig. 7.31 Relationship between the properties of andosols according to WRB classification (**a**), Classification of vitr-andic and non-vitr-andic horizons, according to Shoji et al. (1996) (**b**), Topalović et al. (2004)

Clay fraction (<0.002 mm) content was between 22 and 36%, and it was in accordance with data published for andosols developed in temperate climates. The mineralogy of the clay fraction comprises illite, halloysite (kaolinite) and mixed-layer silicates of illite-smectite (10–14) and chlorite-vermiculite (14–14). Besides these crystalline phases, an abundant volcanic glass phase and cryptocrystalline CT-opal (cristobalite-tridymite opal) are present. The presence of halloysite, allophane and CT-opal is indicative of andosols developed on volcanic tephra of neutral and basic rocks (andesite and basalts) (Wada 1990).

Soils investigated at Mt Radan have a depth of over 30 cm, which satisfies the first criterion for andic and vitric andosols. The values of Al_{ox} + ½ Fe_{ox} of profile 1 range from 0.51 to 1.78%. The lowest value (0.51%) was noted in the A horizon of this soil. In the B horizon, the value of this parameter ranges from 1.72 to 1.78% and is within the range for andic properties. In the B horizon, phosphate retention is >70% (91.5–92%), which satisfies the andic properties for andosols. A volcanic glass percentage in the fine-earth fraction that, when plotted against the Al_{ox} + ½ Fe_{ox} ratio value, lies above the diagonal line (Fig. 7.31a) could be applied as an additional criterion for the determination of vitric andosols (Al_{ox} + ½ Fe_{ox} of 1.2–2%). In profile 1, the content of volcanic glass ranges between 23.34 and 23.44% and complies fully with the framework set out by Shoji et al. (1996), as presented in Fig. 7.31b.

The Al_{py}/Al_{ox} ratio and Si_{ox} content are used as criteria suggested by Shoji et al. (1996) for the separation of andosols into sil-andic and alu-andic variants. An Al_{py}/Al_{ox}

ratio between 0 and 0.5% and a Si_{ox} content >0.6% would indicate the sil-andic variant, while an $Al_{py}/Al_{ox} > 0.5$ and Si_{ox} content <0.6% indicates the alu-andic variant. Profile 1 belongs to the sil-andic variant, as the Al_{py}/Al_{ox} ratio within it ranges from 0.05 to 0.23%, and the Si_{ox} content is between 0.64 and 1.35% (Fig. 7.31a; Table 7.32).

Classification of the soils investigated was carried out according to the World Reference Base (WRB 1998). For the first time in Serbia, processes of andosolization have been detected and described. This is also the second site of andosols on the Balkan Peninsula, after the andosols noted on Santorini Island (Greece). Two types of andosol, with andic (profile 1) and vitric properties (profile 4), are described according to the criteria of the WRB (1998).

Land use. Since Serbia has over 2800 km^2 of volcanic rocks, it is important for their further study to evaluate the effectiveness of the Andosol order of soil taxonomy for providing meaningful classes in Serbia and to acquire fundamental morphological, physical, chemical and mineralogical data on the soils formed on tephra material. Andosols have high potential for agricultural production, but many of them are not used to their full capacity. Andosols are fertile soils, particularly those developed on intermediate or basic volcanic ash and not exposed to excessive leaching. They are easy to cultivate and have good rootability and water storage properties. Andosols are used for a wide variety of crops including wheat, vineyards, potatoes, vegetables, orchard plants, forage crops and tobacco. Their high water holding capacity and good permeability enhance their use for agriculture.

Guidelines for management. Many andosols are intensively cultivated and used for a variety of crops. They are easy to till, with good rootability and water storage, except in strongly hydrated areas. Their major limitation is their considerable PO_4 fixation, caused by active Al and Fe, which render phosphorus unavailable to plants. Ameliorative measures to reduce this effect include remediation by liming and the addition of silica and organic material, and they require the addition of phosphate fertilizer for intensive crop production. Vitric andosols are widespread and in dry environments are successfully used for a variety of crops with minor irrigation. In places, steep topography may be a serious constraint, so andosols on steep slopes are perhaps best kept under forest cover or well-managed pastures.

References

Antić M, Jović N (1965) Genesis and properties of the forest-hunting region of Belje. Deer Belgrade 3:25–36 (Special Edition by the Belje Research Centre Belgrade, in Serbian)

Antić M, Mišić V (1971) Ecology of nut (*Juglans regia* L.) in Đerdapska gorge klisuri. Ekologija Belgrade 6(1):87–92

Antić M, Avdalović V, Jović N (1963) Genesis and properties of soils on the limestone part of Mt. Tara. Bull Fac Forest Belgrade 34/68 (in Serbo-Croatian)

Antić M, Jović N, Avdalović V (1967) Pedological study in some forest communities in Đerdap area. Institute for the Nature Protection of Serbia, Belgrade

Antić M, Jovanović B, Jović N, Munkačević V, Nikolandić S (1972) Forest types in the forest-hunting region of Belje. Current problems facing forestry, the wood industry, and horticulture. Bull Fac Forest Belgrade 15–22 (in Serbo-Croatian)

Antić M, Jović N, Avdalović V (1982) Pedology. Scientific Book, Belgrade (in Serbian)

Avdalović V (1975) Genesis and properties of brown acid soils of the Federal Republic of Serbia. Doctoral dissertation. Bull Fac Forest Belgrade Ser E 47/75 (in Serbo-Croatian)

Bazilevič, NI, Rodin LE (1964) The biological cycle of ash elements and nitrogen for the principal zones of Northern hemispheres. In Proceedings of 8th international conference on soil science, Bucharest, Romania, pp. 833–844

Belanović S, Košanin O (2005) Exchangeable cations in some soils of Stara planina. Bull Fac Forest Belgrade 91:63–76 (in Serbian)

Blackmore LC, Searle PL, Day BK (1987) Methods of chemical analyses of soils. New Zealand Soil Bureau Scientific Report 80, DSIRO, New Zealand

Ćirić M, Aleksandrović D (1959) A view on the genesis of terra rossa. A view on the genesis of terra rossa. In: Proc Fac Agric Belgrade Univ 7:1–12

Čokeša V, Miletić Z, Stajić S (2007) Indicator flora in the mountain beech forest (Fagetum moesiace montanum) on acid soils on gneiss in the MU (Management Unit) of Kacer-Zelenicje (South-eastern Serbia). In: 9th symposium on flora of South-eastern Serbia and neighbouring regions, Niš (Serbia), pp 137–143, 1–3 Sept 2007 (in Serbian)

COST (2002) Soil resources of European volcanic systems. European Cooperation in the Field of Scientific and Technical Research

Cvjetićanin R, Knežević M (2000) Vegetation and soils of Oplenac. Bull Fac Forest Belgrade 83:165–178 (in Serbian)

Dinić A, Pavlović P, Mitrović M, Đurđević L, Kostić O (2001) The effects of the spruce silviculture on the beech forest habitat on Maljen Mt. in Western Serbia. In: Radouglu K (ed) International conference: forest research: a challenge for an integrated European Approach, Thessaloniki, Greece, vo II, pp 559–565

Dinić A, Mišić V, Topalović M, Pavlović P (2002) Forest vegetation of Radan Mt. Area of exquisite characteristics, Institute for the Nature Protection of Serbia, Serbia, pp 77–92

Djurdjević L, Mitrović M, Pavlović P, Gajić G, Kostić O (2006) Phenolic acids as bioindicators of fly ash deposit revegetation. Arch Environ Contam Toxicol 50:488–495

Dokuchaev VV (1883/1948) Russian Chernozem. In: Selected works of VV Dokuchaev. Moscow 1:14–419 (in Russian)

Duchaufour Ph (1965) Précis de pédologie. Masson, France, Paris

Fieldes M, Perrott KW (1966) The nature of allophane in soils. III. Rapid field and laboratory test for allophane. New Zealand J Sci 9:623–629

Filipovski G, Ćirić M (1963) Soils of Yugoslavia. Yugoslav Society of Soil Science, Belgrade

Gajić B, Tapanarova A, Tomić Z, Kresović B, Vujović D, Pejić B (2013a) Land use effects on aggregation and erodibility of Luvisols on undulating slopes. Aust J Crop Sci 7(8):1198–1204

Gajić G, Pavlović P, Kostić O, Jarić S, Djurdjević L, Pavlović D, Mitrović M (2013b) Ecophysiological and biochemical traits of three herbaceous plants growing on the disposed coal combustion fly ash of different weathering stage. Arch Biol Sci Belgrade 65 (4):1651–1667

Galić Z (2014) Monitoring of microclimatic conditions and soil moisture in *Quercetum frainetto-cerris* stand on cambisol. Poplar Novi Sad 193(194):37–45 (in Serbian)

Galić Z, Ivanišević P, Orlović S, Pekeč S (2009) Soil characteristics of the sites in protective forests in the lowland part of Vojvodina. Forest Belgrade 3–4:29–38 (in Serbian)

Jones JB, Segnit ER (1971) The nature of opal. I. Nomenclature and constituent phases. J Geol Soc Australia 18:57–68

Jović N (1965) The characteristics of acid soils and browned humus siliceous soils in the association *Piceetum excelsae serbicum silicicolum* on Kopaonik. Master's thesis. Bull Fac Forest, Belgrade 38 (in Serbian)

Jović N (1966) Podzolic brown soils in Serbia, theit genesis and characteristics. Doctoral dissertation, Faculty of Forestry, Belgrade University

Jović N (1969) Character and group and fractional composition of the humus of the evolutionary series of soils on limestone under Bosnian pine (*Pinetum heldreichii*). Soil Plant Belgrade 18(1–3):59–70 (in Serbian)

Jović N (1973) Brown podzolic soils of Serbia—genesis and properties. Bull Fac Forest, Belgrade Ser E 42:1–88 (in Serbian)

Jović N, Jović D (1972) Soils in virgin forests of Bosnian pine and their productive traits. Soil Plant Belgrade 21(2) (in Serbian)

Jović D, Jovanović B, Jović N, Stefanović V, Burlica Č, Cestar D, Gašperšič F (1976) Using a typology in contemporary forest management in Yugoslavia. Bull Fac Fores Belgrade 53:5–23

Knežević M, Cvjetićanin R (2003) Soils and beech communities in a series of sample plots on the Kučaj Mountains. Forest Belgrade 1–2:113–124 (in Serbian)

Knežević M, Košanin O (2002) Edaphic potential of beech forests on Brezovica. Bull Fac Forest Belgrade 86:135–145 (in Serbian)

Knežević M, Košanin O (2004) Soils in montane beech communities on limestone terrains on Mt. Ozren. Forest Belgrade 3:87–95 (in Serbian)

Knežević M, Košanin O (2006) Characteristics of acric luvisol on the limestones of Mt. Magleš. Bull Fac Forest Belgrade 93:97–104 (in Serbian)

Knežević M, Košanin O (2009) Genesis and characteristics of soils in the A-R stage in forest ecosystems of Tara National Park. Bull Fac Forest Belgrade 99:75–90 (in Serbian)

Knežević M, Košanin O (2010) Rendzinas in the community of beech, fir and spruce with Scots pine in Tara National Park. Bull Fac Forest Belgrade 101:101–112 (in Serbian)

Knežević M, Babić V, Galić Z, Košanin O (2011) Soil properties in sessile oak forests (*Quercetum montanum typicum* Čer. et Jov. 1953) in the area of Fruška Gora. Bull Fac Forest Belgrade 104:97–108

Kolarić LJ, Živanović LJ, Popović V, Ikanović J (2014) Influence of inter-row spacing and cultivar on the yield components of soybean (*Glycine Max.* [L.] Merr.). Agri For 60(2):167–176

Košanin O, Gajić B (2008) Characteristics of some serpentinite soils in Austrian pine stands in the area of Divčibare-Bukovi. Forest Belgrade 4:89–98 (in Serbian)

Košanin O, Knežević M (2002) Characteristics of organic matter of soil at the A-C stage in grass communities at Deliblato Sands. In: 7th symposium on flora of South-Eastern Serbia and neighbouring regions, Dimitrovgrad (Yugoslavia), pp 53–59, June 5–9, 2002 (in Serbian)

Košanin O, Knežević M (2005) Production potential of the soil in sessile oak coppice forests. Bull Fac Forest Belgrade 92:87–97 (in Serbian)

Košanin O, Knežević M (2007) Forest soils in the 'Čezava' Management Unit in 'Đerdap' National Park. Forest Belgrade 1–2:25–38 (in Serbian)

Kostić O (2007) The influence of a monoculture of spruce and Douglas fir on the pedogenesis and properties of soil on Maljen. Master's thesis, Faculty of Forestry, Belgrade University, Belgrade (in Serbian)

Kostić N, Dugalić G, Tomić Z, Jovšić D (2005) Mineral and chemical composition of soils developed on volcanic tephra of Radočelo, Igroš and Takovo. Publication of XI Congress DPZ-SCG, Budva, pp 126–137

Kostić O, Mitrović M, Knežević M, Jarić S, Gajić G, Djurdjević L, Pavlović P (2012a) The potential of four woody species for the revegetation of fly ash deposits of 'Nikola Tesla-A' thermoelectric plant (Obrenovac, Serbia. Arch Biol Sci Belgrade 64(1):145–158

Kostić O, Mitrović M, Jarić S, Đurđević L, Gajić G, Pavlović M, Pavlović P (2012b) The effects of forty years of spruce cultivation in a zone of beech forest on Mt. Maljen (Serbia). Arch Biol Sci Belgrade 64(3):1181–1195

Kovda VA (1973) The principles of soil science. Sci Moscow (in Russian)

Marković M, Cupać S, Đurović R, Milinović J, Kljajić P (2010) Assessment of heavy metal and pesticide levels in soil and plant products from agricultural area of Belgrade, Serbia. Arch Environ Contam Toxicol 58:341–351

Milivojević JŽ, Đalović IG, Jelić MŽ, Trifunović SR, Bogdanović DM, Milošev DS, Nedeljković BD, Bjelić DĐ (2011) Distribution and forms of manganese in vertisols of Serbia. J Serb Chem Soc 76 (8):1177–1190

Mitrović M, Pavlović P, Lakušić D, Stevanović B, Djurdjević L, Gajić G, Kostić O (2008) The potential of *Festuca rubra* and *Calamagrostis epigejos* for the revegetation on fly ash deposits. Sci Tot Environ 407:338–347

Mitrović M, Kostić O, Jarić S, Karadžić B, Djurdjević L, Gajić G, Lj Oberan, Pavlović D, Pavlović M, Pavlović P (2012) Photosynthetic efficiency of four woody species growing on fly ash deposits of a Serbian 'Nikola Tesla—A' thermoelectric plant. Polish J Environ Stud 21(5):1339–1347

Mizota C, van Reeuwijk PL (1989) Clay mineralogy and chemistry of soils formed in volcanic material in diverse climatic regions. In: Soil monograph 2. International Soil Reference and Information Centre (ISRIC), Wageningen, Netherlands

Muckenhausen E (1962) Entstehung. Eigenschaften und Systematik der Boden der Bundesrepublik Deutschland. Frankfurt am Main, DLG-Verlag

Neugebauer V, Zakosek H (1962) Die smonitza, Notizbl. Hess. Bodenforsch, No 90, Wiesenbanden

Parfitt RL, Childs CW, Eden DN (1988) Ferrihydrite and Allophane in Four Andepts from Hawaii and implication for their classification. Geoderma 41:223–241

Pavićević N (1953) Types of soils at Suva planina. Plant Soil, Belgrade, year II, No.1, Belgrade

Pavićević N (1956) Buavice the Montenegrin karst. Agricultural and Forestry Chamber of NR Montenegro, Belgrade

Pavlović P (1992) Analysis of litter and humus condition in forest communities on Mt. Cer. Master's thesis, Faculty of Biology, Belgrade University, Belgrade (in Serbian)

Pavlović P (1998) Pedological components of the metabolism of some forest communities on Mt. Maljen. Doctoral dissertation, Faculty of Biology, Belgrade University, Belgrade (in Serbian)

Pavlović P, Mitrović M (2013) Thermal power plants in Serbia—the impact of ash on soil and plants. In: Anđelković M (ed) Energy and the environment, scientific meetings, Book 4, Serbian Academy of Sciences and Arts (SANU), Belgrade, pp 429–433 (in Serbian)

Pavlović P, Topalović M (2002) Pedological bonity characteristics of Mt Radan. Area of exquisite characteristics (Radan). Institute for the Nature Protection of Serbia, Serbia and Montenegro, pp 38–46

Pavlović P, Mitrović M, Djurdjević L (2004) An ecophysiological study of plants growing on fly ash deposits from the 'Nikola Tesla-A' thermal power station in Serbia. Environ Manage 33:654–663

Pavlović P, Mitrović M, Djurdjević L, Gajić G, Kostić O, Bojović S (2007) Ecological potential of *Spiraea van-hauttei* (Briot) Zabel for urban (Belgrade city) and fly ash deposit (Obrenovac) landscaping in Serbia. Polish J Environ Stud 16(3):427–431

Pivić et al (2008) Research project: Study into the degree of threat to soil and water from hazardous and harmful materials in the neighbourhood beside the ash deposit site in the municipality of Obrenovac. Soil Science Institute, Belgrade (in Serbian)

Pivić et al (2014) Research project: Soil amelioration in the municipality of Veliko Gradište aimed at high-value food production. Soil Science Institute, Belgrade (in Serbian)

Shoji S, Nanzyo M, Dahlgren AR, Quantin P (1996) Evaluation and proposed revision of criteria for andosol in WRB for soil resources. Soil Sci 161(9):604–615

Sibirtsev NM (1895) Genetische Bodenklassifikation. Zap. Nova-Alexandr. Agr. Inst, pp 1–23 (in Russian, quoted by Kubiena 1953)

Sibirtsev NM (1896) Soil classification as applied to Russia. St. Petersburg (in Russian)

Sibirtsev NM (1953) Selected works. Sel'khozizdat, Moscow 3:396–420

Škorić A, Filipovski G, Ćirić M (1985) Classification of Yugoslav soils. p 72, Academy of Sciences and Arts of Bosnia and Herzegovina. Special edition, Book 78, Sarajevo (in Serbo-Croatian)

Stebut AI (1923) Types of land in the Shumadie area. Bull Serb R Acad, SU, Belgrade (in Serbian)

Stebut A (1924) Soils of the Drina/Sava/Morava regions and their agricultural value. Belgrade (in Serbo-Croatian)

Stebut IA (1953) Agropedology. Part 3. Scientific Book, Belgrade (in Serbo-Croatian)

Tomić Z, Kostić N, Nešić Lj, Belić M (2005) Mineral and chemical composition of soils developed on quartz-latite in Fruška Gora (in Serbian). In: XI Congress DPZ-SCG, Budva, Proceedngs, pp 198–207

Tomić Z, Kostić N, Jovšić D, Lj Nešić, Belić M (2008) Mineral composition of soil developed on vulcanites of Fruška Gora (Serbia). Acta biologica Iugoslavica - Series A: Soil Plant, Belgrade 57(1):35–44

Topalović M, Kostić NM, Pavlović P (2001) Contribution to the study of soils on vulcanites on Mt. Radan (Serbia). In: X Yugoslav Society of Soil Science Congress, 22–26 Oct 2001, Vrnjačka Banja, Serbia, Abstract 9 (in Serbian)

Topalović M, Kostić NM, Pavlović P (2004) The first record of andosols in Serbia (Radan Mt.). Acta biologica Iugoslavica - series A: Soil Plant, Belgrade 53(2):147–158

Wada K (1990) Minerals and mineral formation in soils derived from volcanic ash. Proc. In: 9th international clay conference, Strasbourg, pp 69–78

World Reference Base (WRB) (1998) Diagnostic horizons, properties and materials. World Reference Base for Soil Resources, FAO, ISSS-AISS-IBG, IRSIC, Rome, Italy

Živković M (1955) The genesis and the most important traits of eutric cambisol in Serbia. Soil Plant Belgrade 4(1–3):5–85 (in Serbo-Croatian)

Živković M (1965) Contribution to the study of the genesis of humus-siliceous soils in Serbia. Soil Plant Belgrade 14(2):211–225 (in Serbo-Croatian)

Živković BM, Neugebauer VK, Tanasijević Đ, Miljković N, Stojković L, Drezgić P (1972) Soils of Vojvodina. Institute for Agricultural Research, Novi Sad

8 Order of Hydromorphic Soils

Soils of this order are characterized by temporary or permanent excessive wetting in part of or through the entirety of the profile. The genesis and properties of hydromorphic soils mainly depend on the hydrologic regime which includes: sources of water and the manner of excessive wetting; the zone in the profile affected by this wetting; and the duration of the excessive wetting. Soils of this order cover approximately 20,897.15 km^2 or 23.7% of the land mass of Serbia.

8.1 Class I: Soils with Stagnation of Rainwater with A-Eg-Bg-C Soil Horizon Sequence

8.1.1 Pseudogley

The name of the soil type. This name was introduced by the German soil scientist Kubiëna (1953), wanting to distinguish this soil, in which wet and dry periods alternate, from soils with a permanent stagnation of groundwater and a G horizon. Subtypes of this soil according to the classification system used in Serbia are as follows: plateau (on flat ground) and slope (on gentle slopes). They differ in terms of the duration of the phase when they are covered in water. In the past, in ex-Yugoslavia, despite there being no similarities with the processes of podzolization, this soil type was also classified as podzol (due to the upper part of the profile being grey and light in texture, while the lower part is darker and has a heavier texture).

The WRB name. Stagnosol.

Genesis and sequence of horizons. In the conditions found in Serbia, pseudogley forms in regions with annual precipitation of over 700 mm, with distinct wet and dry periods. The parent substrate has different textures, and hence beneath a relatively permeable layer, there is a layer that is impermeable to water. This can be the result of intensive lessivage processes, leading to the formation of secondary pseudogley. Primarily, pseudogley forms on substrates with good water-holding properties, through the deposition of layers with different granulometric compositions (alluvial deposits), or the deposition of silty aeolian deposits over a clayey layer. Primary pseudogley has a general A-Ig-IIg-C profile, while pseudogley which has formed as secondary soil from luvisol during intensive lessivage has an A-Eg-Bg-C profile. In forests, the humus horizon has a thickness of 5–10 cm, while in meadows, it is 15–25 cm and has the properties of ochric (Aohr) humus (Fig. 8.1).

Beneath this horizon, there is a **g** horizon, which is grey with the occurrence of mottling and/or concretions, black, dark brown or rusty red in colour, from several mm to 1–2 cm in size, as a consequence of redox processes, caused by the alternation of dry and wet phases in the profile. If pseudogley evolves from luvisol, then the **g** horizon coincides with the E horizon. The morphological properties of the impermeable layer can differ, from the B horizon of luvisol to the IIg horizon in layered profiles depending on the conditions for deposition. What they do have in common is their compactness and clayey composition, which causes impermeability to water, and signs of surface gleization in the upper parts of the impermeable layer. Classification at the level of varieties and forms is made according to the criteria set out in Table 6.1 (Chap. 6 Soil Classification).

Position in an evolutionary series. On Tertiary sediments, pseudogley is the last member of the evolutionary series, which started with regosol, continuing through black soil on loess and eutric cambisol, to luvisol on loess.

Distribution. Pseudogley is linked to flat terrain, most often old alluvial, diluvial and lacustrine terraces. It is also found on gentle slopes of undulating terrain, where standing water drains away more quickly. As already mentioned, this soil is divided into two subtypes according to the classification system used in Serbia depending on the place where it occurs. In lower lying regions of Serbia, it is linked to oak forests, from xerophytic oak forests of Italian and Turkey oak and mesophytic forests of sessile oak and European

P. Pavlović et al., *The Soils of Serbia*, World Soils Book Series,
DOI 10.1007/978-94-017-8660-7_8

Fig. 8.1 Pseudogley (*photo* Čakmak, D.)

hornbeam, to pedunculate oak forests. With the removal of forest vegetation, the pseudogleization process intensifies due to the lack of tree roots which affects the drainage, and instead of forest vegetation, meadows form. They are widespread in western Serbia and Pocerina, Metohija and around Kladovo, as well as on the old river terraces of many rivers (the Kolubara, the Drina, etc.), Fig. 8.2. Pseudogley in Serbia covers about 5.6% of the land mass, i.e. approximately 4922.19 km^2.

Physical and chemical properties. The most important characteristic when it comes to the physical properties of pseudogley is the marked textural differentiation of the profile. The surface horizons (A and g) are usually silt loam, with more than 40% silt, while the impermeable layer is clayey loam with more clay than in the surface horizons. The water regime of pseudogley is connected to this texture and the alternation of wet and dry phases, in between which there is a longer or shorter damp phase.

The high silt content and poorly defined structure in the surface parts of pseudogley means that during the wet phase, this part of the profile turns into a pulpy mass, while during dessication, it becomes hard and compact. The humus horizon has the greatest porosity (up to 50%) and the air capacity is from 10 to 15%. However, in the **g** horizon, these

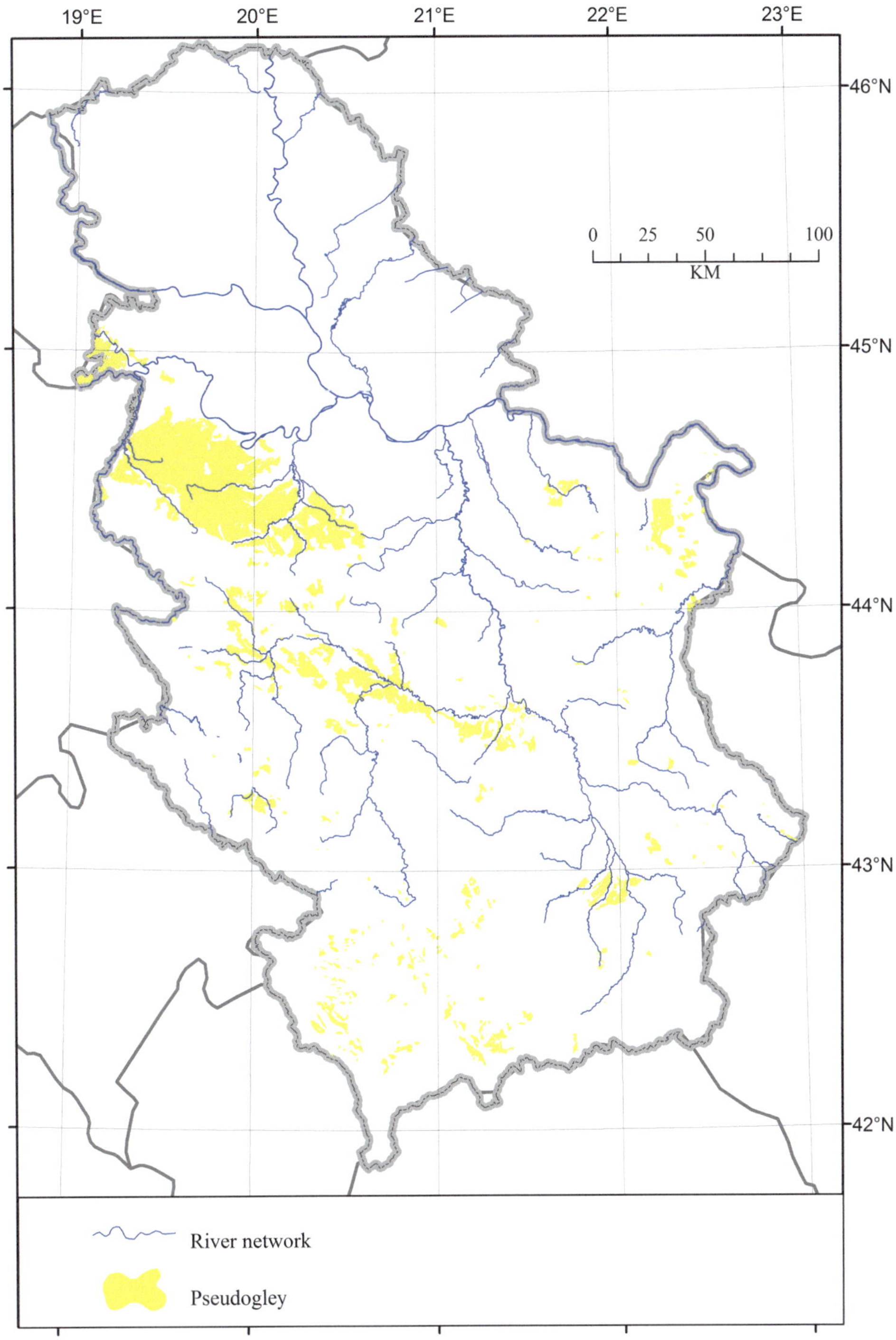

Fig. 8.2 Distribution of pseudogley

values decrease and become so low in the B (or IIg) horizon that it is practically impermeable to water and very poorly aerated (Antić et al. 1982). The soil has a poor temperature regime. Pseudogley has 3–5% humus in forests and meadows and 2–3% in arable fields, although this drops off sharply as depth increases. Physico-chemical characteristics of pseudogley are presented in Tables 8.1 and 8.2.

Land use. Pseudogley is a soil with low productivity, which lacks oxygen during the wet phase and readily available water during the dry phase. The production capability of forest pseudogleys is variable, but those which are deep with a long damp phase are productive soils.

Guidelines for management. Amelioration procedures are required in order to eliminate the excess moisture. It is necessary to introduce organic material and readily available phosphorus and potassium in order to increase the soil's productivity. Pseudogley is prone to erosion due to its silty texture and its low ability to absorb water, which results in

Table 8.1 Physical characteristics of pseudogley

Horizon	Granulometric composition (%)				
	2.0–0.02 mm	0.02–0.002 mm	<0.002 mm	Total sand	Clay + silt
Ap					
X	41.03	34.70	25.85	41.03	60.55
min	29.90	17.40	15.50	29.90	32.90
max	67.10	42.20	34.30	67.10	76.50
Eg					
X	37.22	31.56	33.15	37.22	64.70
min	21.10	16.50	14.90	21.10	31.40
max	68.60	40.10	44.90	68.60	83.50
Btg					
X	29.98	32.14	39.23	29.98	71.37
min	15.80	15.20	21.30	15.80	36.50
max	61.40	42.60	49.50	61.40	87.60

Source Mrvić et al. (2007) and Pivić et al. (2008)

Table 8.2 Chemical characteristics of pseudogley

Horizon	pH		Y1 ccm n/10 NaOH	Adsorption complex				Humus (%)	N (%)	Al (mg/kg)	P_2O_5 (mg/100 g)	K_2O (mg/100 g)
	H_2O	$CaCl_2$/KCl		T-S (cmol kg^{-1})	S (cmol kg^{-1})	T (%)	V (%)					
Ap												
X	5.85	4.92	10.28	6.78	17.10	23.88	63.25	2.67	0.17	191.15	19.03	15.15
min	4.75	3.45	1.30	1.31	4.83	15.42	18.80	1.17	0.15	35.80	7.23	9.70
max	7.30	6.80	20.88	10.59	36.52	37.83	96.55	4.13	0.18	346.50	30.83	20.60
Eg												
X	5.60	4.50	10.15	8.10	16.70	24.80	65.30	1.39	0.08	152.10	6.52	13.25
min	4.85	3.60	2.27	5.60	8.04	13.64	31.34	0.12	0.06	1.60	6.33	11.40
max	6.40	5.70	17.62	11.84	22.84	34.68	88.22	3.51	0.09	404.80	6.71	15.10
Btg												
X	5.94	4.61	9.04	6.73	22.04	28.77	74.31	0.76	0.06	219.60	3.19	12.88
min	4.70	3.20	2.57	1.79	10.05	15.35	38.12	0.01	0.05	0.00	1.30	10.40
max	7.20	5.80	18.92	11.30	31.41	36.59	94.62	1.60	0.08	560.70	5.87	15.00

Source Mrvić et al. (2007) and Pivić et al. (2008)

surface runoff. When deciding on species for afforestation, those with deep roots (oak, hornbeam, and in some cases Scots pine) should be chosen in order to decrease the risk of erosion.

8.2 Class II: Initial Hydromorphic Soils with I-II-III-IV…, or (A)-G, or (A)-C Horizon Sequence

8.2.1 Fluvial or Alluvial (Fluvisol)

The name of the soil type. The name comes from the Latin word *fluvius*, meaning river, which clearly highlights the origin of this soil. Other names include recent river deposits, alluvial deposits, alluvial soil, or simply alluvium.

The WRB name. Fluvisol.

Genesis and sequence of horizons. These are undeveloped soils with an excess wetting of the profile due to groundwater and flood waters. The genesis of fluvisol is conditioned by the hydrological regime of the river and by the nature of the material which accumulates. In terms of the river's course, the coarsest material (proluvium) is deposited on the aggradation plain at the river's source, while the finest material is deposited in its delta. In terms of the distance from the riverbed, the heaviest material accumulates immediately next to the riverbank, while the lightest material is deposited furthest away, beside the river terrace. The conditions for deposition change when the course of the river bed moves, and hence the deposition zone also moves. In this way, very different material can be deposited on top of previously deposited material, and deposition of new material can occur on top of already formed soils with a humus horizon. Hence, the deposition process is a permanent one, with the pedogenesis of these soils often interrupted, meaning there are young soil formations on floodplains. All these soil formations are categorized as fluvisol, as they originate from river deposits in terms of their pedogenesis (Fig. 8.3). The morphology of their profile is characterized by a varying number of different textured layers, which are distinct from one another to varying extents and are denoted using Roman numerals (I, II, III, etc.), starting with the surface layer as the youngest. In terms of genetic horizons, there is an (A) horizon and maybe a G or C horizon. Classification at the level of varieties and forms is made according to the criteria set out in Table 6.1 (Chap. 6 Soil Classification).

Position in an evolutionary series. The position of these soils is a specific one as they are found in the foreshore regions of rivers and are often flooded. The light texture enables the quick withdrawal of water from the profile and the penetration of air. As a result, slight greyish-blue mottling is often visible in the profile, even up to a depth of 2 m. If protected from flooding, the genesis of this soil type begins with an (A)-C profile (regosol on fluvial deposits), with only a very small amount of organic material in the form of tiny black grains. The next stage is the A-C class of automorphic soils, (leptosol, eutric or dystric) where the humus-accumulative horizon can be up to 20 cm deep and finishes with cambisol. In Serbia, this pattern of genesis was first described by Tanasijević et al. (1951).

Fig. 8.3 Fluvisol (*photo* Mrvić, V.)

Distribution. Fluvisols are found in the valleys of the Danube, Sava, Morava and Tisa rivers (Fig. 8.4). Even though they are undeveloped soils, they usually are highly fertile and are populated by mixed forests of poplar and willow and grassy vegetation. In Serbia, these soils cover approximately 7.9% of its land mass, i.e. about 7020.54 km^2.

Physical and chemical properties. Fluvisols are most commonly sandy, and more rarely sandy loam, which determines their air–water properties, and they are structureless. Physical characteristics of fluvisols are presented in Table 8.3. Although the water-holding capacity is low, the air capacity is relatively high. Plants are mainly supplied with water from groundwater, which is often at depths of up

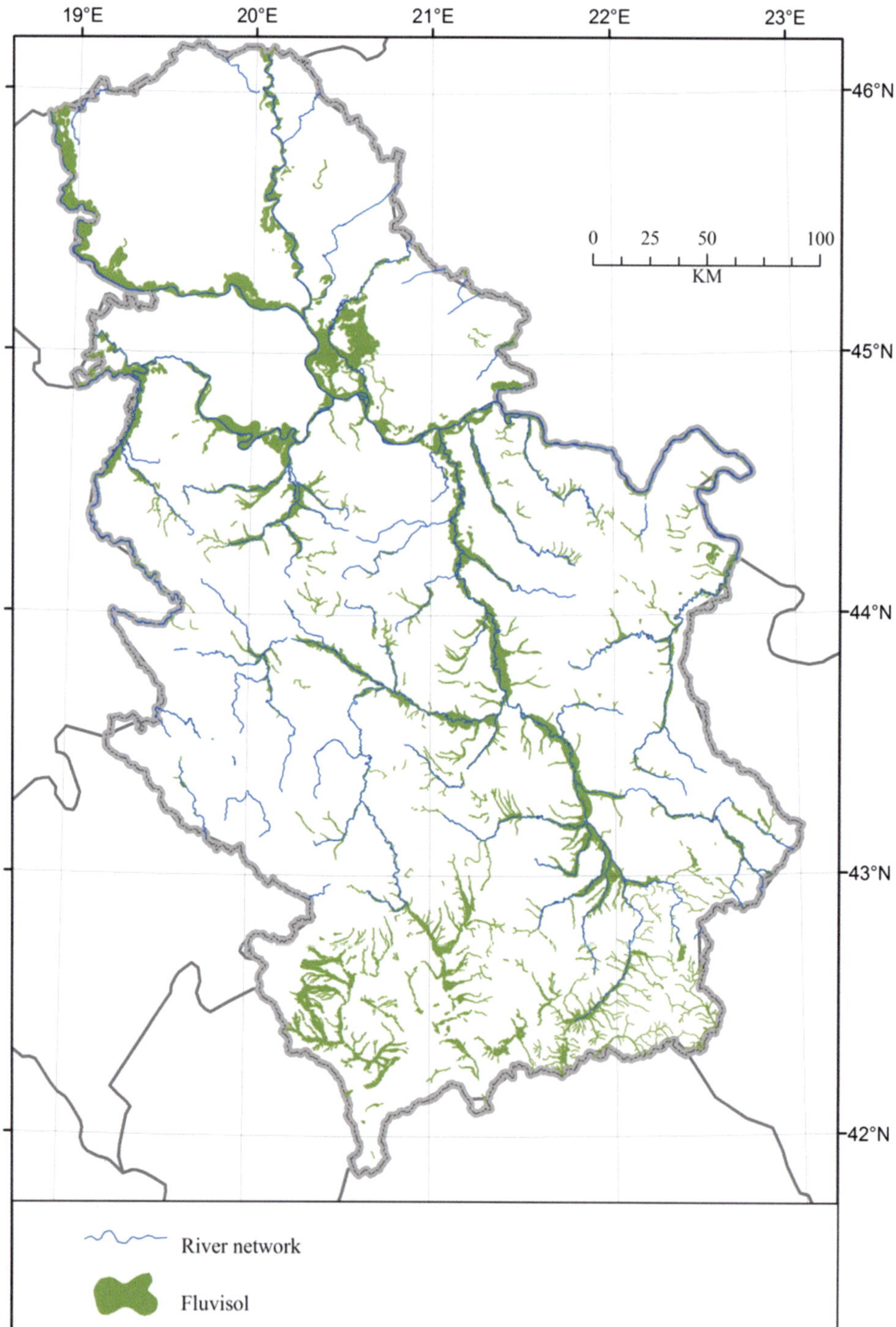

Fig. 8.4 Distribution of fluvisol

to 2 m. The majority of fluvisols in Serbia are calcareous and contain 5–30% carbonates, but there is no regularity in terms of their distribution in the profile. They contain a small amount of humus, 1–2%. In sandy varieties, this level falls to below 1%. Phosphorus levels in fluvisol vary, while they are well supplied with potassium, apart from the sandy varieties (Table 8.4).

Land use. Fluvisols mainly have quite favourable physical and chemical properties. The only exceptions are gravelly deposits and pure sand. In such places, purple willow grows. The ecological production value depends on the flooding and groundwater regimes. These habits are used for poplar plantations and, with irrigation and fertilization, for vegetable cultivation.

Guidelines for management. If used for vegetable cultivation, in addition to irrigation and fertilization, flood defence measures are required.

8.3 Class III: Semi-gleyic Soils with A-C-G Soil Horizon Sequence

8.3.1 Fluvial Meadow (Humofluvisol)

The name of the soil type. The pedogenetic processes in this soil are affected by the high biogenic accumulation beneath the mesophytic meadow vegetation and due to wetting through atmospheric, ground and occasionally flood waters. Forest and meadow (which here is secondary, occurring after deforestation) vegetation has an equal impact on these processes, and the addition of the adjective 'fluvial' highlights the fact that these soils originate on river terraces. In Serbia, these soils are also called 'meadow black soil', 'humus alluvium' and 'alluvial pararendzina'.

The WRB name. Fluvisol, humic.

Genesis and sequence of horizons. This soil forms in the central parts of river terraces, where a humus horizon forms due to the reduced intensity of deposition. These areas are less subject to flooding and the process of pedogenesis proceeds more smoothly. Flooding lasts for less than 30 days, while groundwater is at a depth below 1 m, meaning that beneath the humus A horizon and the gley G horizon, there is a C horizon without any significant signs of gleization (A-C-G), Fig. 8.5. The depth of the humus A horizon is 10–20 cm, with calcareous forms having hydromollic humus and non-calcareous ochric humus. This horizon mainly has a marked granular structure, while the C horizon may have a weak polyhedral structure. The G horizon is at a depth below 100 cm. If groundwater is mineralized, then salinization and alkalinization can occur. Classification at the level of varieties and forms is made according to the criteria set out in Table 6.1 (Chap. 6 Soil Classification).

Position in an evolutionary series. Humofluvisols are the next stage in the evolution of fluvisol. They can evolve in two directions. On higher ground, they become terrestrial (automorphic) soils and pedoevolution progresses towards eutric cambisol. In other cases, with more intense surface wetting, they are subject to processes of lessivation and pedoevolution progresses towards pseudogleization.

Distribution. In Serbia, this soil is found in the valleys of the larger rivers, occupying the central parts of river terraces (Fig. 8.6). Natural vegetation is forests of pedunculate oak, narrow-leaf ash and elm, elm and narrow-leaf ash, then alder and poplar, and forests of silver poplar and black poplar. Secondary vegetation is that of wet meadows. In Serbia, humofluvisol covers approximately 5.1% of its land mass, i.e. about 4485.52 km^2.

Physical and chemical properties. Humofluvisols have a mainly loamy granulometric composition (Table 8.5). Their water–air properties are favourable, but are not always in harmony. The water-holding capacity is sometimes higher at the expense of the air capacity. The humus content ranges from 2 to 5%. pH is neutral to slightly alkaline as

Table 8.3 Physical characteristics of fluvisol

Horizon	Granulometric composition (%)					
	2.0–0.2 mm	0.2–0.02 mm	0.02–0.002 mm	<0.002 mm	Total sand	Clay + silt
A						
X	6.70	44.58	35.24	13.49	51.28	48.72
min	0.52	10.12	12.00	3.28	10.64	16.00
max	22.14	82.70	54.98	34.38	84.00	89.36
Gso						
X	10.33	60.21	18.33	10.36	71.15	28.85
min	0.00	6.18	0.04	0.04	6.33	0.32
max	89.52	95.20	55.52	40.42	99.68	93.67

Source Galić et al. (2009b, 2011), Pekeč et al. (2012), Pivić et al. (2014), Pekeč and Novčić (2014) and Kostić (2014)

Table 8.4 Chemical characteristics of fluvisol

Horizon	pH		Y1 ccm n/10 NaOH	Adsorption complex				Humus (%)	C (%)	N (%)	C/N	$CaCO_3$ (%)	P_2O_5 (mg/100 g)	K_2O (mg/100 g)
	H_2O	$CaCl_2$/ KCl		T-S (cmol kg^{-1})	S (cmol kg^{-1})	T (%)	V (%)							
A														
Average	7.51	7.21	2.97	0.89	43.02	43.91	97.97	7.17	3.12	3.37	9.71	16.75	11.09	30.96
min	5.30	6.95	2.97	0.89	43.02	43.91	97.97	1.05	1.60	0.08	9.71	0.42	5.00	26.40
max	7.90	7.47	2.97	0.89	43.02	43.91	97.97	44.70	4.08	12.10	9.71	21.51	19.30	38.80
Gso														
Average	8.03	6.90	3.08	1.74	37.04	38.78	94.73	2.01	1.37	1.67	10.49	14.95	3.88	10.21
min	6.80	5.85	1.19	0.36	25.18	28.30	88.62	0.02	0.94	0.00	6.24	0.00	0.30	2.60
max	8.80	7.51	6.70	4.35	49.35	49.77	99.27	18.60	2.12	11.00	17.67	25.81	11.40	20.00

Source Gajić et al. (2006), Delić et al. (2011), Pekeč et al. (2012), Kostić (2014) and Pivić et al. (2014)

they generally have a high content of carbonates (Table 8.6).

In forests, they are usually non-calcareous throughout the whole profile, and then, they have neutral to slightly acidic chemistry. Phosphorus supply is moderate or poor as a result of the alternating dry and wet phrases in the profile. They are better supplied with potassium if they have a heavier texture. The properties of these soils show wide spatial variations depending on microrelief, which in the central part of alluvial terraces is in the form of elevations and depressions.

Land use. These soils are most suitable for poplar plantations.

Guidelines for management. On clayey varieties, with a high groundwater level, ameliorative measures should be aimed at reducing this water level.

8.4 Class IV: Hypogleyic Soils with A-G Soil Horizon Sequence

8.4.1 Wetland Black Soil (Humogley)

The name of the soil type. This soil is found on fenland. Its name highlights the extremely pronounced accumulation of humus, which is reflected in the black colour of the soil.

The WRB name. Gleysol, humic.

Genesis and sequence of horizons. The presence of groundwater in the profile is characteristic for humogley, but even at its highest level, it never reaches the surface. It most often fluctuates at a depth of about 100 cm. The parent substrate is calcareous loess. Groundwater, together with flood waters, leads to the excessive wetting of the whole profile for the majority of the year (Fig. 8.7).

Despite this, at times when the water table is low (the late summer), the humus horizon dries out, so the transformation of organic matter becomes aerobic rather than anaerobic, leading towards the formation of mature humus. This is more marked in those areas of Serbia where there are drainage canals. Groundwater levels can fluctuate by over 100 cm, and so gleization processes in the Gso subhorizon coincide with this zone of fluctuation, having a specific impact on gleization in humogley. Humogley often has a pronounced horizon of forest litter. The humus-accumulative A horizon is from 40 to 70 cm in depth and there is often a clear transitional A/Gso horizon, too, where varying levels of $CaCO_3$ concretions can be found, while signs of gleization are always present. The Gso subhorizon extends to a depth of over 100 cm, and beneath it a Gr subhorizon can be found. Sometimes, this subhorizon is connected to the parent substrate (Gso/C). The occurrence of calcareous and non-calcareous varieties depends on the impact ratio of flood water to groundwater. Varieties where flood waters have a greater impact are non-calcareous. Classification at the level of varieties and forms is made according to the criteria set out in Table 6.1 (Chap. 6 Soil Classification).

Position in an evolutionary series. Humogley can evolve in two directions. When groundwater is deep enough and when there is no flooding, lessivation can occur, and if this process continues, pseudogley-gley results. The other path is the direct formation of pseudogley when the texture of humogley is heavy. This occurs on old river terraces.

Distribution. This soil is widespread in the river valleys of the Pannonian plain—areas of the Pančevo, Negotin, Godomin, Vršac and Makiš wetlands. They are also found in the Morava river valley (Fig. 8.8). Vegetation on humogley is moist forests of narrow-leaf ash and pedunculate oak, willow and alder. In Serbia, humic gleysols and gleysols cover approximately 5.1% of its territory, i.e. about 4468.90 km^2.

Physical and chemical properties. Humogley is mainly clayey with more than 40% clay and has poor filtration capacity. It has an unfavourable water–air regime and

Fig. 8.5 Humofluvisol (*photo* Mrvić, V.)

contains between 3 and 6% humus. The C/N ratio is from 10 to 20. Calcareous varieties of this soil have neutral to slightly alkaline chemistry, while non-calcareous are usually neutral to slightly acidic. Calcareous humogley can also be salinized, and sometimes alkalinized in the deeper reaches of the profile. They are well supplied with readily available potassium and phosphorus.

Guidelines for management. Through hydroamelioration, these soils can become highly productive and can be used for the successful cultivation of arable crops such as wheat, maize, sunflowers, alfalfa and sugar beet.

8.4.2 Wetland gley (Eugley)

The name of the soil type. Wetlands are terrains with a high level of permanently stagnant groundwater. The international name 'eugley' highlights that these are soils where the processes of gleization are most pronounced.

The WRB name. Gleysol.

Genesis and sequence of horizons. These soils are linked to river valleys, where they are found at the lowest points (in depressions) where groundwater is permanently present and hardly fluctuates, most frequently in the range of 0–80 cm.

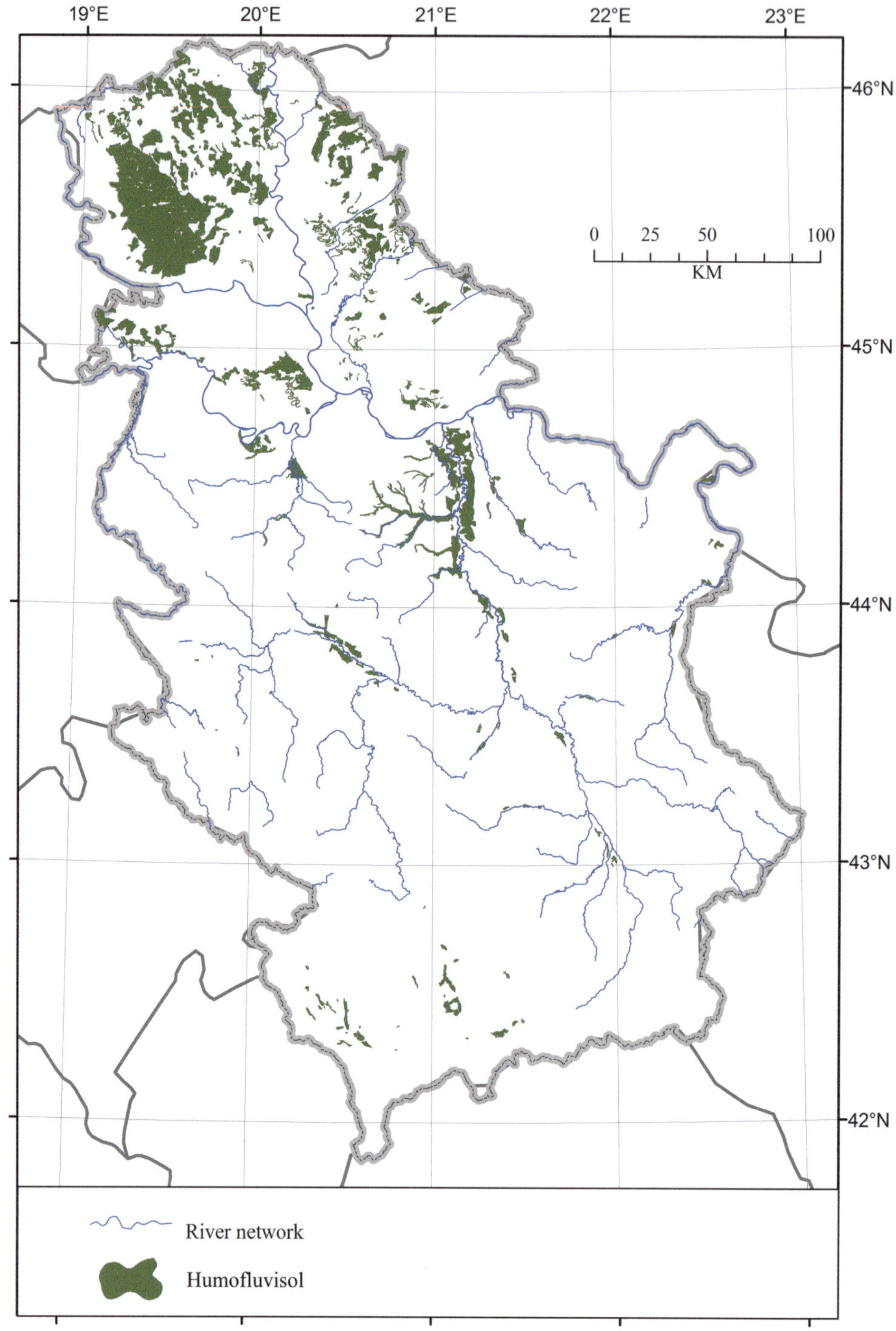

Fig. 8.6 Distribution of humofluvisol

There are two alternating periods in the process of gley formation. The first period is when the water level is at its highest in the profile, air availability is at its lowest, and when the reduction process is most intensive. These conditions are completely anaerobic and all the Fe content is in the form of ferrous hydroxide and ferrocarbonates. The mineral part of the profile takes on a grey–green colour without any reddish spots or concretions. During the summer period, when the groundwater level falls, a period of oxidation begins. The drying of the profile, which begins at the surface, leads to increased biological activity and the release of CO_2, which penetrates the soil profile. Under its influence, solutions of calcium and iron bicarbonates form in the gley horizon, which through evapotranspiration rise above the level of groundwater, oxidize and accumulate in the form of reddish mottling and concretions of $CaCO_3xFe_2(CO_3)_3$. The reductive Gr gley horizon continually takes iron and calcium from the Gso horizon. In wet conditions, a

Table 8.5 Physical characteristics of humofluvisol

Horizon	Granulometric composition (%)					
	2.0–0.2 mm	0.2–0.02 mm	0.02–0.002 mm	<0.002 mm	Total sand	Clay + silt
A						
X	4.60	35.55	39.48	26.45	34.07	65.93
min	2.50	17.30	24.70	10.80	20.60	50.80
max	9.30	43.90	58.00	54.00	49.20	79.40
Gso						
X	9.42	59.75	22.74	15.12	62.14	37.86
min	0.40	29.00	0.80	0.80	21.40	1.60
max	46.10	93.30	54.40	53.10	98.40	78.60

Source Galić et al. (2009b, 2011) and Delić et al. (2011)

Table 8.6 Chemical characteristics of humofluvisol

Horizon	pH		$CaCO_3$ (%)	Humus (%)	N (%)	P_2O_5 (mg/100 g)	K_2O (mg/100 g)
	H_2O	$CaCl_2$/KCl					
A							
Average	7.77	6.86	11.87	4.09	0.15	14.24	17.10
min	7.70	6.82	0.85	1.52	0.08	0.38	11.80
max	7.80	6.89	20.90	5.27	0.22	28.10	22.40
Gso							
Average	8.17	7.27	16.09	1.14	0.06	0.33	9.20
min	7.90	7.21	1.27	0.45	0.06	0.32	9.20
max	8.80	7.32	23.48	2.71	0.06	0.33	9.20

Source Galić et al. (2009b) and Delić et al. (2011)

humus-accumulative horizon forms with hydromull humus, giving the soil an Aa-Gso-Gr-C profile (Fig. 8.9). Classification at the level of varieties and forms is made according to the criteria set out in Table 6.1 (Chap. 6 Soil Classification).

Position in an evolutionary series. If the drying process (either naturally or artificially) of these soils continues, changes occur in the humus itself, with the hydromull humus becoming mull humus similar to that found in chernozems. The process of pedoevolution may proceed towards terrestrial soil if the groundwater level continues to fall and air continues to penetrate the profile. In this case, if aeration is good, the gley horizon can take on the properties of a (B) horizon. If, however, the groundwater level rises, without any fluctuations, the humus layer can turn into peat.

Distribution. In Serbia, these soils are found in parts of Vojvodina, where they occupy plains, plateaus and river valleys which have a constant groundwater level and are not subject to flooding. They are natural habitats for the white willow, *Miosoto-Salicetum albae, Rubeto-Salicetum albae* and *Caricetum-Salicetum albae*. In Serbia, gleysols and humic gleysols cover approximately 5.1% of the landmass, i.e. about 4468.90 km^2.

Physical and chemical properties. Gleysols have adverse physical properties with a clayey texture and poor aeration (Table 8.7). Chemical properties can be extremely heterogeneous, depending on the parent substrate and groundwater level. They can be calcareous and acid soils. Their chemistry is generally neutral to weakly alkaline, while dystric varieties are found sporadically. The degree of base saturation is very high with the domination of absorbed Ca^{2+} ions, while absorbed Na^+ ions dominate in alkaline soils. The humus content can be 7.5% (Table 8.8), but also 10–15% (Antić et al. 1982). The total N, P, and K content is high, but in anaerobic conditions, the mineralization of N is negligible. In anaerobic conditions, biochemical processes lead to the formation of specific reduced compounds (methane, H_2S, and PH_3), which are environmentally harmful.

Land use. In natural conditions, these soils are populated by plant species adapted to the anaerobic conditions in the substrate.

Guidelines for management. It is possible to reduce groundwater levels through hydroamelioration, resulting in good agricultural land, which can be used for several years without the need for fertilization, thanks to the mineralization of the accumulated humus.

Fig. 8.7 Humogley (*photo* Vasin, J.)

8.5 Class V: Peat Soils with T-G Soil Horizon Sequence

The constant stagnation of water throughout the entire profile and on the surface of the soil leads to lush hydrophilic vegetation, whose organic remains break down slowly in the anaerobic conditions, leading to their accumulation in the form of semi-decomposed organic matter—peat. This process of peat formation is known as paludization, from the Latin word *palus* meaning *marsh*. In Europe, peat formation began during the last glacial period 10,000 years ago. If the surface layer of peat is thicker than 30 cm and if it contains more than 30% organic matter, then these soils are classified as peat soils according to the classification system used in Serbia. If not, they belong to other classes of hydromorphic soils as peat varieties. Peat soils differ, both in terms of their physical and chemical characteristics, and also in the place, conditions and method of formation. In the soil classification system used in Serbia, three types of peat soil are defined, depending on the type of wetting, genesis, and properties. Subhorizons are defined according to the degree of decomposition of organic matter: fibric (Tfi), hemic (The), and sapric (Ts). These soils occupy a negligible area of Serbia, but from the palaeobotanical and palaeopedological

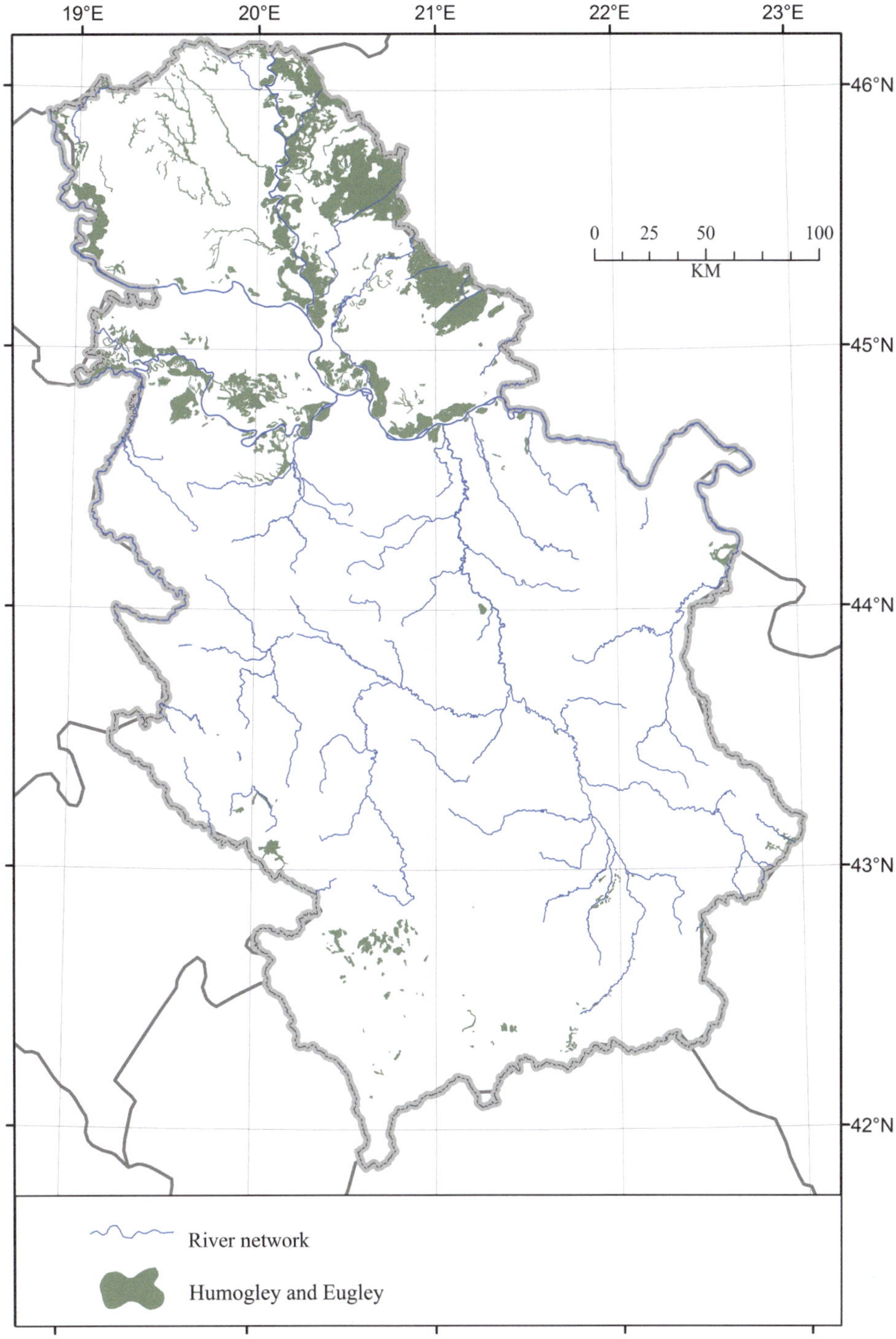

Fig. 8.8 Distribution of humogley and eugley

standpoints, these habitats are of exceptional importance and are often subject to strict protection measures.

8.5.1 Low Moor Peat (Planohistosol)

The name of the soil type. The name planohistosol comes from the Latin word *planus*—flat and the Greek word *histos*—tissue, which highlights the basic characteristic of its structure which is dominated by undecomposed fragments of plant tissue. In Serbia, the local name 'bog peat' is used, which maybe best describes the conditions in which this peat soil forms.

The WRB name. Histosol.

Genesis and sequence of horizons. This soil is a topogenic creation, forming in depressions where water permanently covers the soil surface and where peat forms under the influence of wetland vegetation, mainly in an eutric

Fig. 8.9 Eugley (*photo* Vasin, J.)

environment. Organic remains fall into the water and in such anaerobic conditions, the rate of decomposition is very slow. These depressions are mainly accumulative zones in which mineral suspensions also accrue, meaning the peat can also contain varying amounts of muck. The entire environment is rich in bases and sometimes is also calcareous. In these conditions, a T horizon forms, beneath which there is a gley horizon with extreme reductive characteristics, gyttja or sapropel as a subhydric soil. Subhorizons can be identified which differ in terms of the degree of decomposition (Tfi, The, and Ts) and which can alternate with no real pattern. Classification at the level of varieties and forms is made according to the criteria set out in Table 6.1 (Chap. 6 Soil Classification).

Distribution. In Serbia, low moor peat (bog peat) forms in places where water stagnates in depressions around lakes, former river beds, and the deltas of the major rivers. It is linked to the dying of the plant cover (*Phragmites* sp. *Tipha* sp., *Carex* sp., and more rarely the moss *Hypnum* sp.). The natural vegetation today is reeds, common cat's tail, bulrush,

Table 8.7 Physical characteristics of eugley

Horizon	Granulometric composition (%)					
	2.0–0.2 mm	0.2–0.02 mm	0.02–0.002 mm	<0.002 mm	Total sand	Clay + silt
A						
X	5.62	45.68	30.52	18.16	51.31	48.69
min	0.10	21.59	3.80	3.00	22.32	8.60
max	31.30	89.80	48.36	38.84	91.40	77.68
G						
X	7.74	60.84	19.63	11.81	69.35	30.65
min	0.44	20.60	0.10	0.20	21.20	2.40
max	22.60	95.20	50.40	30.32	97.60	78.80

Source Galić et al. (2009a, b), Belić et al. (2011), Pekeč et al. (2011), Pivić et al. (2014) and Pekeč and Novčić (2014)

Table 8.8 Chemical characteristics of eugley

Horizon	pH		$CaCO_3$ (%)	Humus (%)	N (%)	EC (mS/cm)	P_2O_5(mg/100 g)	K_2O (mg/100 g)
	H_2O	$CaCl_2$/KCl						
A								
Average	8.14	7.70	11.99	3.83	0.20	1.91	5.10	20.68
min	7.16	7.17	8.27	0.29	0.02	0.48	1.10	8.20
max	9.89	8.66	14.17	7.46	0.30	3.33	9.30	33.51
G								
Average	7.50		14.31	1.68				
min	7.33		14.01	0.08				
max	7.68		14.60	7.06				

Source Belić et al. (2011), Pivić et al. (2014) and Pekeč and Novčić (2014)

common sedge, and moss from the genus *Hypnum*, while alder and grey willow can also be found.

Physical and chemical properties. The thickness of low moor peat ranges from 50 cm to several metres. It has a very high water-holding capacity (up to 90 vol.%). In its dry state, the peat horizon is a light, porous mass with a bulk density of only 0.1–0.2 and a total pore volume of 80–90%. It contains 70–80% organic matter. It is a neutral soil, rich in bases with 3–4% nitrogen (C/N ratio of about 20) and calcium (up to 5%). It is characterized by a high adsorption capacity and is mainly saturated with Ca^{2+} and K^{+} ions with a high buffering capacity. The majority of phosphorus is bound to organic matter (Bogdanović et al. 1972, 1982).

Land use. It is applied directly as a fertilizer or can be used in greenhouses with the addition of phosphorus and potassium to achieve optimal ratios.

Guidelines for management. As already mentioned, these soils occupy only a small area of Serbia, but they are of exceptional significance from the palaeobotanical and palaeopedological standpoint, and hence should be subject to strict protection measures. These are often the habitats of endemic flora.

8.5.2 High Moor Peat (Acrohistosol)

The name of the soil type. The name elevated or high moor peat comes from the raised dome formations which occur as this soil develops. The name acrohistosol comes from the Greek words *arcos*—high and *histos*—tissue. The local name in Serbia is moss peat, which points to the origins of this soil (moss).

The WRB name. Histosol.

Genesis and sequence of horizons. This peat soil is conditioned by climate and is linked to atmospheric humidity and deposition. It forms under the influence of moss in dystric environments. Climatic conditions for the formation of this peat soil are characterized by low temperatures, high precipitation and high relative humidity. In conditions where the climate is cold and wet, moss peat soils form on substrates which are poor in bases and populated by mosses from the genus *Sphagnum* sp. The colonization of these mosses can take two forms. The first involves the evolution of extremely acidic podzols, leading to the formation of a thick layer of raw humus, which excludes the already low amount of nutrients from

Fig. 8.10 High moor peat with relict community of Serbian spruce and common alder *Alno glutinosae-Picetum omorikae*, in Mitrovac on Mt. Tara (*photo* Milenković, M)

biological cycling. In wet and cold climates, this highly oligotrophic substrate is colonized by white moss as trees retreat. The other form of colonization occurs in dystric hydromorphic soils, particularly on stagnogley, which is found in depressions. In anaerobic conditions, the poorly decomposed organic remains accumulate, which further increases the oligotrophy of the habitat and leads to mosses replacing hydrophilic vegetation. The peat horizon, which is characterized by recognizable plant tissue, extends from the very surface right down to the parent substrate. The surface layer is composed of the slightly altered remains of the moss and has a light yellow colour, while deeper down, there are the more decomposed and dark coloured layers of the peat horizon, some of which were formed in the period before moss colonization. Beneath the peat layer, there is gley, most often with a developed ortstein. Classification at the level of varieties and forms is made according to the criteria set out in Table 6.1 (Chap. 6 Soil Classification).

Distribution. It is found in the mountain regions of Serbia, on Tara, Kopaonik, Divčibare and Vlasina. These habitats are populated by white moss from the genus *Sphagnum* sp., while tree species include Scots pine, birch, spruce, fir and Serbian spruce.

Physical and chemical properties. The peat horizon is highly porous (a pore volume of up to 95%). However, its permeability to water is low, while it has high water retention (the wilting point is up to 20% water). The organic matter content is from 93 to 95%, but the degree of its decomposition can vary. High moor peat (moss peat) is very acidic (pH 3–4) and poor in bases and nitrogen. The P and K content is very low (Tešić et al. 1969).

Land use. It is used as fuel for heating and as organic fertilizer, in which case it is neutralized through the addition of Ca^{2+} and enriched with N, P, and K.

Guidelines for management. As already mentioned, these soils occupy only small areas in Serbia; however, these habitats are of exceptional significance from the palaeobotanical and palaeopedological standpoint, and hence should be subject to strict protection measures. These are often the habitats of endemic flora. However, in spite of the high level of protection, some endemic species are disappearing from these habitats. One example is the Serbian spruce in the relict community Serbian spruce and common alder *Alno glutinosae-Picetum omorikae*, in Mitrovac on Mt. Tara (Fig. 8.10).

Thick layers of peat, decomposition in anaerobic conditions and the intensive formation of phenolic compounds have an inhibitory effect on the seeds and seedlings of Serbian spruce, which consequently is disappearing from this region. This demands different measures whereby the whole habitat retains its peatland character, the only autochthonous habitat of the Serbian spruce on Earth (Djurdjević et al. 2003).

8.5.3 Transient Peat (Plano-Acrohistosol)

This usually forms as bog peat evolves into moss peat. In terms of the vegetation and properties of this peat soil, elements of both the previously described peat soils are present, with the impact of one or the other predominating. Classification at the level of varieties and forms is made according to the criteria set out in Table 6.1 (Chap. 6 Soil Classification).

The WRB name. Histosol.

8.6 Class VI: Anthropogenic Soils with P-G Horizon Soil Sequence

8.6.1 Rigosol

These are deep ploughed peat or peat gley soils, created for agricultural production. To this aim, the T horizon is mixed with the mineral gley horizon.

8.6.2 Hydroameliorated Soil

This soil originates due to a reduction in the groundwater level and/or hydroameliorative measures, which result in a weakening of gleization in the upper part of the profile. The subtypes of this soil according to the classification system used in Serbia are defined according to the original soil type.

References

Antić M, Jović N, Avdalović V (1982) Pedology. Scientific Book, Belgrade (in Serbian)

Belić M, Lj Nešić, Ćirić V, Vasin J, Milošev D, Šeremešić S (2011) Characteristics and classification of gleyic soils of banat. Field Veg Crop Res Novi Sad 48:375–382

Bogdanović M, Gigov A, Tešić Ž, Todorović M, Stojanović S, Aleksić Ž (1972) The genesis of peat-bog of Veliko Gradište boggy region and its peat. Soil Plant Belgrade 21(2):203–220

Bogdanović M, Gigov A, Tešić Ž, Todorović M, Stojanović S, Antić M (1982) The peat-bogs and peats of Vojvodina with a special reference to the protection of life environment. Soil Plant Belgrade 31(3):334–355

Delić D, Stajković-Srbinović O, Dj Kuzmanović, Rasulić N, Mrvić V, Andjelković S, Knežević-Vukčević J (2011) Effect of bradyrhizobial inoculation on growth and seed yield of mungbean in fluvisol and humofluvisol. Afr J Microb Res 5(23):3946–3957

Djurdjević L, Dinić A, Mitrović M, Pavlović P, Tešević V (2003) Phenolic acids distribution in a peat of the relict community with Serbian spruce in the Tara Mt. forest reserve (Serbia). Eur J Soil Biol 39:97–103

Gajić B, Dugalić G, Đurović N (2006) Comparison of soil organic matter content, aggregate composition and water stability of gleyic fluvisol from adjacent forest and cultivated areas. Agron Res 4 (2):499–508

Galić Z, Ivanišević P, Orlović S, Pekeč S (2009a) Soil characteristics of the sites in protective forests in the lowland part of Vojvodina. Forest Belgrade 3–4:29–38 (in Serbian)

Galić Z, Pekeč P, Pekeč S, Kebert M, Stojnić S (2009b) Characteristics of soil types on small river islands in the Central Danube Basin. Bull Fac Forest Belgrade 100:55–70 (in Serbian)

Galić Z, Orlović S, Vasić V, Galović V, Klašnja B, Stojanović D, Babić V (2011) Phytocenological characteristics in poplar plantations in the protected region of the central Danube basin. Arch Biol Sci Belgrade 63(3):811–817

Kostić O (2014) Ecophysiological characteristics of some woody plant species and their potential for the revitalization of fly ash deposits from a thermoelectric power plant. Doctoral dissertation, Faculty of Forestry, Belgrade University, Belgrade (in Serbian)

Kubiëna W (1953) Bestimmungsbuch und Systematik der Böden Europas. Stuttgart

Mrvić V, Jakovljević M, Stevanović D, Čakmak D (2007) The forms of aluminium in stagnosols in Serbia. Plant Soil Environ 53(11):482–489

Pekeč S, Novčić Z (2014) Possible land use for afforestation of former pond area. Poplar Novi Sad 193(194):1–6 (in Serbian)

Pekeč S, Belić M, Lj Nešić, Orlović S, Ivanišević P (2011) Water physical properties of eugley in protected part of alluvial plains of central Danube Basin. Afr J Agric Res 6(7):1717–1725

Pekeč S, Ivanišević P, Stojanović D, Marković M, Katanić M, Galović V (2012) Properties of different forms of fluvisol in the protected area of the River Danube flood plain in Južna Bačka. Poplar Novi Sad 189(190):19–28 (in Serbian)

Pivić et al (2008) Research project: study into the degree of threat to soil and water from hazardous and harmful materials in the neighbourhood beside the ash deposit site in the municipality of Obrenovac. Soil Science Institute, Belgrade (in Serbian)

Pivić et al (2014) Research project: soil amelioration in the municipality of Veliko Gradište aimed at high-value food production. Soil Science Institute, Belgrade (in Serbian)

Tanasijević Đ, Sandić M, Pavićević N (1951) Types of soils in area of Belgrade and Obrenovac. Arch Agric Sci IV(6):28–37

Tešić Ž, Bogdanović M, Gigov A (1969) Peat-bog Crveni Potok on the Tara Mt. and properties of its Peat. Ekologija Belgrade 4(2):149–163

9 Order of Halomorphic and Subaquatic Soils

9.1 Order of Halomorphic Soils

Halomorphic soils form under the influence of readily soluble salts, of which Na salts are particularly significant. Salts can accumulate in the soil solution and leach from it in the form of free crystals (the class of salt-accumulative soils), or the concentration of salts in the solution may be minimal, but the presence of Na^+ ions in the adsorption complex is high (the class of alkaline eluvial–illuvial soils). Halomorphic soils cover approximately 1140.29 km^2 or 1.3% of the land mass of Serbia. As already mentioned, this order is divided into two classes.

9.2 Class I: Salty Soils with Asa-G or Asa-Cg Soil Horizon Sequence

9.2.1 Solonchak

The name of the soil type. The name 'solonchak' is of Russian origin. It comes from the words *sol*—salt—and the Tartar word *čak*—a lot, meaning soil with a lot of salt.

The WRB name. Solonchak

Genesis and sequence of horizons. These soils form at low altitudes, on salty lacustrine and fluvial sediments, while salts sometimes accumulate through the alternating processes of dissolution and evaporation of water. The process involved in the formation of solonchak is the accumulation of salts through the capillary rise of the soil solution and evaporation from the soil surface. As such, the parent substrate is significant as in some cases, it can be the main source of salts, or their appearance on the surface results from a continuous increase in the concentration of salts in the soil solution as evaporation is greater than precipitation.

Typical solonchaks have a salinized humus-accumulative horizon Asa and parent substrate C (profile Asa-C), or if gleyed, the profile is Asa-CG. The salinized horizon can be thick, even up to 50 cm, while salts can appear on the very surface. They regularly cover aggregates and are found on the walls of cracks. In the dry part of the year, this horizon is grey. Beneath it, there is usually the parent substrate, which is affected by groundwater so it is therefore gleyic. Its colour is greenish-grey with Fe and Mn mottling (Fig. 9.1).

On the basis of the profile morphology, they are divided into black solonchaks (dominated by carbonates and with a black crust from Na humate on the surface) and wet solonchaks (dominated by sulphates). In terms of the kinds of salts, they are divided into pure and mixed: soda, sulphate, chloride, soda-sulphate, sulphate-soda, soda-chloride, chloride-soda, sulphate-chloride, chloride-sulphate and mixed with three salts. Classification at the level of varieties and forms is made according to the criteria set out in Table 6.1 (Chap. 6, Soil Classification).

Distribution. In Serbia, they are widespread in Vojvodina, mainly in Bačka, followed by Banat and Srem (Fig. 9.2). Solonchaks are populated by native halophytic flora (*Salicornia herbacea*, *Champhorosma ovata*, *Hordeum gusoneanum*, *Atropis* sp., *Suaeda maritima*, *Staticea gmelini*, *Salsola* sp., *Canphorosa ovata, Gypsophila muralis, Portulaca* sp. and *Cynodon dactilon*, etc.). In Serbia, solonchak and solonetz cover approximately 1.3% of the land mass, i.e. about 1140.29 km^2.

Physical and chemical properties. Solonchaks have relatively poor physical properties, although this is not so much due to the presence of salts, as the texture and pronounced hydrophilic character of montmorillonite clay and the level of Na present in the adsorption complex. As a rule, solonchaks are rich in clay throughout almost the entire profile. If there are textural differences within the profile, they mainly stem from the different character of the parent substrate (Živković et al. 1972; Antić et al. 1982).

The chemical properties of solonchak are determined mainly by the presence of salts, on the basis of which solonchaks are classified. These are salinized soils and can contain >1% of soluble salts, which are easily soluble and harmful to the development of most plants. Solonchaks can be classified according to their salt content into the following: typical (>1% salts), highly salinized (0.5–1% salts),

P. Pavlović et al., *The Soils of Serbia*, World Soils Book Series,
DOI 10.1007/978-94-017-8660-7_9

Fig. 9.1 Solonchak (*photo* Vasin, J.)

moderately salinized (0.5–0.7% salts) and weakly salinized (0.25–0.5%). Generally, typical solonchaks are rich in $CaCO_3$ and also contain sodium chlorides, nitrates, sulphates and carbonates. The proportion of total salts ranges from 1 to 2%, but in extreme cases can be up to 20%. The degree of alkalization depends on the Na(Ca + Mg) ratio. The wider this ratio, the more intensive alkalization is. In addition to the degree of salinity, the cation composition of soluble salts is important, with Na predominating. Generally, its levels are higher than the total content of all the other cations. The reaction of solonchaks in water ranges from pH 7.8 in moderate solonchaks to 10.45 in lime-soda solonchaks. As a rule, solonchaks are poor in organic material, with the humus content ranging from 1 to 2%, and only rarely reaching 4% (Živković et al. 1972; Antić et al. 1982).

Land use. The production capabilities depend on the total salt content, their composition and the cation composition in the water solution and adsorption complex. Furthermore, they also depend on the physical properties. Moderate solonchaks are covered by impoverished meadows mainly populated by halophytic species. As a rule, woody vegetation is not found on this soil. In the summer, during the dry period, due to the high accumulation of salts on the surface, solonchaks resemble white spots due to the plant cover. The worst are typical solonchaks (pH around 10), which have virtually no vegetation.

Guidelines for management. The use of solonchaks for plant cultivation without ameliorative measures is almost impossible. In their natural state, solonchaks can only be used for cultivating chamomile (Antić et al. 1982), e.g. in Vojvodina. Amelioration is aimed at removing salts from the profile with approaches varying, depending on whether the soils are chloride-sulphate solonchaks or soda solonchaks. It is believed that chemical amelioration is not necessary for the former, while it is a must for the latter group and takes the form of adding gypsum. The main ameliorative measure for these soils is desalinization, which removes salts from the soil through irrigation, and is repeated several times with care being taken not to reintroduce salts into the profile.

9.3 Class II: Alkalized Soil with A/E-Btna-C Soil Horizon Sequence

9.3.1 Solonetz

The name of the soil type. The name 'solonetz' is of Russian origin and arises from adding the Russian diminutive form

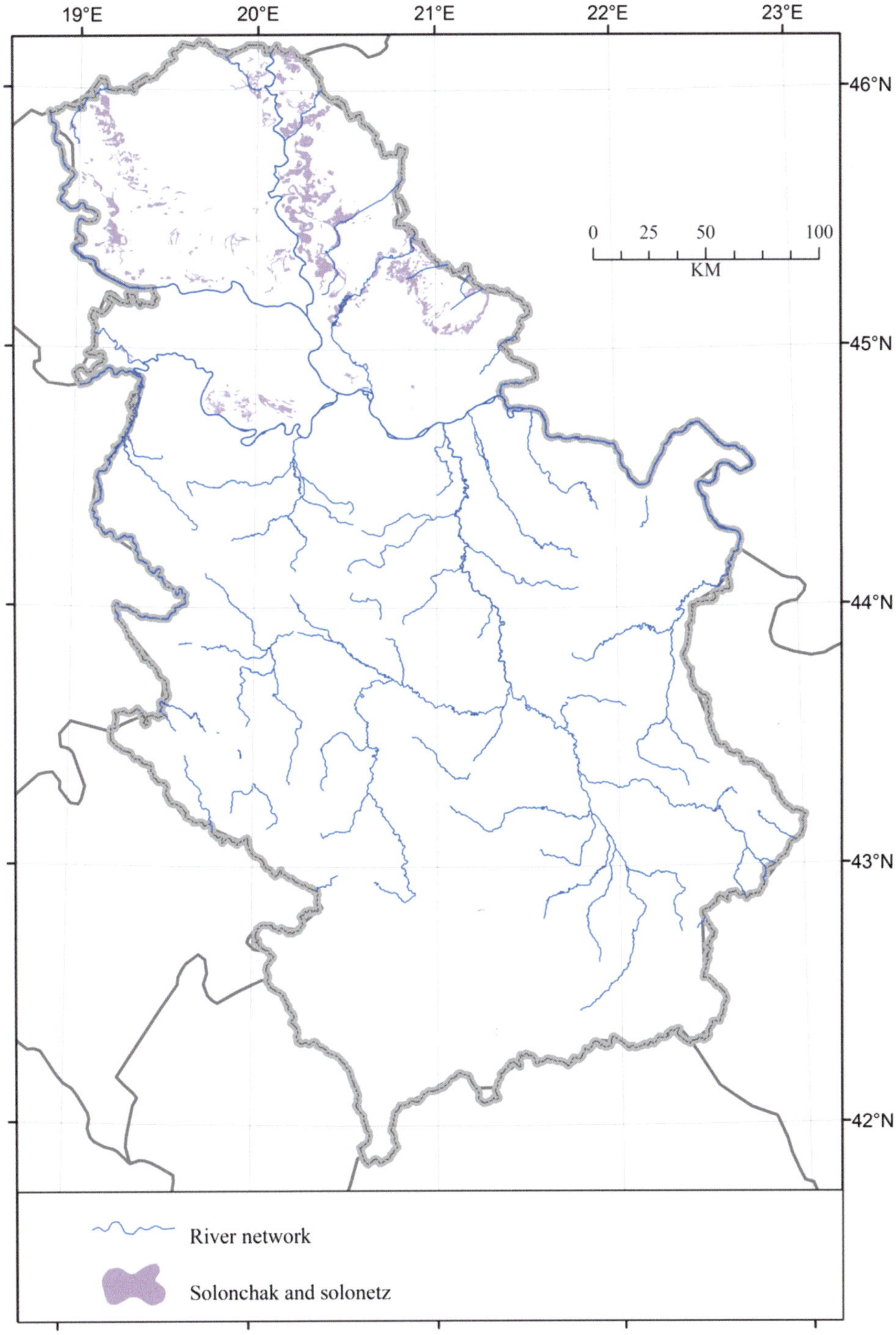

Fig. 9.2 Distribution of solonchak and solonetz

'etz' to the basic word 'solon', thus highlighting that this is soil with a moderate salt content. Solonetz is also an Na soil. It differs from solonchak in that there is an absence of free salts, particularly in the upper layer of the profile. Its profile is more complex, stemming from the complexity of the processes occurring.

The WRB name. Solonetz

Genesis and sequence of horizons. The formation of solonetz is accompanied by a series of complex reactions. The basic process is the introduction of Na into the adsorption complex and its relatively high presence. Solonetz evolves from solonchak, through a process which occurs in three stages: desalinization, formation of soda and the introduction of Na, and the lessivation of clay (Giedrojc 1935), or as a biogenic process requiring organic matter (Antipov-Karataev and Manaeva 1958). As a biochemical process, the formation of soda is linked to a reduction in sodium sulphate in the presence of organic matter and anaerobic sulphate-reducing bacteria. During hydrolysis, the highly basic and weakly acidic salt Na_2CO_3 causes an extremely alkaline reaction of the soil solution, whose pH is >8 and can even reach 11. The presence of Na ions causes the intensive peptization of organic and mineral colloids and their translocation in the profile. Having reached the lower layers where there are free salts, the colloids coagulate, forming an illuvial horizon, which is called 'solonetz', from which this soil gets its name. It is thought that changes in the adsorption complex are far-reaching and that the liberation of sesquioxides and SiO_2 results. As it dries, the soil material in the illuvial horizon disperses, leading to the columnar structure characteristic for this soil type. There are three clear horizons in the solonetz profile. The remaining part of the A horizon usually mixes with the eluvial horizon, denoted as A/E. The ochric humus horizon has a relatively clear crumbly structure, and there is a sharp transition into the B horizon, which is usually divided into Btna and Btsa subhorizons. The B horizon is clearly marked by its specific columnar or prismatic structure. It is characterized by a heavy texture and high compactness. In addition, the majority of the organic and mineral colloids accumulate there. This is why it is quite plastic and shiny, and sometimes, a grey film of SiO_2 appears on the surface. The parent substrate has varying levels of salt. Beneath this horizon, there is generally groundwater. Solonetz has a A/E-Btna-C profile (Fig. 9.3). Classification at the level of varieties and forms is made according to the criteria set out in Table 6.1 (Chap. 6, Soil Classification).

Distribution. It is found at even lower altitudes than solonchak, most often on sedimentary substrates– loess-like, loamy and clayey. Vegetation mainly consists of *Camphorosma annua* Pall. and *Artemisia maritima* L. It is found among solonchaks, but does not generally occupy a large continuous area, instead appearing in spots. These soils are more widespread than solonchaks. In Serbia, they are found in Vojvodina (in Banat and to a lesser extent in Bačka and even less so in Srem), as in Fig. 9.2. In Serbia, solonetz and solonchak occupy approximately 1.3% of its land mass, i.e. about 1140.29 km^2.

Physical and chemical properties. The properties of solonetz are determined by the salinized horizon, which as a rule is characterized by the presence of highly dispersed clay particles. On the other hand, there is the eluvial horizon, poor in clay, almost structureless and relatively light in texture. The salinized horizon is generally alkaline, while the eluvial horizon has neutral chemistry. The water–air properties differ greatly, with the eluvial horizon being relatively permeable to water, while the salinized horizon swells and prevents water percolating through the profile. In its dry state, both horizons, but in particular the salinized horizon, are a highly coherent mass that cracks in the form of columns or prisms and is virtually impenetrable to root systems. The alkaline chemistry, on the one hand, and the poor water–air properties, on the other, almost completely prevent the development of vegetation, which is shown by the scarcity of natural vegetation.

Land use. The cultivation of solonetz is practically impossible. In its dry state, not even the heaviest tools can till it, while, when wet, trying to move across it is futile. Hence, without ameliorative measures, solonetz cannot be used for agricultural production.

Guidelines for management. In order to be used for agriculture, solonetz requires major changes to its physical and chemical properties. In terms of physical properties, changes should involve improving its water–air regime and increasing the filtration capacity. Ploughing must reach depths of 60–70 cm, i.e. the whole of the salinized horizon, bringing it up to the surface. For solonetz where the parent substrate is rich in $CaCO_3$ (loess or loess-like material), deep ploughing brings the parent substrate to the surface and the excavated $CaCO_3$ leads to a change in the adsorption complex of the soil. This measure is applied to steppe solonetz and is known as self-amelioration.

Improving the chemical properties should be aimed at replacing the Na ions in the adsorption complex with Ca ions and reducing pH values. This is achieved through adding gypsum, with not only the Ca ion being important, but also the anion (SO_4'') itself. Of the three basic anions (CO_3'', Cl' and SO_4''), the sulphate anion has the most positive impact on forming structural aggregates (Antić et al. 1982).

9.4 Order of Subaquatic Soils

These soils form at the bottom of water bodies (lakes and wetlands). The dominant process is that of sedimentation, in which water biocoenoses participate. Vegetation consists of associations of *Myriophylla-Potametum* Soo, while in standing water and slow moving water, there are associations of water lily, club moss, etc. These submersed communities accelerate the processes of soil formation and represent

Fig. 9.3 Solonetz (*photo* Vasin, J.)

pioneer associations in the populating of water bodies. Thanks to them, at the bottom of water bodies, subaquatic soils form, with limnic, eutrophic and calcareous varieties. In Serbia, these soils have no economic significance and only when dried out can they be used for plant cultivation, at which point these soils, in the new conditions, change drastically and lose their previous characteristics. In Serbia, they occupy very small areas. According to the classification system used in Serbia (Škorić et al. 1985), they are divided into three classes, which we will only briefly mention:

9.5 Class I: Initial Subaquatic Soils with (A)-CG Soil Horizon Sequence

9.5.1 Protopedon

This soil type is the initial stage of development of subaquatic soils on loose sediments. There is no developed humus-accumulative horizon. Under the influence of lower plants (algae) or higher plants, an initial subaquatic, poorly developed (A) horizon forms. In the (A) horizon, raw humus forms. Beneath the (A) horizon, there can be non-gleyed C or gleyed G sediment. It evolves towards gyttja.

9.6 Class II: Humic Subaquatic Soils with A-C or A-G Soil Horizon Sequence

9.6.1 Gyttja

This is the Swedish name for this soil type, and it is the next member of the evolutionary series of subaquatic soils after protopedon. In the A horizon, a clay-humus complex forms. The humus is well mixed with the mineral part of the soil, while organic matter is well decomposed and humified. There is not the rotting or unpleasant smell that sapropel has. The colour of the humus-accumulative horizon is grey to grey-black. When dry, it is good soil under agriculture.

9.6.2 Dy

This name also comes from Swedish. Unlike gyttja, this soil is acidic. It has low biological activity. In the A horizon, the coagulation of brown humic acid occurs from the water solution coming from nearby high peat soils rich in plant and animal remains. It is of no significance for forestry and agriculture.

9.6.3 Sapropel

This soil forms in anaerobic conditions in standing water which is poor in oxygen. In such conditions, organic matter rots and hydrogen sulphide (H_2S) is released along with other gases with an unpleasant odour. The soil is reddish, which comes from the presence of iron sulphide. When dry, if the sediment is non-calcareous, an extremely acidic soil results due to the iron sulphide which oxidizes into sulphuric acid.

9.7 Class III: Anthropogenic Subaquatic Soils with P-C or P-G Soil Horizon Sequence

9.7.1 Drained Gyttja

This is the first stage after the drying of gyttja. Once drained, quite a good soil results. The further evolution of these soils proceeds as per the development model for terrestrial soils.

9.7.2 Drained Sapropel

This is the first stage after the drying of sapropel. With agronomic measures, it is possible to use these soils, once dry, for plant cultivation. The further evolution of these soils proceeds as per the development model for terrestrial soils.

References

Antić M, Jović N, Avdalović V (1982) Pedology. Scientific Book, Belgrade (in Serbian)

Antipov-Karataev IM, Manaeva Ly (1958) The role of magnesium in Solonetz properties. Agrokem Talajt 7:1–14 (in Russian)

Giedrojc KK (1935) Chemical soil analysis. CET. Ed. 4th stereotypical. I. - L. Symptoms - 1935, at 533 S

Škorić A, Filipovski G, Ćirić M (1985) Classification of Yugoslav soils. p 72, Academy of Sciences and Arts of Bosnia and Herzegovina. Special edition, Book 78, Sarajevo (in Serbo-Croatian)

Živković BM, Neugebauer VK, Tanasijević Đ, Miljković N, Stojković L, Drezgić P (1972) Soils of Vojvodina. Institute for Agricultural Research, Novi Sad

Land Use

10

A proportion of the agricultural land undergoes a change in use every year. In second half of twentieth century, Serbia lost irrecoverably approximately 220,000 ha of agricultural land due to construction of industrial, mining, energy and traffic infrastructure. Projections based on existing data on the annual reduction in the amount of fertile land suggest that the loss on this basis could amount to as much as 180,000 ha over the next 30 years.

10.1 Current State and Trends of Land Use

To assess the current state and trends of land use change in Serbia, CORINE Land Cover (CLC) databases were used. The European Commission has been using the CORINE (Coordination of information on the environment) programme since its implementation in 1985 and the land cover project within this programme is intended to provide consistent localized geographical information on the land cover of European countries. As part of the Regional Environmental Reconstruction Programme (REReP) for south-eastern Europe (2001–2006), the CORINE land cover project was extended to the countries of the western Balkans, including Serbia. The CLC nomenclature described in the CORINE Land Cover Technical Guide (European Commission 1994) and a related document (European Environment Agency 1999) divides land use categories into five main classes (artificial surfaces, agriculture areas, forests and semi-natural herbaceous vegetation, wetlands and water bodies). Within each of these classes, there are numerous subclasses. A map of the main CLC classes in Serbia is presented in Fig. 10.1.

Wetlands and water bodies cover 1,103,831 ha. Inland aquatic ecosystems may be grouped into riverine (river/stream), lacustrine (lake/pond) and palustrine ecosystems, covering freshwater springs, peat bogs, fens, marshes, swamps, shrub-dominated swamps and swamp forests (Scott and Jones 1995; Karadžić et al. 2015). These ecosystems represent large transitional zones (ecotones), supporting high biodiversity. The Convention on Wetlands of International Importance was adopted in the Iranian town of Ramsar in order to protect the biodiversity of wetland habitats. Unfortunately, the percentage of such habitats is decreasing at the global level as well as in Serbia. Inland wetlands in Serbia cover 25,877 ha (Table 10.1). The greatest proportion of these sites are found in Vojvodina, along large rivers (the River Danube, Tisza and Sava), Fig. 10.2. Protected wetland habitats of special biodiversity importance include Lake Palić, Lake Ludaško, Lake Vlasina, Zasavica, Carska Bara, the forests of Gornje Podunavlje and the forests of Morović-Bosut.

Watercourses and water bodies are of great commercial importance as important water transport routes and forage fish ponds and for the irrigation of agriculture fields. Serbia's irrigation system covers 180,000 ha. The Danube-Tisza-Danube system of irrigation channels in Vojvodina enables the intensive cropping of corn, wheat and other agricultural production.

Soils along the watercourses and wetlands are very heterogeneous. Different forms of colluvial soils with coarse unconsolidated erosional deposits dominate along the banks of fast-flowing rivers in mountainous regions, while the highly productive soils such as fluvisol and humofluvisol (WRB, Fluvisol, humic) along large lowland rivers are characterized

P. Pavlović et al., *The Soils of Serbia*, World Soils Book Series,
DOI 10.1007/978-94-017-8660-7_10

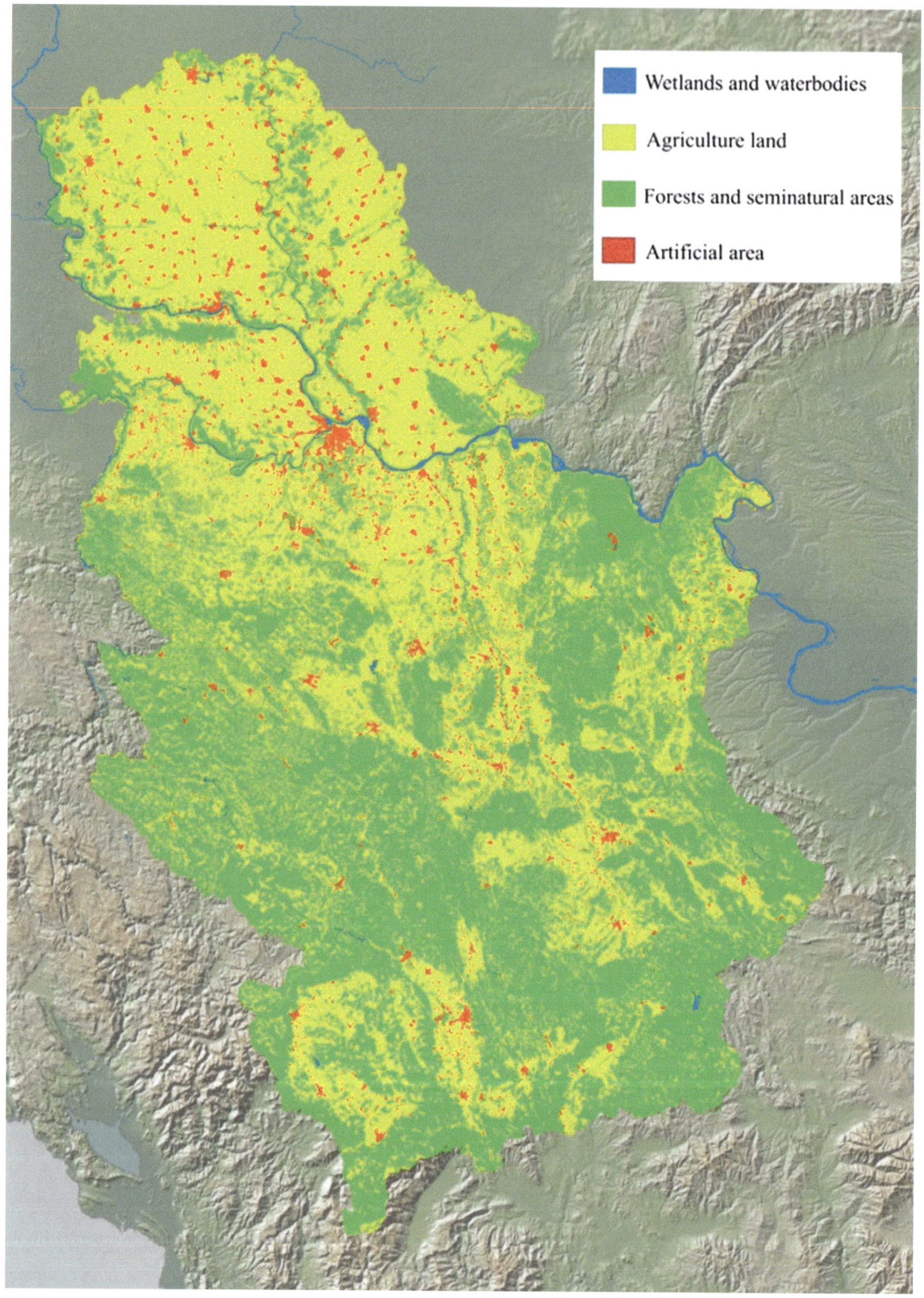

Fig. 10.1 Land use (Map of the main CLC classes)

Table 10.1 Area of wetlands and water bodies (km^2)

CLC code	Land cover type	Total
4.1	Inland wetlands	25,877
5.11	Water courses	80,301.2
5.12	Water bodies	27,494.2

by fine siltous alluvial deposits. Various types of marsh gley such as humogley (WRB, Gleysol, humic) and eugley (WRB, Gleysol) dominate in lowland and wetland areas exposed to temporary flooding, and low moore peat (WRB, Histosol) is distributed locally in swamp and lacustrine ecosystems (Carska Bara, Lake Ludaško, Lake Vlasina, Zasavica, etc.).

Forests and semi-natural herbaceous vegetation cover a relatively high proportion of Serbia (Table 10.2). Forestry is an important commercial branch, enabling economic and social development, especially in economically poor regions. Serbian forests are both state-owned and privately owned. The majority of the state-owned forests are managed by the public enterprise 'Srbijašume', while privately owned forests, outside areas under conservation protection, are generally used for commercial purposes. The commercial

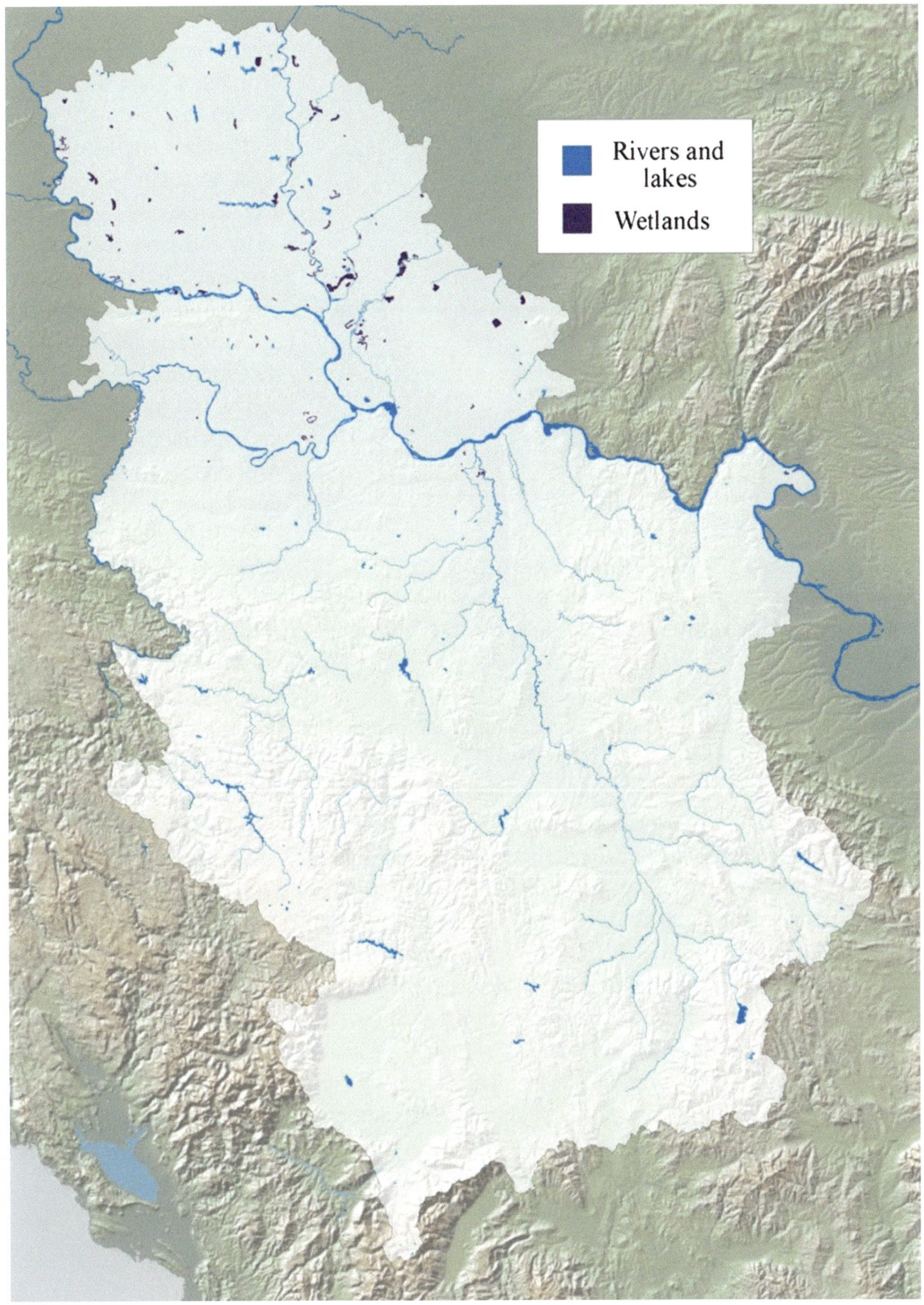

Fig. 10.2 Water bodies and wetlands

products from forestry include the production of paper and paperboard, industrial timber, construction timber and firewood. Besides a significant role in socio-economic development, forests are important 'reservoirs' of biodiversity, affecting hydrological cycles and preventing soil erosion.

A large proportion of riparian broad-leaved forests are distributed across the Pannonian Plain, along large rivers. However, the greatest proportion of broad-leaved forests is distributed in hilly and mountainous regions (Fig. 10.3). Mixed forests cover 142,369.3 ha and are distributed at higher altitudes, while diverse coniferous forests are also distributed in mountainous and alti-montane regions. A relatively large area of the Pannonian Plain is covered by planted coniferous and mixed forests. These lowland coniferous forests have been planted at Deliblato Sands and Subotica Sands to prevent aeolian erosion of arenosols.

Table 10.2 Forests and other types of semi-natural vegetation in Serbia (km^2)

CLC code	Land cover type	Area
3.1	**Forests**	
3.11	Broad-leaved forest	2,611,544.4
3.12	Coniferous forest	121,694.4
3.13	Mixed forest	142,369.3
3.2	**Shrub and herbaceous vegetation**	581,786.6
3.21	Natural grasslands	306,477.2
3.24	Transitional woodland-shrub	581,786.6
3.3	**Habitats with sparse vegetation**	
3.31	Dunes, sands	1569.8
3.32	Bare rocks	283
3.33	Sparsely vegetated area	38,721.2

Herbaceous vegetation in Serbia is represented by anthropogenic meadows, pastures and natural grasslands (Fig. 10.4).

Anthropogenic meadows are distributed in hilly and mountainous regions, and these ecosystems are primarily affected by the frequency and intensity of mowing and/or grazing. Natural herbaceous vegetation in Serbia is represented by chasmophytic, steppe and subalpine grassland communities. Chasmophytic azonal vegetation is distributed in ravine habitats, covering bare rocks, screes, lithosol (WRB, Leptosol), rendzinas (WRB, Leptosol, calcaric) and kalkomelanosol (WRB, Leptosol), while a large area of steppic vegetation is fragmentarily distributed across the Pannonian lowland. These fragments are developed on chernozem (Pannonic steppic grasslands), arenosols (Psammophytic steppic vegetation on unconsolidated sands) and halomorphic soils (salt steppes). Cryophilic grasslands are represented by alpine vegetation, growing on either shallow calcareous rendzina or acid rankers.

Agriculture areas dominate in Serbia. They are distributed mainly in the province of Vojvodina, on the large Pannonian Plain. A significant percentage of these areas are found along large rivers (the Morava river valley and the Danube valley in eastern Serbia) and on the Kosovo plateau.

Soil under agriculture (Arable land) covers 2,166,699.6 ha (Table 10.3). Different cereals (corn, wheat and oats) are planted on chernozem (in Vojvodina) as well on fluvisols and vertisols (in hilly regions and along the valley of the River Morava). Compared to cereals, plantations of both industrial crops (sugar beet, sunflower and soybean) and vegetables cover a much smaller area.

Permanent crops (vineyards, orchards and berry plantations) are distributed in hilly regions. Apples, pears, plums and sour cherries are the main fruit crops in Serbia. A relatively large percentage of vineyards are distributed in Vojvodina (arenosols in the municipalities of Subotica and Vršac).

Pastures are distributed both in mountain and alpine areas with a small percentage also distributed in Vojvodina. Heterogeneous agriculture areas (complex cultivation patterns and land occupied by agriculture with significant natural areas and abandoned crop fields) are distributed in hilly and mountainous regions (Fig. 10.5).

Artificial surfaces are the result of urbanization, the development of agriculture, industry, mining, the energy sector and transport infrastructures, which are necessary for socio-economic growth. The province of Vojvodina and the Morava river valley are regions with the most intense development of urban areas, agriculture and industry (Fig. 10.6). Continuous and discontinuous urban fabric covers more than 2,200,000 ha in Serbia (Table 10.4).

While important for socio-economic progress, these anthropogenic activities inevitably create serious environmental problems (air, water and soil pollution, a reduction in biodiversity and climate change). Industrial, commercial and transport units are distributed within large urban conglomerations and mineral extraction sites represent a particularly serious environmental problem. The largest dumps of mineral deposits are located near the Obrenovac thermoelectric power plant and within the Bor and Majdanpek, Kostolac and Kosovska Mitrovica industrial basins.

The process of biodiversity loss is marked, not only in Serbia but also in the wider region of the Balkan Peninsula. According to the Red Data Book of the Flora of Serbia (Stevanović 1999), four local endemic taxa have disappeared from Serbia, and consequently, from the global gene pool (*Althea kragujevacensis* Pančić, *A. vranjensis* Diklić and Nikolić, *Scabiosa achates* Vis. and Pančić, and *Trapa annosa* Janković). Furthermore, more than 40 vascular plants have disappeared from Serbia, but not from other parts of the Balkan Peninsula, while populations of numerous plant species have permanently declined over the past 50 years.

10.2 Changes in the Use of Agricultural Soil

Beside biodiversity loss, anthropogenic activities significantly contribute to soil degradation (induced soil erosion, reduction of soil quality, soil contamination). Between 1957 and 1993, Serbia lost irrecoverably approximately 220,000 ha of agricultural land to the construction of industrial, mining, energy and traffic infrastructure. The expansion of urban and industrial zones (urban settlements,

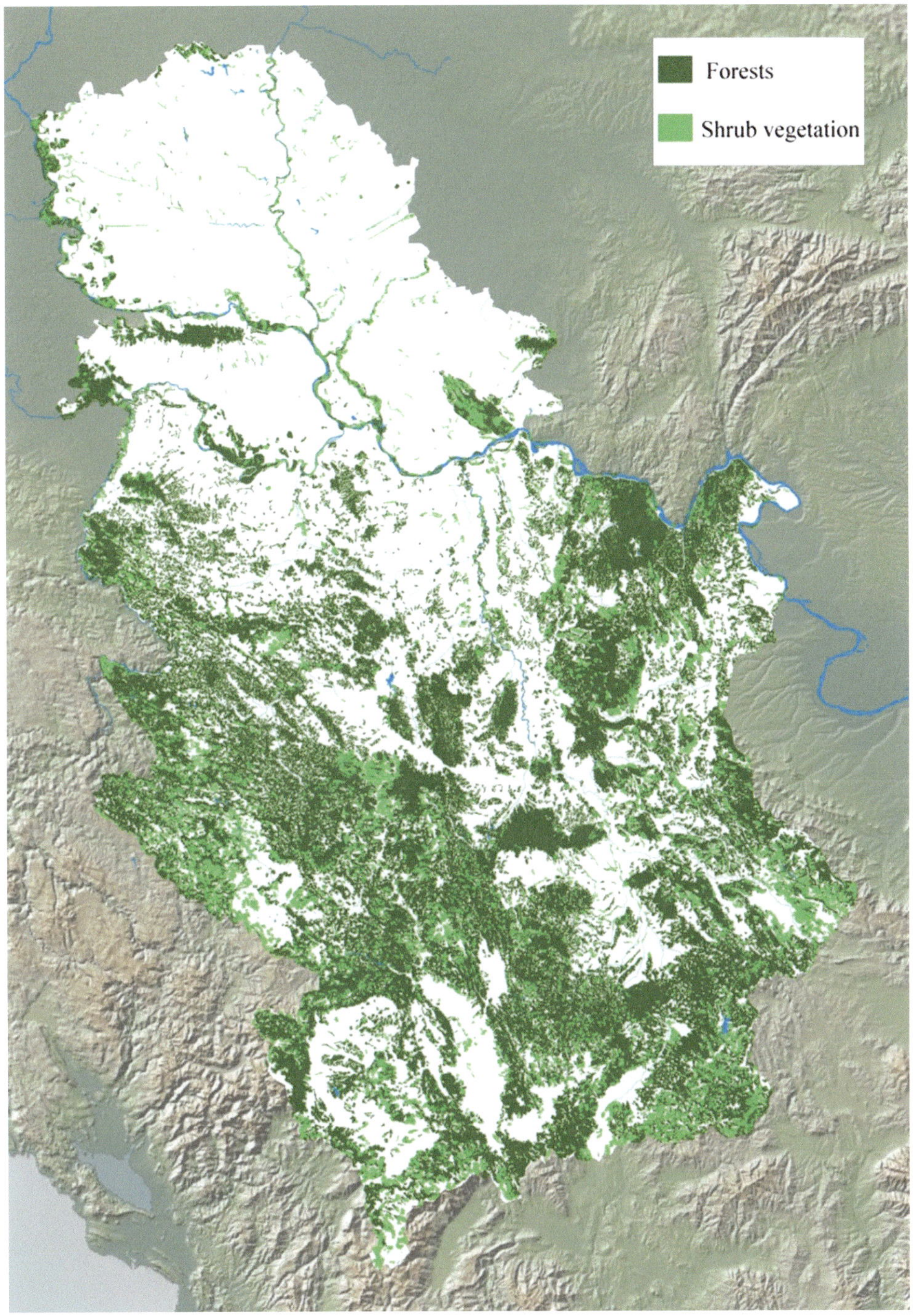

Fig. 10.3 Forest and shrub vegetation

roads, factories, mines, hydroelectric power plants, airports, borrow pits, etc.) usually incorporates flat, fertile land. In recent years, the construction of residential blocks in larger towns and the building of industrial zones and industrial parks in the immediate vicinity of towns have intensified, particularly in Vojvodina. The construction or widening of major roads, such as motorways and regional highways, takes up yet more land. For example, as a part of the pan-European Corridor X, the E-75 motorway occupies an additional 566 ha of land and the E-80 motorway an additional 1022 ha (http://www.putevi-srbije.rs/pdf/koridorxsrp.pdf 2009). The result of these processes is a further reduction in the percentage of agricultural land in the Republic of Serbia. For example, in 1990 it totalled 66.2%, but in 2010, it had been reduced to 65.7% (without the Autonomous Province of Kosovo and Metohija). The loss of agricultural

Fig. 10.4 Grassland areas

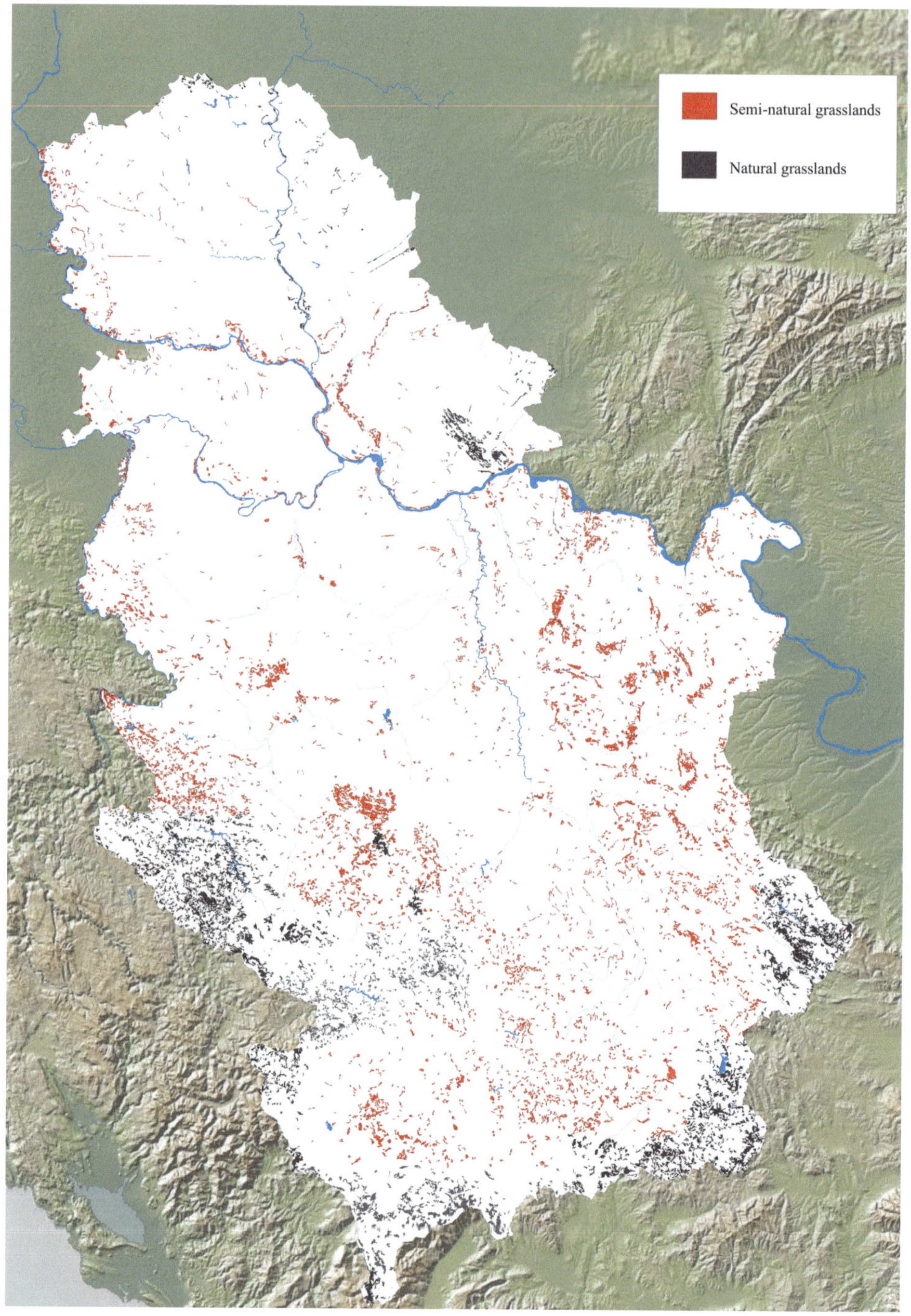

Table 10.3 Soils under agriculture

CLC code	Land cover type	Area (km^2)
2.1	**Arable land (Soil under agriculture)**	2,166,699.6
2.2	**Permanent crops**	
2.21	Vineyards	17,652.9
2.22	Fruit trees and berry plantations	9602.8
2.3	**Pastures**	187,598.1
2.4	**Heterogeneous agriculture areas**	
2.42	Complex cultivation patterns	1,432,080.5
2.43	Land occupied by agriculture with significant areas of natural vegetation	1,162,615.8

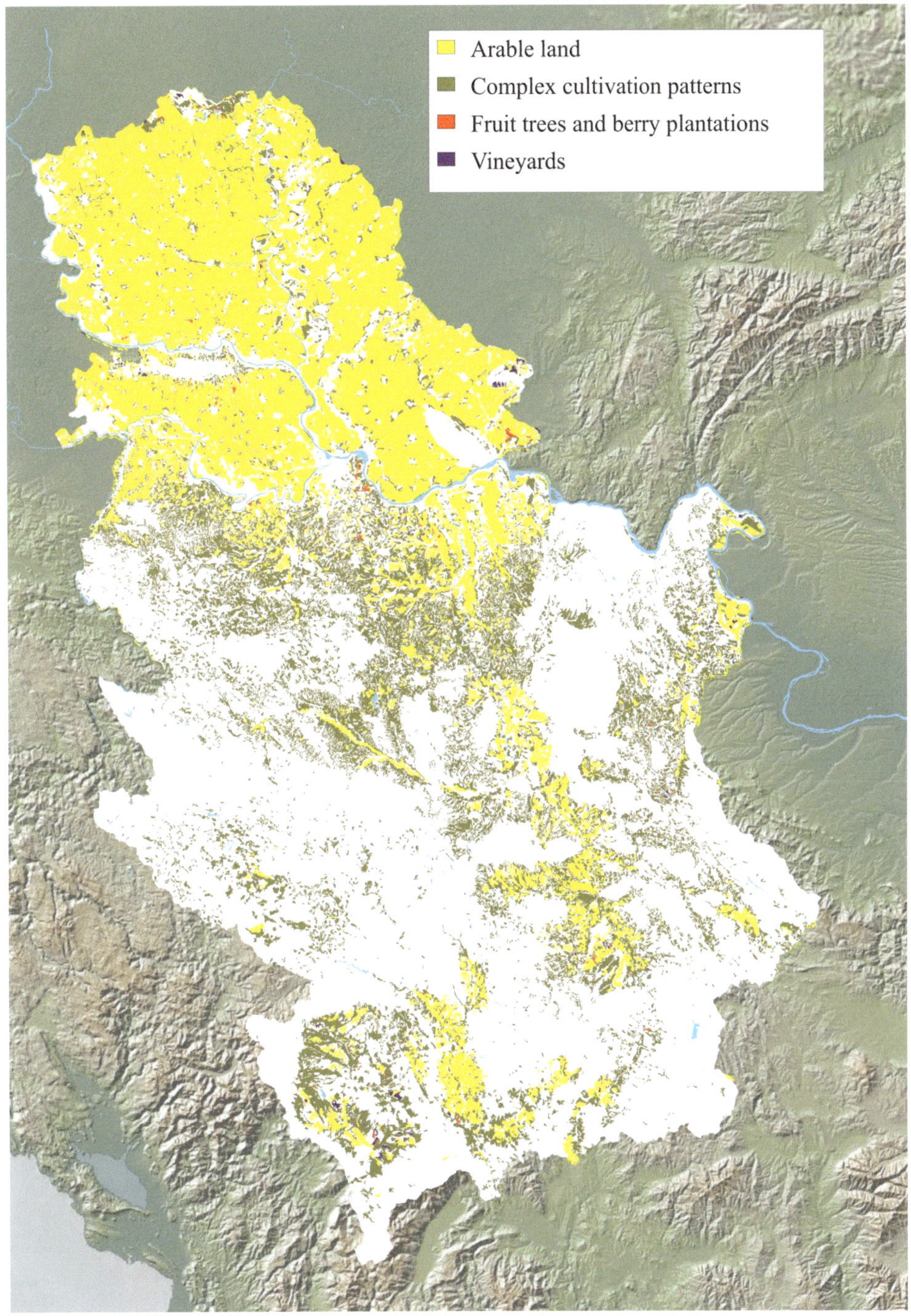

Fig. 10.5 Agricultural areas

land on a daily basis amounts to approximately 16.1 ha or around 5876.5 ha annually. Projections based on existing data on the annual reduction in the amount of fertile land suggest that the loss on this basis could amount to as much as 180,000 ha over the next 30 years (National Strategy of the Sustainable Use of Natural Resources 2012).

When it comes to changes in land use, current trends suggest that by 2020 agricultural land in Serbia will have been reduced by 50% (Kadović 2013). The latest data show that in the last twenty years, the amount of agricultural land has decreased by as much as 39,074 ha. Since 1990, pastures have disappeared to the greatest extent (164,583 ha),

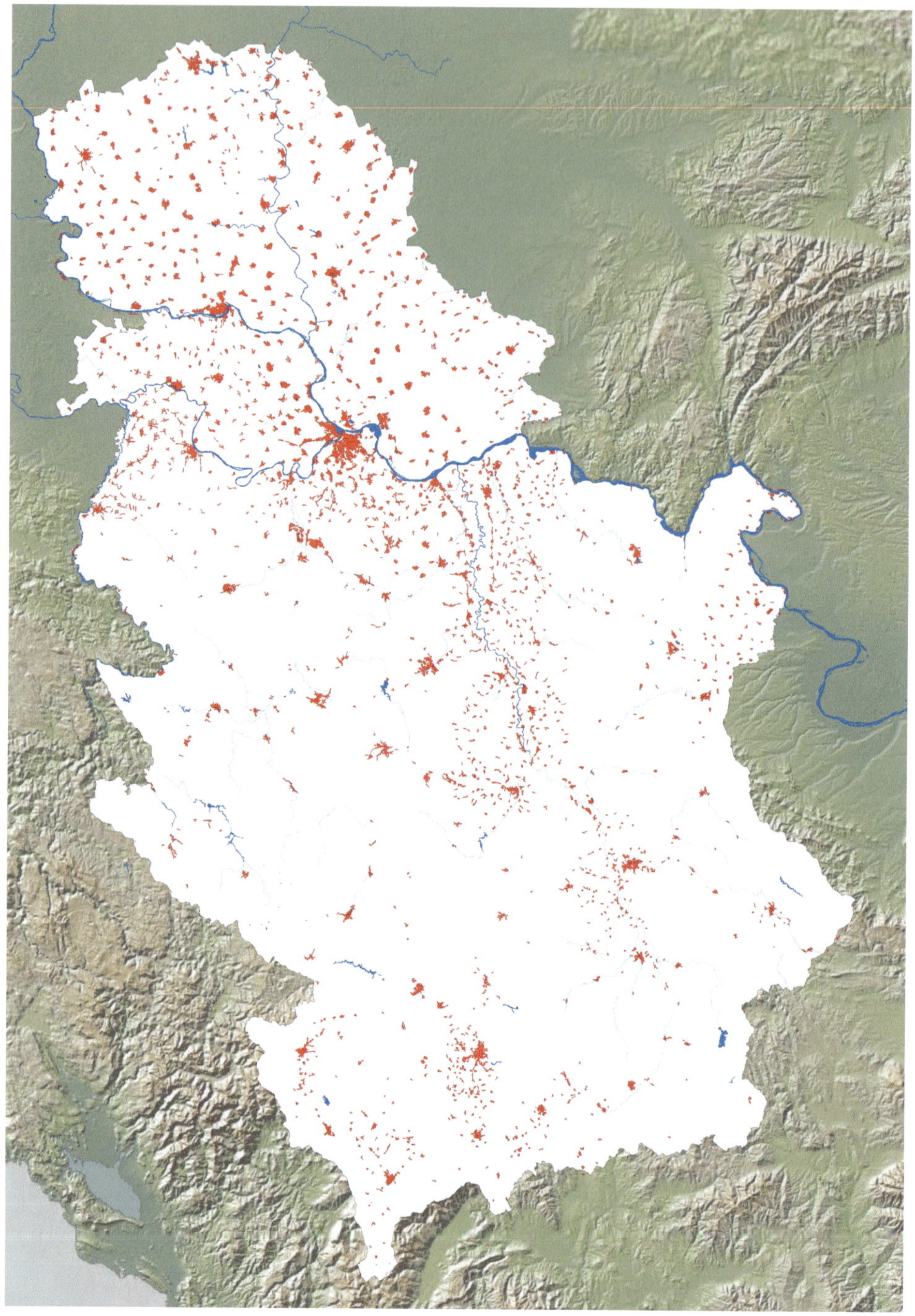

Fig. 10.6 Artificial areas

followed by vineyards (29,878 ha) and orchards (18,529 ha) (Statistical Office of the Republic of Serbia 2013). At the same time, forested areas and semi-natural regions have increased by 1975 ha, which is the result of the afforestation of agricultural land and the expansion of areas covered by coniferous and mixed forests (Kadović 2013). Furthermore, the area of wetlands, characterized by the category of inland marshes, has decreased by 119 ha, while the area of water bodies has risen by 2343 ha, mainly due to the construction of new, artificial lakes.

Table 10.4 Artificial surfaces

CLC code	Land cover type	Area
1.1	**Urban fabric**	
1.11	Continuous urban fabric	239.8
1.12	Discontinuous urban fabric	2,241,870.2
1.2	**Industrial, commercial and transport units**	
1.21	Industrial or commercial units	21,024.5
1.22	Road and rail networks	1205.8
1.23	Port areas	363
1.24	Airports	1901.8
1.3	**Mine, dump and construction sites**	
1.31	Mineral extraction sites	96,236.4
1.32	Dump sites	2096.9
1.34	Construction sites	59
1.4	**Artificial, non-agricultural vegetated areas**	
1.41	Green urban areas	34,737.6
1.42	Sport and leisure facilities	2345

References

European Commission (1994) CORINE land cover technical guide, EUR 12585 EN, OPOCE Luxembourg

European Environment Agency (1999) CORINE land cover—a key database for European integrated environmental assessment. European Environment Agency, Belgium

Kadović R (2013) The potential of Serbian forests for climate change mitigation and environmental protection. In: Andjelković M (ed) Energy and environment, Scientific meetings, Book CXLIII, Serbian Academy of Science and Arts (SANU), pp 493–514

Karadžić B, Jarić S, Pavlović P, Mitrović M (2015) Aquatic and wetland vegetation along the Sava River. In: Milačič R, Ščančar J, Paunović M (eds) The Sava River, Springer, pp 249–316

National Strategy of the Sustainable Use of Natural Resources (2012) Official Gazette of the Republic of Serbia, no. 33/2012 (in Serbian)

Scott DA, Jones TA (1995) Classification and inventory of the world's wetlands. Vegetatio 118:3–16

Statistical Office of the Republic of Serbia (2013) Census of agriculture 2012—Agriculture in the Republic of Serbia—Labour force and activities of agricultural holdings in the Republic of Serbia, Belgrade (in Serbian)

Stevanović V (ed) (1999) The red data book of the flora of Serbia 1. Serbian Ministry of Environmental Protection, Faculty of Biology, University of Belgrade and Serbian Environmental Protection Agency (in Serbian)

11 Environmental Issues

A significant proportion of the land area of the Republic of Serbia occupied by areas prone to erosion, landslides, surface exploitation of mineral resources and dump sites can be classified as degraded. The basic natural processes of soil degradation in Serbia are as follows: water and wind erosion; the loss of organic material, compaction due to an increase in volumetric weight and a decrease in soil porosity; salinization through the accumulation of soluble salts in soil; landslides, i.e. the moderately fast or fast movement of a mass of soil and rock down slopes. However, various human activities connected to the exploitation of natural resources result in its immediate loss as well as pollution and degradation, such as mineral extraction activities, industry, borrow pits for industrial requirements and the extraction of gravel and sand at the river banks.

To a great extent, the problems of environmental issues in the Republic of Serbia are the results of outdated technology and equipment, as well as low energy and raw material efficiency, which is a consequence of a lack of financial resources for improving the existing industrial infrastructure dating back several decades.

The main sources of soil pollution due to pollutant gasses (SO_2 and NO_x) and heavy metals in Serbia are the processes involving the combustion of fossil fuels, in particular at thermoelectric power plants, followed by the processing of sulphide ores, lead, zinc and copper ores, the production of sulphuric acid and paper, oil refineries, and of course transport (internal combustion engines) (Figs. 11.1 and 11.2).

Inappropriate agricultural practices also have a negative effect on soil quality, including the uncontrolled and inadequate application of artificial fertilizers and pesticides.

Various applications of coal combustion, production and application of phosphorus fertilizers, mining industry and formation of radioactive waste dumps contribute to redistribution of natural radioactivity. A military operation, undertaken by the North Atlantic Treaty Organization (NATO) against Yugoslavia in 1999, caused environmental contamination with depleted uranium of southern parts of Serbia (UNEP 2002).

11.1 Soil Degradation

The basic natural processes of soil degradation in Serbia are as follows: water and wind erosion; the loss of organic material, compaction due to an increase in volumetric weight and a decrease in soil porosity; salinization through the accumulation of soluble salts in soil; landslides, i.e. the moderately fast or fast movement of a mass of soil and rock down slopes. However, various human activities connected to the exploitation of natural resources from the land result in its immediate loss, such as mineral extraction activities, industry, borrow pits for industrial requirements and the exploitation of gravel and sand at the edge of rivers.

Approximately 20% of the land area of the Republic of Serbia occupied by areas is prone to erosion, landslides and surface exploitation of mineral resources, and dump sites can be classified as degraded. During 2008 alone, approximately 122,100 ha of land was eroded (Statistical Office of the Republic of Serbia 2008). According to current estimates, 37,300 m^3 of eroded materials produced in Serbia each year, 9350 m^3 (25%) of which is deposited in reservoirs or other water ecosystems (National Program of Environmental Protection 2010).

11.1.1 Water and Eolian Erosion

Soil erosion processes relate to its permanent loss, its physical degradation, or deterioration in its chemical quality

P. Pavlović et al., *The Soils of Serbia*, World Soils Book Series,
DOI 10.1007/978-94-017-8660-7_11

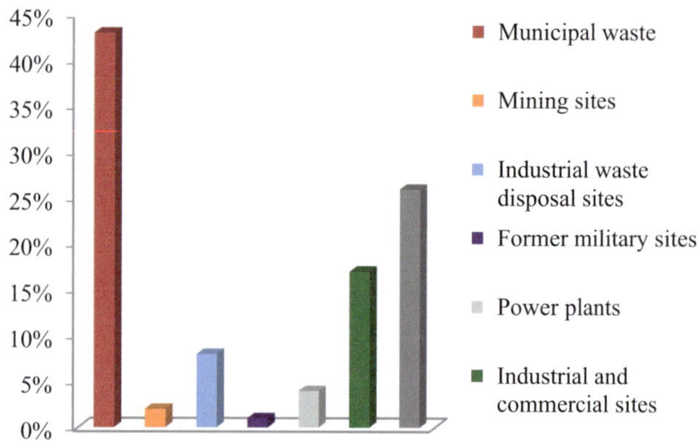

Fig. 11.1 Soil polluting activities from localized sources (% total sites; Vidojević and Manojlović 2006)

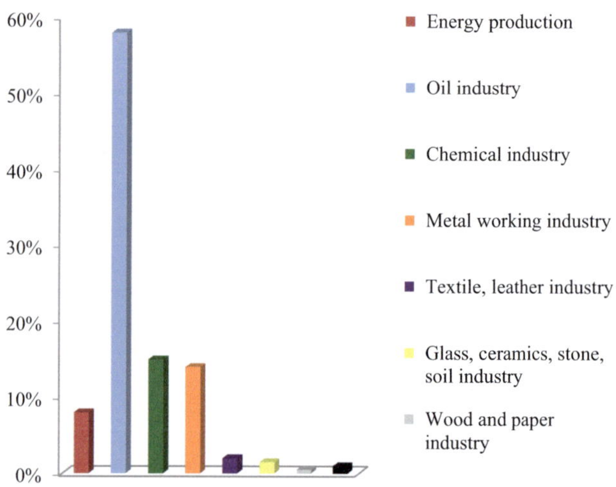

Fig. 11.2 Industrial and commercial branches responsible for local soil contamination (% of total; Vidojević and Manojlović 2006)

due to the activity of water, glaciers or wind (water, glacial, eolian erosion). In many European countries, including Serbia, with the marked extremes in terms of relief and climate, erosion is one of the basic and most frequent causes of soil loss and degradation, making it the primary factor of those which threaten soil as a natural resource.

Serbia is predisposed to erosion processes considering that approximately 70% of its total surface area is found in hill and mountain regions (Zlatić et al. 2008; Ristić et al. 2012), where soil degradation mainly occurs due to rockfalls, rockslides, and debris flows. About 25–30% of land in Serbia is potentially unstable, while approximately 8–10% is subject to active slide processes. In terms of surface area of degraded soil, by far the most important factor in soil degradation in Serbia is water erosion (Table 11.1), followed by eolian erosion.

According to the Erosion Map of Serbia compiled between 1966 and 1971 (1983), 76,354.43 km^2 or 86.39% of the land area of Serbia is affected by varying the intensities of erosion, while the zone of accumulation covers 12,024.71 km^2 (13.61%).

Soil erosion affects up to 80% of soils under agriculture in Serbia. Water erosion is predominant in the central regions and the hilly and mountainous regions, while wind erosion is the primary cause of soil degradation in Vojvodina, affecting 85% of soils under agriculture with an annual loss of over 0.9 tonnes of material per hectare (Cooper et al. 2010). In fact, this soil is not lost, but is relocated (transferred) to faraway areas and the degradation is reflected in the displacement of parts of fertile surface horizons which are affluent in nutrients (Figs. 11.3 and 11.4).

11.1.2 Soil Acidification

Acid soils lead to many unfavourable soil characteristics when it comes to plant nutrition. Because of increased soil acidity, the violation of soil buffering properties occurs due to the leaching of Ca and Mg ions, which can also cause physical soil degradation via the peptization of colloids. Together with increasing soil acidity, the content of mobile Al increases, which can be toxic for plants. Readily available nutrients transform into forms with poor availability (for example, phosphorous). On the other hand, the higher solubility of some microelements (Zn and B) can bring about their accelerated leaching from the root zone and, therefore, results in their deficiency for plant nutrition (Mrvić 2010).

Table 11.1 Water erosion (category I = highest intensity, category V = lowest intensity of erosion)

Region	Area (km^2)	Category of erosion				
		I	II	III	IV	V
Vojvodina	21,206	48	336	947	15,193	4982
Western Serbia	14,902	578	2149	4870	5394	1911
Central Serbia	11,180	110	1407	2935	4504	2224
Eastern Serbia	15,009	629	1789	4448	6840	1303
Southern Serbia	14,877	1050	2060	3132	7403	1232
Kosovo and Metohija	10,887	473	1426	2972	4711	1305
Total	88,361	2888	9138	19,386	43,914	13,035

Fig. 11.3 Erosion of agricultural soil near Krupanj, Western Serbia (*photo* Ristić, R.)

Natural acid soils are widespread, accounting for over 60% of the total arable land, mostly in central Serbia. Approximately, 43% of the area can be classed as very acid to acid soils, while acid to slightly acid soils account for 20%, with only 35% being categorized as slightly acid and neutral soils (National Strategy of Sustainable Use of Natural Resources 2012). Over the last few decades, the decreased input of organic matter and the application of only mineral (nitrogen) fertilizers have accelerated soil acidification processes. Particularly endangered are the areas found in south-eastern Serbia, Šumadija, the Kolubara Basin, the Jadar valley and Pocerina, as well as areas around Leskovac. These are mostly lowland or hillside types of pseudogley or their leached variants, acid vertisols, podzolic eutric cambisols, and diluvial, brown or leached brown soils of mountainous regions, which are rather poor in bases, moderately to heavily acidic, with very poor texture and low organic content. They are all suitable for the cultivation of most cereal crops (Jakovljević et al. 2005; Kresović et al. 2009; Đalović et al. 2010).

In terms of forest soils, they are also found in those areas of central and western Serbia already mentioned. Beech forests dominate in the forest reserves of Serbia, accounting for 60% of the volume of trees and 31% of the surface area they occupy. Coppice forests and shrubs comprise 52.56% of the total area populated by beech (Medarević et al. 2003). Such a high proportion of coppice and degraded forests, which are often unable to make use of the production potential of the habitat, meant that the need arose for the melioration of these forests in order to improve their quality. For this reason, the mid-twentieth century witnessed the start of a program of intensive afforestation in western Serbia with the planting of forest plantations (silviculture) which had unsatisfactory results. According to data from the National Inventory of Forests, the surface area of silviculture in Serbia today totals 174,800 ha, 71.4% of which comprises coniferous plantations (European black pine, spruce, Douglas fir, eastern white pine, etc.) (Banković et al. 2008). When it came to choosing coniferous species for the melioration of degraded forest areas and for the afforestation of barren land, their productive capabilities were of the greatest importance. However, they were also planted in localities that, due to ecological conditions, would have been inhabited naturally by deciduous species. Hence, for example, Norway spruce (*Picea abies* (L.) Karst.), which was planted at mid and low altitudes, found itself completely

Fig. 11.4 Erosion of forest soil near Krupanj, Western Serbia (*photo* Ristić, R.)

outside its natural distribution area, occupying some of the best habitats for deciduous species in Serbia. The long-term cultivation of conifers in a deciduous habitat has led to a reduction in biodiversity, as well as changes in the soil, most pronounced in the top soil layer. There were found to be lower soil moisture levels, lower active and substitutional acidity, the depletion of the adsorption complex in base cations, and lower levels of N, P and K in the spruce stand in relation to the beech stand. The higher C/N ratio of spruce litter caused its lower decomposition rate in comparison with beech litter. All these changes have led to degradation and a reduction in this ecosystem's productivity (Kostić 2008; Kostić et al. 2012a).

In addition, industrial activity causes localised soil acidification in the vicinity of large industrial complexes with the major anthropogenic sources being thermoelectric power plants (combustion of fossil fuels—coal, oil and natural gas), smelters for metal production and traffic. For example, the thermoelectric power plants in Obrenovac, Lazarevac, Kostolac and Obilić, as well as the smelters in Bor, Smederevo and Trepča, emit aerosols and pollutant gases, above all CO_2, SO_2 and NO_x, which contribute to the occurrence of localized acid rain. In Bor and the surrounding areas, acid rainfall has a very low pH (pH 2–3). The emission of pollutant gasses and particles will continue to be one of the most significant forms of soil degradation in the future, too. The areas in Serbia where a reduction in pH is predicted in the future are those areas with the greatest excesses of sulphur deposition between 1980 and 2009, dominated by dystric, shallow and skeletoid soils, from rankers to dystric cambisol. Previous research identified these regions as areas vulnerable to the deposition of pollutants, leading to possible ecosystem damage (Kadović and Belanović 2009; Belanović et al. 2012). By comparing pH values measured in the field during research and those simulated by the VSD model, model sensitivity was noted for pH values between pH 5.6–6.7 (Perović et al. 2012). Acid rain has an impact on the Ca and Mg contents in soil, which leads to them being leached from the soil and a cation exchange imbalance in the soil. Soils with a pH 5.0–6.0 are potentially the most sensitive to the process of acidification due to their poor buffering capacity, particularly those with a pH between 5.5 and 6.5 with a low cation exchange capacity. Particularly hazardous is Ca deficiency in soils, since with an insufficiency of Ca, plants resorb Al ions which are normally present in the insoluble non-toxic form of aluminium hydroxide. It appears that when soil pH is <5, Al ions dissolve in water and become toxic to plants. Aluminium ions cause a stunting of root growth and prevent roots from taking up calcium. The result may be the overall retardation of plant growth (Mrvić et al. 2009).

11.1.3 Depletion of Organic Matter in Soils

The content of organic matter in soil is very sensitive to changes in land use, yet recently changes have increasingly been linked to climatic changes too. Soil organic matter (SOM) contributes to overall soil fertility and is a source and reserve for plant nutrients. As the most important component of SOM, humus is stable, but its reduction in soil is the result of the effects of various ecological and anthropogenic factors.

In Serbia, the content of organic matter in soils varies. It is low in the soils of dry regions, high in soils of the moderate climatic region and extremely high in organogenic and peat soils. Anthropogenic degradation also occurs and is most often a consequence of agricultural production methods which involve the irregular use of organic fertilizers, the application of only mineral (nitrogen) fertilizers which may be insufficient or even excessive, the burning of crop residues and the cultivation of plants with high biological potential, which generally leads to the depletion of SOM and plant nutrients, while the effect of climatic changes on humus content is significantly less.

Over the last few decades due to economic deterioration, mainly because of the high costs associated with fertilization (fuel and equipment) as well as the reduction in the number of livestock, the importance of applying organic fertilizers, primarily manure, has been marginalized. Based on the example of chernozem in Vojvodina, where soil analyses were carried out in the 1990s (a project undertaken by the Serbian Ministry of Agriculture, Forestry and Water Management, entitled 'Control of soil fertility and levels of harmful and hazardous substances in the soils of Serbia') followed by research from 2002 to 2004 at the same locations, it was found that organic matter had fallen by 0.05–0.2%. Admittedly, this reduction in the organic matter in chernozem is certainly negligible when compared to that in the acid soils of central Serbia (ranker, ranker and eutric cambisol, dystric cambisol and pseudogley). These soils have considerably less organic matter (on average 1–2% humus) as they are found in areas characterized by a wetter climate. The higher levels of precipitation contribute to intensified leaching of bases from the organic complex, increased acidity and decreased stability of the organic complex, which is hence more prone to mineralization. The Serbian Ministry of Agriculture and Environmental Protection monitors the state and quality of soils across the country. Research into soil fertility in Serbia showed the following: the content of organic carbon (OC) in the surface layer of soil was between 1 and 2% (45.13%), between 2 and 5% (29.01%) and between 0 and 1% (19.41%) (Environmental Protection Agency 2007). The most recent data suggest that fertility is improving: the majority of soil samples have an average 1–2% humus (62.78%), fewer samples contain 2–5% humus (29.22%), while only 6.59% of soils have humus content <1% (National Strategy of Sustainable Use of Natural Resources 2012).

11.1.4 Soil Salinization

Natural or primary salinity results from the accumulation of soluble salts in soils or groundwater over long geological periods, mainly by the weathering of parent minerals, which releases salts of various types, such as chlorides, sulphates, carbonates and bicarbonates of sodium, magnesium and calcium, whereas the occurrence of so-called secondary salt-affected soils is becoming ever more visible due to the application of different agricultural practices, mainly irrigation (Fig. 11.5). However, there are human influences besides irrigation that lead to the adverse effects of secondary salinization, such as overgrazing, deforestation in semi-humid and semi-arid areas, contamination by chemicals and the accumulation of airborne or waterborne salts (Dajić 2006; Pessarakli and Szabolcs 1994). In Serbia, salinized and alkaline soils are mainly localized in Vojvodina (233,000 ha), and the majority of these soils are cultivated apart from two types (solontchak and solonetz) with exceptionally low production value due to the presence of salts (National Strategy of the Sustainable Use of Natural Resources 2012).

Only 2% of the soils of central Serbia are classified as alkaline soils. Unfortunately, the problem of the salinization of other types of soils under agriculture in Vojvodina, like chernozem, is becoming more and more common due to intensive watering. However, there is also the risk of secondary salinization of hydromorphic soils through groundwater, which, when at its highest during winter and spring, rises into the surface layers (into the adsorption zone of plants' root systems) and deposits salts, hence salinizing them permanently. With salinized soils occupying a huge area in Serbia, this problem is a significant one, particularly as it concerns large cultivable areas of naturally fertile land, which are often the only source of income from agricultural production for the population of these parts. Hence, this problem deserves attention and salt-tolerant cultures should be grown on such soils rather than salt-sensitive species. Salt-tolerant plant cultures include *Brassica napus* L. (canola), *Asparagus officinalis* L. (asparagus), *Hordeum vulgare* L. (barley), *Avena sativa* L. (oats), *Secale cereale* L. (rye) and *Beta vulgaris* L. (sugar beet), whereas salt-sensitive cultures such as *Phaseolus vulgaris* L. (bean), *Daucus carota* L. (carrot), *Allium cepa* L. (onion), *Pastinaca sativa* L. (parsnip), *Fragaria vesca* L. (strawberry), *Malus sylvestris* L. (apple), *Prunus domestica* L. (plum), *Pyrus communis* L. (pear), *Prunus persica* L. (peach) and *Rubus* spp. should be avoided (Maas 1990; Francois and Maas 1994; Dajić 2006).

Vegetation on salty soils with poor physical characteristics such as solonetz and solontchak which have no agronomic potential or whose amelioration, is very expensive and occurs in moderately wet, flat areas which flood in spring and dry out in summer, meaning they are extremely dry habitats. They are also affected by the micro-relief of salt marshes, occurring in the best drained areas of salt marsh pastures in Vojvodina. These salty steppes and salt marshes are of great importance for maintaining the biodiversity, as they are unique habitats for halophytic species and endemic associations. In the past, these habitats were used as rich pastures for livestock grazing. Therefore, the Pannonian salt steppes and salt marshes are highly endangered and according to EEC directives (Habitats Directive 92/43/EEC within the Natura 2000 network) have been designated as priority

Fig. 11.5 Negative effects of irrigation (*photo* http://www.zdravasrbija.com/lat/Zemlja/Ratarstvo/1622-Negativni-efekti-navodnjavanja.php)

conservation habitat types. Some of them are special nature reserves such as Okanj Lake in the municipality of Bečej (protected by 1st, 2nd and 3rd degree protection regimes) and the 'Slano Kopovo' Special Nature Reserve, which is located north-east of Novi Bečej and the River Tisa.

11.1.5 Mining Activity and Industry Effects on Soils

With the extraction of mineral resources, whether at underground or surface sites, the soil becomes completely degraded. Surface coal mining leads to the immediate loss of soil, and with the excavation of overburden (the layer of material lying above the coal) deposol is formed, which is deposited on the soil surface or in parts of the excavated area. This form of degradation is particularly pronounced in the Kolubara and Kostolac basins where the layers of lignite lying below a surface layer of the best quality lowland soils are excavated, as well as in Bor and Majdanpek, where the underground extraction of mineral-rich ore, primarily copper and gold, takes place.

In the Bor Basin, geological research dates back to 1897 and ore extraction to 1902 in several of the mines located in this Basin: the underground mine known as 'Jama', which has been in continuous operation since 1902 with record production levels between 1996 and 1998 when up to 1.9 million tons of ore was extracted annually, the 'Veliki Krivelj' mine, where from 1983 to the present day approximately 194.6 million tons of ore has been extracted with around 178.8 million tons of waste and the Mali Krivelj—Cerovo ore field where approximately 20 million tons of ore was excavated between 1993 and 2002 (http://rtb.rs/).

The first reserves of copper ore in Majdanpek were discovered at the end of 1953 and from then until 2010, 331,108,859 t of copper ore and 776,697,772 t of waste were extracted from the 'Južni revir' (Southern Mining District) open pit and 37,946,339 t of copper ore and 141,783,047 t of waste from the 'Severni revir' (Northern Mining District) pit (http://rtb.rs/rtb). The mining activity resulted in the degradation of large areas of agricultural and forest land (Fig. 11.6).

In the Kolubara Coal Basin, underground coal mining began in 1896 with opencast mining beginning in 1950. From 1950 to the present day, a total of 1,011,624,912 t of lignite has been extracted with 2,176,323,550 m^3 of overburden (http://www.rbkolubara.rs). The Kostolac Basin encompasses the right bank of the River Danube, from the confluence with the Velika Morava to the Ram-Gorica hills in eastern Serbia. Coal exploration dates back to 1870 and excavation to 1873 when the first 7500 t were produced. Up to 2010, the opencast mines in the Kostolac Basin had produced 9,000,000 t of coal.

Opencast coal mining currently occupies approximately 12,000 ha in Serbia, with excavation sites swallowing up a further 200 ha each year. The resulting deposol is mainly found in areas which were covered by fertile alluvial soils.

Today, both regions can be considered ecological black spots since recultivation of these areas has been virtually non-existent since the 1990s (Fig. 11.7). Fertile, agricultural land has been virtually destroyed, and areas of deposol have no production potential.

The newly excavated areas appear to be greatly degraded, and the spontaneous vegetation which appears after a period of several years only partially improves the picture. Therefore, without the revitalization of deposol and the fly ash deposits of thermoelectric power plants, the technology which is currently used for the excavation and exploitation of lignite will lead to numerous ecological, geomorphological and hydrological changes.

Approximately 40,000 ha of land has been directly polluted and degraded by open-pit mines and spoil tips in the major mining basins in Serbia. Excavation of non-metal and non-ferrous ores occupies around 53% (13,500 ha) of the total area destroyed by such activities in Serbia, with mining in Majdanpek and Bor accounting for the greatest proportion of this. Namely, the area of damaged and destroyed agricultural land in the municipality of Bor is estimated to total approximately 60.6% of the total agricultural land, i.e. around 1110 ha in Bor and 12,060 ha in Majdanpek (The Privatization Agency, Republic of Serbia 2006).

Opencast coal mines in Lazarevac and Kostolac occupy approximately 11,000 ha, while the deposits of ash and slag resulting from coal combustion take up around 1500 ha of fertile agricultural land. The largest of these sites are at the 'Nikola Tesla A and B' thermoelectric power plant in Obrenovac (more than 800 ha, Fig. 11.8), while the ash deposits in Kostolac spread across approximately 250 ha. In addition, in Vojvodina there are also temporary disposal pits for drilling fluid from oil wells—600,000 m^3 of drilling mud has been deposited at the disposal site in Novo Miloševo (Waste Management Strategy 2010–2019, 2010).

A specific form of degradation took place during the NATO bombing campaign when soils were mechanically damaged by powerful bombs exploding and polluted by depleted uranium, oil derivatives and toxic substances released from destroyed industrial plants (heavy metals, above all mercury and cadmium, ammonia, hydrochloric acid, raw materials and finished products containing polyethylene and polyvinyl chloride, transformer PCB and pyralene oil, polychlorinated dibenzodioxins, furans and polychlorinated biphenyls).

Fig. 11.6 Bor copper mine (*photo* http://rtb.rs/)

11.1.6 Disposal of Industrial and Municipal Waste

In opencast and underground coal mines, a major problem are the deposits of ash and slag, and mine, metallurgical and flotation tailings, which unfortunately are frequently found on fertile agricultural land. Moreover, the discharge of waste water from flotation and tailing processing plants degrades the soil in industrial zones and the valleys of rivers which flow next to industrial deposit sites (for example, the Veliki Timok River next to Bor). Waste water from flotation at the copper mine in Majdanpek pollutes the Veliki Pek River, while the waste water from the crushing plant and heavy vehicle repair workshops pollute the Mali Pek River (The Privatization Agency, Republic of Serbia 2006).

Soil quality is threatened by the inadequate disposal of industrial waste, but also by the complete absence of planned recultivation in the industrial basins. The problem lies in the fact that during ore excavation the amount of overburden is several dozen times greater than the ore produced, and hence, compared to coal mining, soil degradation is considerably higher as areas of deposol are for the most part devastated and without vegetation.

The inadequate management of waste and chemicals causes soil degradation by occupying areas of land and through the emission of harmful and toxic materials which infiltrate the soil profile. A large amount of waste material is generated by mining and the technological preparation of the ore for metallurgical and chemical processing, with the highest proportion produced by the chemical industry

Fig. 11.7 Kostolac coal basin (*photo* http://www.te-ko.rs)

Fig. 11.8 Fly ash deposit at 'Nikola Tesla A' thermoelectric power plant in Obrenovac (*photo* Pavlović, P.)

(37.6%) and primary metallurgy (29.1%) (National Strategy of the Sustainable Use of Natural Resources 2012).

In addition, the deposition of large amounts of industrial waste originating during the combustion of lignite in thermoelectric power plants is also a significant cause of soil degradation. Namely, in Serbia, 8 power plants ('Nikola Tesla A', 'Nikola Tesla B', 'Kolubara', 'Kostolac A', 'Kostolac B', 'Morava', 'Kosovo A' and 'Kosovo B') with a total capacity of 5171 MW provide 70–75% of Serbia's annual electricity requirements, burn low calorie lignite and produce approximately 8 Mt of ash per year (Electric Power Industry of Serbia, Technical Report 2010). The 'Nikola Tesla A' power plant (which comprises six generation units with a combined capacity of 1652 MW) and 'Nikola Tesla B' (which comprises two generation units with a combined capacity of 1240 MW) produce between 15 and 16 billion KWh of electrical energy annually, which is a third of the total capacity of Serbia's electric power system. In doing so, they burn around 25 Mt of lignite and produce 4–4.7 Mt of ash, which is deposited at ash deposit sites covering a total area of approximately 800 ha (Popović et al. 2013). Considering the fact that these two power plants are in the immediate vicinity of Belgrade, their operations are both directly and indirectly endangering the lives of almost 2 million people.

If not used, ash is treated as waste. Unlike in developed industrial countries where the utilization of ash varies from 84% in Japan, 70% in China and 56% in the European Union, to 32% in the USA (ACAA 2001; Cao et al. 2008), only 2.7% of the ash is used in Serbia, mainly in the cement industry, while the majority of the ash is deposited in the open air. As such, to date 250–300 Mt of ash has been deposited at sites which occupy over 1500 ha of cultivable land (Životić et al. 2012; Pavlović and Mitrović 2013; Popović et al. 2013).

The total amount of non-hazardous waste generated in the Republic of Serbia in 2009 was 2,374,374 t (Table 11.2), while in 2012 it rose to 7,968,517.44 t, with 238,480.72 t of hazardous waste, totalling 8,206,998.16 t (Central European Development Forum, CEDEF 2014). The highest contributor to the total waste produced is fly ash from coal, accounting for 67%, but other forms of waste originating from thermal processes are also present: bottom ash, slag and sludge, etc. Then comes the waste from the physical and chemical processing of minerals in non-ferrous extractive metallurgy, waste phosphoric and phosphorous acid, metals which contain iron, cables, concrete materials, ceramic waste and soil from the cleaning and washing of sugar beet, as well as the sludge produced during the washing, cleaning, peeling, centrifugation and separation processes in the agriculture and the food processing industries. When it comes to municipal waste, data from the Environmental Protection Agency show that there are 164 landfill sites in the Republic of Serbia which are used by the local utility companies for waste disposal. Often they do not meet the technical or sanitary requirements set out by EU regulations and there are still a very large number of 'unauthorized tips', which contribute significantly to soil degradation.

Table 11.2 Sources and levels of industrial and municipal waste in Serbia, 2008–2009

Industrial activity	Total amount of waste (t)
Mining operations	15,009,969
Manufacturing industry	1,961,276
Supply of electricity, gas, steam, etc.	5,699,875
Municipal waste	2,374,374

Approximately 40% of the municipal waste generated in Serbia is disposed of at these unauthorized tips, which are not regulated by public utilities companies. According to a report from 2009, there are 4481 such sites (Statistical Office of the Republic of Serbia 2012). Hence, the uncontrolled disposal of industrial and often municipal waste in large industrial centres such as Bor, Pančevo, Novi Sad, Smederevo, Obrenovac and Kragujevac degrades large areas of cultivable land. Less than 20% of these areas, degraded by mining and other industrial operations, are subject to natural and artificial recultivation, i.e. revegetation. Without these measures, we have ecological black spots, which result from coal and ore excavation, soil borrow pits, the localization of municipal waste dumps, and disposal sites and drainage areas for waste water.

When these activities cease, major ameliorative measures are required to return the degraded soils (deposols) to their original state or to turn them into a productive natural resource. However, there is no systematic implementation of such measures on deposols as it is rare that they are restored to agricultural production. One positive example is the abandoned mines of the Kolubara Coal Basin in western Serbia, which have been afforested or even turned into agricultural land. Of the approximate 120 km^2 on which the four 'Kolubara' coal mines are situated, around 1200 ha of recultivated land is currently being used (approximately 850 ha are forested, while maize, wheat and clover are cultivated on around 300 ha). Besides the abandoned mines, the overburden and tailing dumps are also being restored—agricultural cultures will be grown for the production of biomass as ecological fuel.

Deposition of ash at the deposit sites of the 'Nikola Tesla A' thermoelectric power plant is performed by sluicing it into three ash ponds, two of which are inactive at any one time. Biological recultivation is carried out at the inactive ash ponds through the sowing of a grass–legume mixture and the planting of shrubs and woody plant species. With the periodic sluicing of ash into the ash ponds with the formed plant cover (in 5- to 13-year cycles), the succession of vegetation spanning several years is destroyed. Once again, a biologically barren area is formed, and the revegetation process starts afresh (Pavlović et al. 2004; Mitrović et al. 2008).

The recultivation of ash deposit sites is undertaken as a conservation measure so as to prevent ash dispersal by wind. In fact, one of the best solutions for maintaining ash deposit sites is revegetation as it allows for the physical and chemical stabilization of this mobile substrate (Haynes 2009;

Pavlović et al. 2004; Mitrović et al. 2008). The biological restoration of ash deposits can lead to effective stabilization, the protection of deposits from wind erosion and an improvement in the quality of the habitat, and in both the microclimate and aesthetic appearance.

In Serbia, the biological restoration of ash deposit sites (recultivation) is a regular and permanent activity. It takes the form of forestation (the planting of tamarisk, poplar, black pine, Siberian elm, linden, black locust, etc.), as well as the sowing of various grass species (Kostić et al. 2012b; Mitrović et al. 2012). A sprinkler system is set up to spray the perimeter banks of the deposit site and to water the grasses, as is a system of water cannons to water the dry, flat areas of the sites. Furthermore, there have been attempts to bind the ash with cultures of a variety of forest trees, various grain species and industrial and medicinal plants (Dželetović and Filipović 1995). However, ecological research at the ash deposit sites of the 'Nikola Tesla A' power plant has shown that of the approximate 100 different plant species on the ash deposits, more than 80 have naturally colonized the site from the surrounding area (Project reports 2001, 2002; Fig. 11.9). The majority are anemochorous, annual and biennial, weed and ruderal species. It is these weed species and elements of ruderal flora that naturally colonize the deposit sites which have great potential for the revegetation process. It has been established that only the sown xero-mesophyllic grass *Festuca rubra* L. and the planted tamarisk remain in any great number several years after planting.

All the other species—the mesophyllic grasses, grains and legumes—only remain in traces, mainly on the banks. In the central areas of the ash ponds, where the ash fractions are finest, there are almost no plants at all apart from the remains of withered tamarisk plantlets. What differentiates the species that naturally colonize the deposit sites from those that are sown or planted is their high adaptability to different ecological conditions, which over a period of years gives them the upper hand when it comes to colonizing the ash deposit sites (Pavlović and Mitrović 2013). They do not require fertilization and most do not need watering.

11.2 Soil Pollution

11.2.1 Heavy Metals

Heavy metals are permanently present in soil, but their level and behaviour, and as such, their mobility and availability to plants depends on the parent substrate, the soil pH, the organic matter and clay content in the soil, the physical composition of the soil, moisture content, calcium carbonate content, etc. If the concentration of heavy metals in the soil is significantly higher than normal, it points to contamination due to the natural geochemical origin of the soil or from anthropogenic sources.

The results of Mrvić et al. (2009) showed that most soils are not contaminated with heavy metals. Increased concentrations of elements occur mainly in the hilly and mountainous areas. The exception is the elevated Cu and Ni concentrations in Vojvodina, where a higher Cu content has been found in the vicinity of Sombor, Vršac, Senta, Novi Sad and Pančevo in plants at vineyards due to the long-term application of copper-based compounds for plant protection, while Ni exceeded the maximum allowable concentrations (MAC) only in the vicinity of Vršac. Ni and Cr are often over the MAC (>100 mg/kg), as is As (>25 mg/kg), compared to other elements (Pb, Cu, Cd, Zn and Hg). The origin of heavy metals is mainly geochemical, as is the case with the soils across the world (Kabata-Pendias and Pendias 2001; Table 11.3).

The situation in central Serbia is as follows: the elevated heavy metal content across large areas is not the result of anthropogenic activity, but is connected to the geochemical origin and genesis of some soil types. The average Ni content in the soils of central Serbia is 58 mg/g (Mrvić et al. 2009) with approximately 80% of soil having Ni concentrations <50 mg/g. Elevated Ni occurs in soils formed on ultrabasic or basic rocks which have a high natural content of this element as well as Cr, Mg, Fe, Co, etc., but also on sedimentary and metamorphic rocks originating from the transformation of ultrabasic rocks (Protić et al. 2004), which are mainly localized in areas of central and western Serbia.

In western Serbia, these soils are found in the region of the Zlatibor peridotite massif with varying degrees of serpentinization than those in the region of Mt. Maljen and Suvobor, and across the massif of the mountains of Troglav, Stolovi, Čemerno and Željin all the way to Golija and Kopaonik. They are also in parts of central Serbia comprising the Rudnik massif to the Gledić mountains, following the valleys of the Velika Morava and Kolubara rivers where the distribution of serpentinite is connected to the deluvial and alluvial processes along the watercourses (Jakovljević et al. 1997). Soils formed on ultrabasic and basic rocks include ranker (WRB, Leptosol, eutric) and occasionally regosol (WRB, Leptosol) and cambisols. They are populated by forests, pastures and meadows, where the Ni concentration is >100 mg/g, but can reach levels as high as 1900 mg/g (Mrvić et al. 2009) (Fig. 11.10). In terms of Cr, this element is found in the same soil types as Ni.

The average Cr concentration in soils is 48 mg/g (Mrvić et al. 2009) with approximately 80% of the soils having a Cr concentration of around 50 mg/g (Fig. 11.10). Chromium is less mobile than Ni so its translocation is minimal—as much as 99% of the chromium taken up by plants remains in the roots. Anthropogenic sources of Ni include the application

Fig. 11.9 Stages in the revegetation of the fly ash deposits at 'Nikola Tesla A' thermoelectric power plant (Obrenovac) (*photo* Pavlović, P., Mitrović, M.)

of phosphorous fertilizers and emissions from metallurgical plants.

Cooper concentrations in soils range from 1 to 1219 mg/g with the average concentration in central Serbia being 27 mg/g (Mrvić et al. 2009), which is somewhat lower than the average value for soils in the world (Adriano 1986). Natural zones with Cu concentrations between 50 and 100 mg/g are characteristics for eastern Serbia due to the parent substrate (a massif of andesite and dacite rich in copper) (Fig. 11.10). Anthropogenic copper contamination of soil is highest in the Bor Ore Basin. However, there are also soils with low Cu content

Table 11.3 The content of elements in soil types—median values (mg/kg) (Mrvić et al. 2009)

Soil type Serbian class./WRB	As	Cd	Cr	Cu	Hg	Ni	Pb	Zn
Arenosol and Regosol on sand/Leptosol	4.60	1.11	11.95	21.29	0.09	22.41	24.44	36.61
Colluvium/Colluvial soil	6.10	0.43	23.30	19.10	0.05	22.71	27.73	36.00
Kalkomelanosol/Leptosol, Lithosol/Leptosol, Regosol/Leptosol	8.40	1.20	24.95	20.66	0.10	20.45	33.80	40.55
Rendzina/Leptosol, Regosol/ Leptosol Lithosol/Leptosol	7.50	1.10	24.40	17.00	0.09	18.00	25.20	35.00
Ranker on granite, granodiorite, Quartz latite/Leptosol dystric	5.30	0.30	12.00	13.00	0.08	11.00	23.00	35.80
Ranker on andesite, dacite, tuff/Leptosol dystric	8.00	0.35	19.75	21.00	0.07	13.63	46.15	38.50
Ranker on serpentine and basic rocks/Leptosol eutric	4.20	0.70	194.50	20.56	0.08	164.00	44.70	38.25
Ranker on sandstones, flysch and cherts/Leptosol dystric	7.70	0.65	29.30	19.00	0.08	25.80	38.50	42.40
Ranker on schist and gneiss/Leptosol dystric	7.70	0.35	21.00	19.00	0.08	18.00	32.50	41.00
Chernozem	4.70	0.20	29.95	20.35	0.04	32.00	23.00	43.02
Vertisol	6.80	0.70	32.05	21.90	0.06	22.21	29.53	37.00
Eutric cambisol/Cambisol, eutric	6.20	0.40	30.05	18.40	0.07	22.94	27.50	40.00
Dystric cambisol/Cambisol, dystric	6.40	0.65	15.10	19.48	0.08	14.64	26.66	42.58
Kalkocambisol/No adequate name and kalkomelanosol/Leptosol	7.80	0.78	29.05	19.65	0.12	29.18	37.58	47.88
Luvisol	5.50	0.71	25.00	17.50	0.05	26.77	26.10	42.63
Pseudogley/Stagnosol	7.40	0.65	22.65	15.40	0.07	30.83	36.08	47.50
Fluvisol	8.00	0.50	30.05	20.90	0.08	32.00	30.00	44.55
Humofluvisol/Fluvisol, humic	7.15	0.30	47.50	21.86	0.05	58.22	33.72	44.84
Humogley/Gleysol, humic and eugley/Gleysol	6.10	0.20	33.40	22.23	0.07	30.79	28.71	42.00

(<3 mg/g) such as regosol, ranker (WRB, Leptosol) and dystric cambisol. The mass application of copper-based compounds for the conservation of fruit and vegetable crops, and particularly grapevines, results in elevated concentrations of this element in winemaking regions such as Župa, Negotinska krajina and the area around Kruševac and Niš, as well as in the orchards of Kosmaj, Jošanička banja and Krupanj (Mrvić et al. 2009).

The average cadmium concentration in the soils of central Serbia is 0.805 mg/g (Mrvić et al. 2009) with 1 mg/g being measured in the majority of samples (Fig. 11.10). Elevated Cd concentrations occur in alluvial soils and pseudogleys (WRB, Stagnosol) in western Serbia (Loznica), in kalkomelanosols (WRB, Leptosol) in south-western Serbia (Pešter plateau), and in rendzinas (WRB, Leptosol, calcaric) in eastern (Nišava river valley and around Bor) and south-western Serbia (Mokra Gora mountain). Cadmium may be geochemical in origin in which case it is found in soils rich in zinc ores, carbonates, phosphorites, black schists and clays. Cadmium originating anthropogenically can be the result of the combustion of fossil fuels and various processes in the metal, paint and plastics industries. It can also occur from processes in the production of batteries, transport (found in engine oils and tyres), and as the result of the intensive use of fungicides and phosphorous fertilizers on agricultural land.

The average lead content in the soils of central Serbia is 40 mg/g (Mrvić et al. 2009) (Fig. 11.11). In the majority of soil samples, the average Pb concentration is 50 mg/g. Elevated Pb concentrations have been recorded in the valleys of the Velika Morava and Zapadna Morava rivers, where a natural source of lead are those soils formed on rocks rich in metals in western Serbia (around Zvornik and Bajina Bašta, Kopaonik Mt., and the mountains around Raška), and to some extent in the south of Serbia (Surdulica) or along busy traffic arteries.

Zinc is a rare soil pollutant in Serbia with average concentrations being 48 mg/g (Mrvić et al. 2009). The majority of soil samples had 50–62 mg/g Zn (Fig. 11.11), although higher concentrations occurred locally in those soils lying on reserves of sulphide ores (around Medveđa). Sources of zinc contamination include mining activities, energy plants and the use of chemicals in agriculture.

The average mercury content in soils is 0.120 mg/g (Mrvić et al. 2009), with the majority of samples containing 0.1 mg/g. Elevated levels of mercury can be found in forest soils, kalkomelanosol (WRB, Leptosol) at Jelica Mt., dystric cambisol at Goč Mt. and lithosol (WRB, Leptosol) at the 'Oštra stena' barite mine near Ljubovija (Fig. 11.11).

The average arsenic content in soils is 11 mg/g, but in the majority of samples As concentrations are <10 mg/g (Mrvić et al. 2009), Fig. 11.11. Elevated levels of this element have

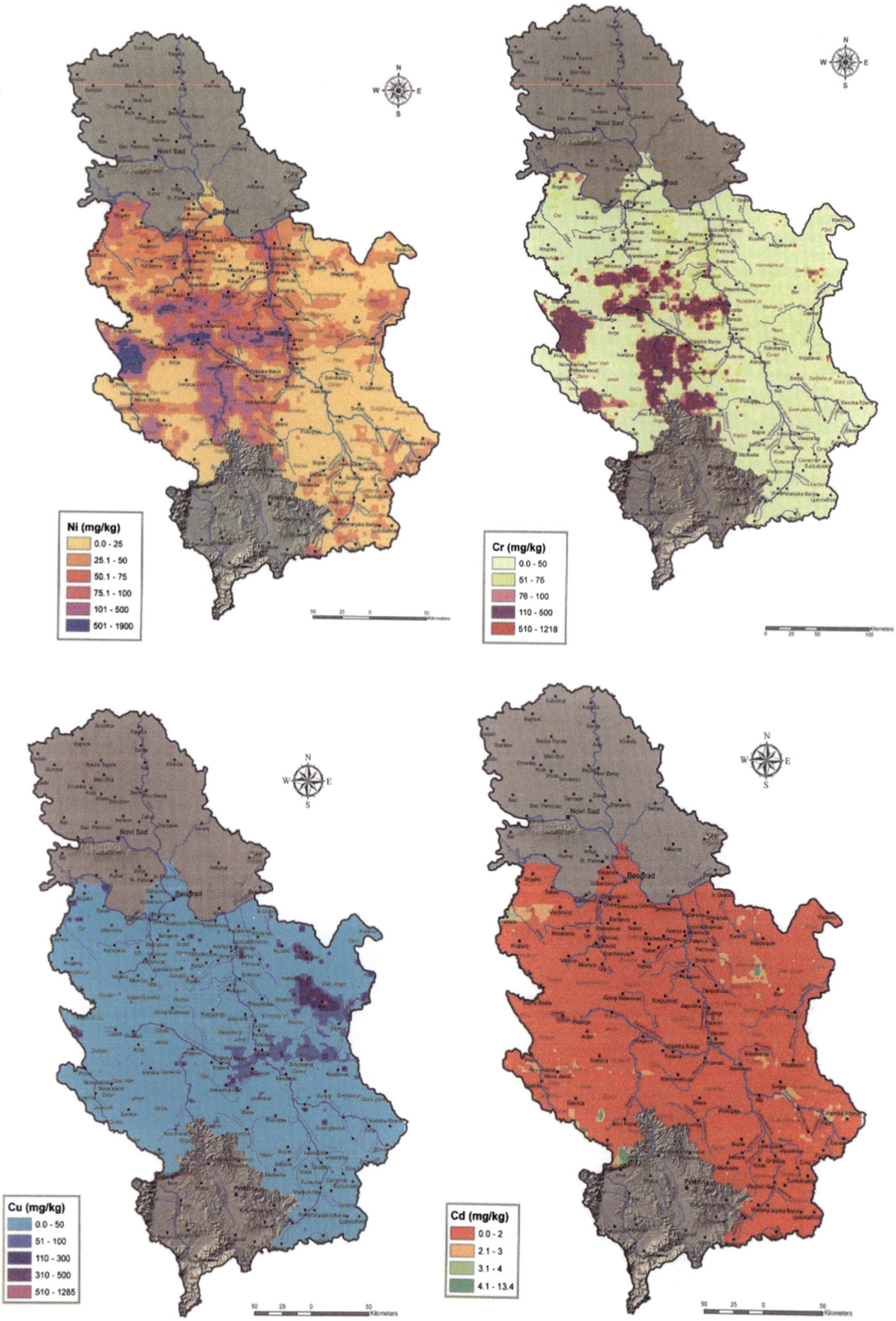

Fig. 11.10 Distribution of nickel, chromium, copper and cadmium in the soils of Serbia (Mrvić et al. 2009)

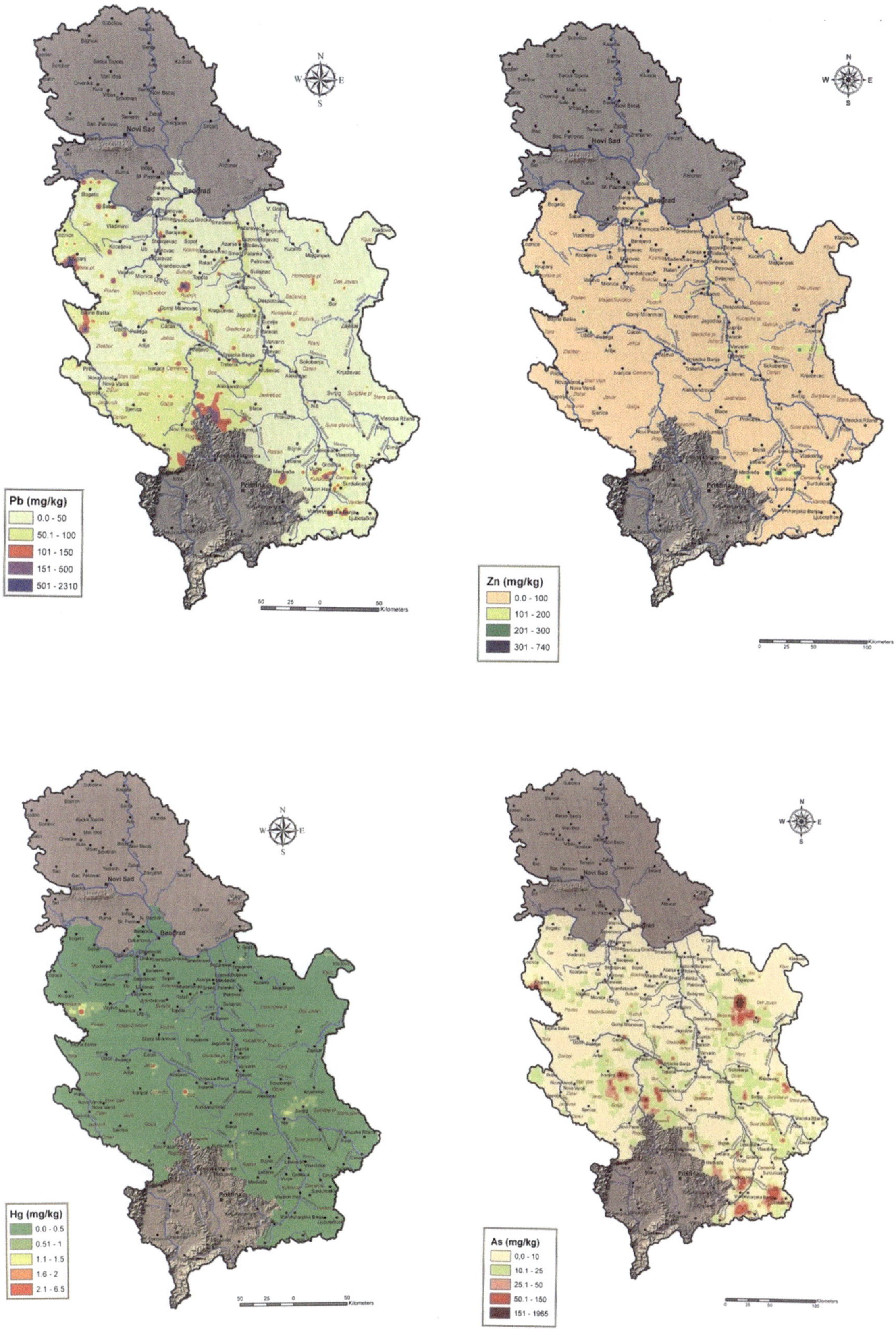

Fig. 11.11 Distribution of lead, zinc, mercury and arsenic in the soils of Serbia (Mrvić et al. 2009)

Fig. 11.12 Beech mountain forests of Crni Vrh, and Beljanica Mt., eastern Serbia (*photo* Pavlović, P.)

been recorded around the Bor Basin as a result of copper ore processing. Naturally acidic forest soils with elevated As levels are found in western Serbia (Ivanjica) and rankers (WRB, Leptosols) in south-eastern Serbia (Vranje, Bosilegrad).

Over the past few decades, forest soils have been also exposed to the atmospheric deposition of heavy metals, which, besides having a direct impact on the trees, also accumulate in the soils and have a negative effect on the chemical processes in them. In fact, the highest concentrations of metals have been measured in the layers of forest litter and the surface organo-mineral horizons (Kadović and Knežević 2002, 2004).

The results of Belanović et al. (2003), who analysed the content and distribution of Pb, Zn and Cd in soils under beech forests in the Đerdap region, serve to illustrate this problem. Their results revealed the following: Pb, Zn and Cd concentrations decrease significantly with soil depth and are below critical levels (Vanmechelen et al. 1997; Kadović and Knežević 2002). Later, extensive research into the content of heavy metals (Zn, Mn, Cu, Fe, Cd, Pb, Ni and Cr) in the organic horizon (forest litter) in beech forests in Serbia carried out by Kadović et al. (2005) showed that Cd concentrations range from 0.19 to 1.50 mg/kg, with an average of 0.98 mg/kg, and that all the concentrations measured are lower than the average content for European soils (Rademacher 2001), Fig. 11.12. The content of Zn, ranging from 2.45 to 108.04 mg/kg, with an average concentration of 53.78 mg/kg, is also lower (Vanmechelen et al. 1997).

In the litter of the beech forests of Serbia, concentrations of copper were found to be between 5.99 and 81.91 mg/g, with an average level of 16.85 mg/kg, which is somewhat higher than the average for European soils (Rademacher 2001). Concentrations of Mn measured in the soils vary widely, ranging from 512.64 to 5200.78 mg/kg with an average concentration of 1627.83 mg/kg, which is above the average level for European soils (Kadović et al. 2005). Iron in beech litter in Serbia is found in concentrations between 588.66 and 25,909 mg/kg, with an average level of 3448 mg/kg (Kadović et al. 2005). In the organic soil layers, concentrations of Fe are below 10,000 mg/kg, apart from in dystric cambisol in the beech mountain forests of Crni Vrh in eastern Serbia (Fig. 11.13), where levels are over 20,000 mg/kg (Kadović and Knežević 2002).

The levels of Pb in forest litter range from 3.99 to 108.11 mg/kg, with average concentrations of 25.53 mg/kg

Fig. 11.13 Beech mountain forest along the Dojkinačka reka at Stara planina Mt., eastern Serbia (*photo* Mihailović, M.)

(Kadović et al. 2005). Adriano (1986) believes that Pb concentrations in soils >11.0 mg/kg are of anthropogenic origin, which is why the surface horizons (the organic layers) are richest in this element. Concentrations of Ni in the organic layers of beech range from 1.00 to 110.04 mg/kg, with an average concentration of 21.48 mg/kg, while Cr levels are between 3.4 and 120.2 mg/kg, with an average of 19.2 mg/kg. According to Kadović et al. (2005), the concentrations of heavy metals measured in the organic soil layers of Serbia's most significant forest ecosystems clearly show that atmospheric deposition is the main source of these elements in forest soils. This is of particular significance when it comes to Pb and Cd because their deposition in Serbia is either in line with or above average levels for other European countries.

Anthropogenic pollution caused by heavy metals is particularly pronounced in the vicinity of major industrial plants with lead and zinc most commonly the problem. Tailings are dumped in the surrounding areas, from where they are dispersed by the wind onto soil and plants. It is estimated that production processes at the Bor Mining and Smelting Combine result annually in the release of 300–500 t of sulphuric acid, 300–500 t of arsenic, 30–100 t of lead and 10–35 t of zinc. Wastewater from the mine and production processes, which is exceptionally acidic and has a high heavy metal content such as Cu, As, Pb and Zn, is discharged into the Bor and Krivelj rivers, which carry it into the Timok and Danube rivers (National Strategy of Sustainable Use of Natural Resources 2012). The potential for arsenic contamination is present in the area around Bor mine (mainly to the west of the town of Bor) and in the Timok valley, as well as adjacent to the Žagubica—Bor highway, which are all contaminated with other pollutants, too. In addition, elevated levels of this element have been recorded in some soils in the vicinity of the Resava coal mines.

Extreme forms of soil pollution have been noted around thermoelectric power plants due to the emission of aerosols and gases containing SO_2 and NO_x, which are blown by the wind in the form of ash onto surrounding soils and vegetation. The huge areas of ash deposits from thermoelectric plants in Serbia are an environmental risk due to the leaching of chemical substances (salts and heavy metals) into the soil and groundwater, while the fine ash particles are scattered from the dry surfaces of the ash deposit sites across the surrounding areas, hence endangering the health of the local inhabitants (Pavlović and Mitrović 2016) (Fig. 11.14).

The physical, chemical, morphological and radioactive characteristics of the coal combustion residues (CCR) at thermoelectric plants vary depending on the quality of the coal, the combustion process, the length of exposure to atmospheric conditions, the morphological characteristics of ash particles and the age of the ash. Ash contains high concentrations of chemical elements such as As, Ba, Be, B, Cd, Cr, Co, Cu, Pb, Mn, Ni, Se, Sc, Ta, V and Zn. Fine ash particles carried by the wind can also contaminate soil and vegetation in areas around the ash deposit sites, but also further afield. The consequences of contamination can be far-reaching and long-term (Pavlović et al. 2004, 2007).

In Belgrade alone, there are three thermoelectric power plants ('Nikola Tesla A' and 'Nikola Tesla B' in the municipality of Obrenovac and 'Kolubara' in the municipality of Lazarevac), and their operations endanger the health of nearly 2 million inhabitants both directly and indirectly. They are continually operational while intensive agricultural production takes place nearby. Hence, Dragović et al. (2013) found elevated concentrations of Cd, Co, Cr, Cu, Hg, Mn, Ni, Pb and Zn in the soils around the 'Nikola Tesla A' thermoelectric power plant, with concentrations of Cr, Hg and Ni being above the maximum permitted levels for agricultural soils as prescribed by local regulations. Preliminary results of research into total concentrations of Al, As, B, Ba, Cd, Co, Cr, Cu, Fe, Hg, K, Li, Mg, Mn, Mo, Na, Ni, Se, Pb, Sr and Zn in soil samples from gardens and fields in the vicinity of the open-cast mine and the 'Kolubara' thermoelectric power plant in Lazarevac and the fly ash deposit sites at the 'Nikola Tesla A' plant in Obrenovac showed that the majority of the elements analysed fell within the permitted range for concentrations in soils.

However, there were elevated concentrations of Al, As and Fe, and the content of B and Ni was above the maximum permitted concentrations for agricultural soils (Regulations on permitted levels of hazardous and harmful substances in soil and irrigation water and methods for their testing, 'Official Gazette of the Republic of Serbia', no. 23/94), measured in garden soil in Lazarevac. This is the result of atmospheric deposition from emissions and the dispersal of fly ash deposits from the 'Kolubara' thermoelectric power plant onto surrounding land. It was noted that the Kolubara and Sava rivers bursting their banks in 2014 contributed to an increase in Fe, As, Cr and Ni concentrations in cultivable soil (Kostić et al. 2015).

In terms of industrial areas, research by Relić et al. (2011) at the industrial zone of the town of Pančevo revealed elevated concentrations of Cd, Zn and Pb in surface layers of soil in the vicinity of a petrochemical plant, Cd and Zn in the vicinity of a nitrogen fertilizer factory, and Pb in the vicinity of the Messer plant for the production of gas and gas mixtures. Likewise, high concentrations of Cd, Co, Cu, Ni, Pb and Zn have been measured in soils close to the iron and steel works in Smederevo, exceeding the average levels for these elements in the soils of Europe (Dragović et al. 2014).

When it comes to watercourses, heavy metals and other toxic elements accumulate in the flood zone of rivers and in the areas around confluences, whether their presence in the soil be the result of the natural disintegration of the parent substrate or of anthropogenic origin (numerous industrial plants such as thermoelectric power plants are built in riparian zones). The testing of sediment, soil and vegetation in the riparian zone of the Danube at 43 sites, from the river's source to its mouth at the Black Sea, revealed a significant positive correlation between concentrations of these elements in sediment and soils. Analyses of soil from the floodplains of the stretch of the Danube which passes through Serbia showed very high concentrations of Cu (>500 mg/kg) at the point where the River Timok flows into the Danube. Elevated concentrations were also measured in other parts of the Serbian stretch of the Danube: of Cd where the Tisa flows into the Sava at Belegiš; of Cr at Banatska Palanka/Bazias; of Ni and Zn downstream from Pančevo; and the presence of Hg at sites upstream from Pančevo/downstream from the Sava, and downstream from Pančevo (Pavlović et al. 2016). The contamination at certain sites is naturally occurring, originating from the rock in the river basin, while at other locations anthropogenic pollution from industrial plants dominates, responsible for 'point source pollution'. Previous analyses of Ca, Cd, Co, Cr, Cu, Fe, Mn, Pb, Ni, Zn, Be and V in sediments of the Tisa, Sava and Danube rivers, and their tributaries revealed that Fe, Mn, Be and V are mainly of lithogenic origin, while Cd, Cr, Pb, Ni, Zn and Cu were recognised as elements of anthropogenic origin, with Co partly lithogenic and partly anthropogenic in origin (Sakan et al. 2011). Furthermore, elevated levels of Cd, Pb, Cu and Zn were measured, which could pose a potential risk for living organisms in these aquatic ecosystems (Fig. 11.15).

Municipal landfills, localities with spilt oil derivatives, industrial waste dumps, industrial and commercial zones, mines, oil wells, oil depots and power stations also contribute to soil pollution.

For all these reasons, a regulation on permitted levels of hazardous and harmful substances in soil and irrigation water and methods for their testing ('Official Gazette of the Republic of Serbia', no.23/94) (Table 11.4). The programme for the systematic monitoring of soil quality includes the establishment of a state and local network of localities for monitoring soil quality. The state network of

Fig. 11.14 Ash disposal site in the vicinity of 'Nikola Tesla A' thermoelectric power plant in Obrenovac (*photo* Pavlović, P.)

Fig. 11.15 Riparian zone of Danube river (*photo* Pavlović, P.)

localities is established for monitoring soil quality at the level of the Republic of Serbia at localities where there is or there could potentially be soil pollution and which are of particular significance for the country as a whole.

When deciding on parameters, those which could reach levels that present a significant risk to human health and the environment are taken into consideration. The time schedule for soil sampling, analysing samples, and data processing and presentation depends on the stability of the analysed parameters, their forms and their concentrations in the environment, so measurements are taken at one-, three- or five-year intervals. Contaminated areas are identified so remediation programmes can be devised based on the confirmed presence of soil pollutants. These are areas where degradation processes are exhibited, such as waste disposal sites, business and industrial sites (i.e. locations where pollution is caused by active or inactive industries or operations in the vicinity of which hazardous materials are deposited), accident sites or sites contaminated due to emergency events, including breakdowns, and brownfield land where activities once took place which could have contaminated the soils. A database of contaminated locations is an integral part of the environmental protection information system, which is run by the Environmental Protection Agency of the Republic of Serbia. In those cases where threshold values and the threshold level for concentrations of contaminants in the soil have been exceeded, further investigations are carried out at the contaminated locations so as to determine the degree of contamination and to devise remediation programmes for the clean-up of contaminated soils.

11.2.1.1 Chemical and Physical Speciation of Metals in Soil

In order to establish heavy metal availability to plants, chemical speciation analyses were carried out on 40 selected samples representing the major soil types in Serbia. The results of the sequential dissolution analysis of soil materials are presented in Fig. 11.16 and their distribution in Tables 11.5 and 11.6. Chemical speciation was carried out at Macaulay Land Use Research Institute, Scotland, by using the method proposed by Quevauville et al. (1994).

The investigated soils contained large residual chromium and iron fractions. The limited mobility of Cr and Fe is due to resistance to the weathering of primary oxide minerals

Table 11.4 Threshold and remediation levels for concentrations of heavy metals and levels which may suggest significant soil contamination (Official Gazette of the Republic of Serbia, no. 23/94)

Metals	Soil (mg/kg of absolutely dry matter)	
	Threshold level	Remediation level
Cadmium (Cd)	0.8	12
Chromium (Cr)	100	380
Bakar (Cu)	36	190
Nickel (Ni)	35	210
Lead (Pb)	85	530
Zinc (Zn)	140	720
Mercury (Hg)	0.3	10
Arsenic (As)	29	55
Barium (Ba)	160	625
Cobalt (Co)	9	240
Molybdenum (Mo)	3	200
Antimony (Sb)	3	15
	Threshold level	Levels which may suggest significant contamination
Beryllium (Be)	1.1	30
Selenium (Se)	0.7	100
Tellurium (Te)	–	600
Thallium (Tl)	1	15
Tin (Sn)	–	900

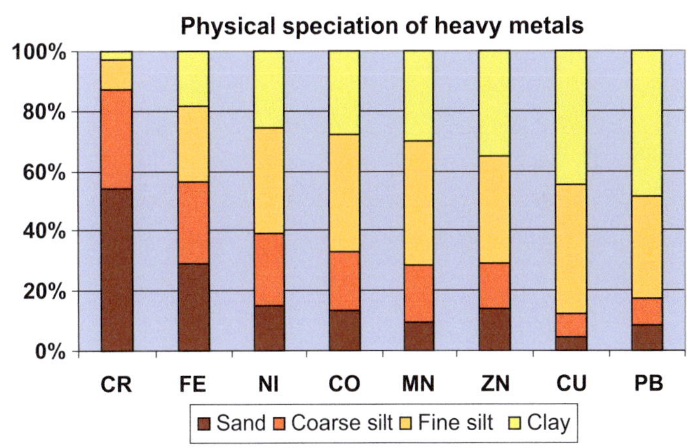

Fig. 11.16 Physical speciation of heavy metals

(chromite and magnetite) where chromium is often substituted for iron. Iron is a macroelement in the Earth's crust and between 0.73 and 8.86% Fe is present in the investigated soils, whilst chromium is found in concentrations between 16 and 776 mg/kg. Maximum concentrations are found in ranker (WRB, Leptosol, eutric) on serpentinite, humic fluvisol (WRB, Fluvisol, humic) and brownized chernozem (WRB, Chernozem), in some eutric cambisol, luvisol and vertisol, as well as in fluvisols derived from their erosion, and some rendzina (WRB, Leptosol, calcaric) from central Serbia. They are the least mobile heavy metals with relatively small variations, between 80 and 98% Cr and 70–90% Fe, in the residual fraction.

Iron ions comprise 6–24% of ions combined into oxides and carbonates. The iron oxide fraction gradually increases with an increase in rainfall, from chernozem, humic fluvisol (WRB, Fluvisol, humic) and some eutric cambisol (6–8%) in the north, through pseudogley (WRB, Stagnosol), fluvisol and vertisol, to luvisol and dystric cambisol (16–26%) in the south. Organic-bound chromium (up to 15%) is found in vertisol, humogley (WRB, Gleysol, humic) and humic fluvisol (WRB, Fluvisol, humic), where it is combined in organo-mineral complexes with kerogen derived from bituminous marls or clays (Kostić et al. 2002).

The investigated soils contain 124–2844 mg/kg manganese. Maximum amounts are found in ranker on

Table 11.5 Chemical speciation of heavy metals

Heavy metal		Mean	Min	Max	St. dev.
Fe	Exch	0.50	0.09	1.92	0.54
	Ox	13.81	6.02	24.67	6.38
	Org	2.91	1.42	4.04	1.49
	Res	82.78	69.68	92.21	7.14
Mn	Exch	17.76	10.34	27.78	4.49
	Ox	56.80	43.83	72.26	7.19
	Org	5.06	1.72	9.25	2.12
	Res	20.37	13.31	28.51	4.89
Cr	Exch	1.98	0.00	4.55	1.71
	Ox	1.37	0.01	2.66	0.81
	Org	9.14	1.19	14.77	3.30
	Res	87.50	80.11	98.64	4.53
Ni	Exch	9.75	5.75	39.52	7.03
	Ox	26.93	11.71	44.95	8.00
	Org	9.28	1.00	17.16	3.90
	Res	54.04	39.74	72.65	8.44
Co	Exch	9.92	2.78	17.43	4.52
	Ox	39.55	30.50	52.00	5.58
	Org	3.30	0.47	6.89	2.06
	Res	47.22	26.22	59.83	8.91
Cu	Exch	6.90	0.93	28.75	6.96
	Ox	9.59	0.00	39.55	9.03
	Org	15.58	5.55	40.16	9.19
	Res	67.92	37.50	83.92	12.89
Zn	Exch	15.12	1.44	43.37	12.45
	Ox	22.49	12.53	42.00	9.02
	Org	7.39	0.54	14.00	3.40
	Res	55.00	25.36	83.24	16.69
Pb	Exch	1.37	0.00	7.92	1.26
	Ox	41.80	11.11	78.82	17.82
	Org	37.62	2.70	56.43	13.63
	Res	17.56	2.60	84.21	20.10

Table 11.6 Averages of the physical speciation of heavy metals in soils (%)

	Cr	Fe	Ni	Co	Mn	Zn	Cu	Pb
Clay	2.81	18.60	25.85	28.06	30.37	35.40	44.82	48.84
Fine silt	10.12	25.06	35.25	38.76	40.88	35.38	42.70	33.88
Coarse silt	32.75	27.10	24.00	19.78	19.04	15.38	8.17	8.98
Sand	54.32	29.24	14.90	13.40	9.71	13.84	4.31	8.30

serpentinite (WRB, Leptosol, eutric) and luvisol, whilst minimum concentrations are found in arenosol, chernozem, eugley and humogley (WRB, Gleysol, humic). Manganese is most common in the oxide/hydroxide fraction at 44–72%, then in the residual fraction (13–28%) and as exchangeable cations or carbonate-bound (10–25%), and least so in the organic fraction (2–9%). The oxide fraction of pseudogley (WRB, Stagnosol) and humogley (WRB, Gleysol, humic) indicates that a change in underground water levels is favourable for the formation of Mn/Fe oxide/hydroxides as the crucial minerals in the fraction. The highest concentrations of exchangeable or carbonate-bound Mn are found in calcareous humic fluvisol (WRB, Fluvisol, humic) developed on alluvial terraces, eutric cambisol on sandy carbonates, and luvisol and dystric cambisol on marly flysch (Kostić et al. 1998a).

All soils contain almost equal amounts of cobalt (16–20 mg/kg), with the exception of ranker (WRB, Leptosol, eutric) developed on serpentinite and Arenosol from sand plains. The manganese oxide/hydroxide mineral wad/psilomelane is a strong scavenger for cobalt (30–52%) in the surface weathering zone. Soluble cobalt is depleted from the soil surface horizon where an average of 10% is found in extractable or carbonate forms. Residual forms of cobalt (26–60%) are found in soils developed on greenschist and Fe–Mg mineral-rich rocks. The organic form (up to 7%) of cobalt is mainly found in ranker WRB (Leptosol, eutric) developed on peridotite and serpentinite and humic fluvisol (WRB, Fluvisol, humic) and humogley (WRB, Gleysol, humic) in oxbows and marshes on alluvial terraces.

Nickel and zinc (6–662 mg/kg Ni and 11–514 mg/kg Zn) demonstrate similar behaviour in the soils. They are found in ranker (WRB, Leptosol, eutric) and rendzina (WRB, Leptosol, calcaric), eutric cambisol and humic fluvisol (WRB, Fluvisol, humic) in central Serbia. They are most common in the residual fraction (25–98%), then in the oxide/hydroxide fraction (12–52%) and as exchangeable cations or carbonate-bound (3–43%), and least so in the organic fraction (0.4–17%). The oxide/hydroxide fraction is most often found in pseudogley (WRB, Stagnosol), humogley (WRB, Gleysol, humic) and luvisol, whilst they are found as exchangeable cations or carbonate-bound in calcareous humic fluvisol (WRB, Fluvisol, humic) and chernozem developed on loess terraces, as well as in luvisol and dystric cambisol developed on Cretaceous flysch. The organic fraction is most common in ranker (WRB, Leptosol, eutric), vertisol, humogley (WRB, Gleysol, humic) and fluvisol, where it is bound in organo-mineral complexes. Higher concentrations of Ni and Cr are found on the alluvial plains of the River Velika Morava, near the confluence with Jasenica, which drains an area rich in serpentinite. Further analyses have shown that in spite of increased Ni and Cr contents in humic fluvisol (WRB, Fluvisol, humic), there is not a significant heavy metal uptake by plants (Antić et al. 2005).

Copper (5–340 mg/kg Cu) exhibits different behaviour from the other elements. The majority of copper (37–84%) remains in residual forms, mainly bound to aluminosilicates in chernozem on loess plateaus, and eutric cambisol and vertisol developed on Neogene lake sediments. Up to 40% of copper is found in the organic fraction, strongly suggesting biophille cycling, particularly in ranker (WRB, Leptosol, eutric), humic fluvisol (WRB, Fluvisol, humic), pseudogley (WRB, Stagnosol) and dystric cambisol. The organic fraction of copper correlates negatively with that found in the oxide/hydroxide fraction. The manganese minerals wad and psilomelane can absorb up to 40% of copper in cambisols and brownized chernozem (WRB, Chernozem). The exchangeable copper ions are found mainly in calcareous humic fluvisol (WRB, Fluvisol, humic) and humogley (WRB, Gleysol, humic) developed on alluvial terraces in marshy environments and Vertisol developed on lake deposits, suggesting that copper could react with calcite and form soluble copper carbonate.

Maximum values of lead are found in ranker (WRB, Leptosol, eutric) (264 mg/kg Pb) on serpentinite, while the levels range from 56 to 225 mg/kg Pb in eutric cambisol, fluvisol, humogley (WRB, Gleysol, humic), humic fluvisol (WRB, Fluvisol, humic) and vertisol developed on the alluvial plains of the Morava and Kolubara rivers and their tributaries, which drain the area around known Pb, Zn and Cu mines in central Serbia. Analyses have shown that Pb is leached and migrates to oxide (up to 79%) and organic (up to 56%) fractions, with only 0–8% remaining in exchangeable forms, whilst about 22% of lead is found in the residual form associated with aluminosilicates.

Exchangeable Pb is found in chernozem on loess plateaus, in eutric cambisol and humic fluvisol developed on loess terraces in Šumadija, in Vertisol on Neogene lake deposits and in dystric cambisol on marly flysch. The Pb organic fraction is common in the vertisol, humogley (WRB, Gleysol, humic), pseudogley (WRB, Stagnosol) and fluvisol of Šumadija. The lead oxide fraction is found in eutric cambisol developed on flysch and chernozem on loess plateaus (Kostić 2001).

In order to investigate the physical speciation of heavy metal concentrations, particle size analysis was carried out on soil samples. The samples were split into four fractions as follows: sand (2–0.2 mm), coarse silt (0.2–0.063 mm), fine silt (0.063–0.002 mm) and clay (<0.002 mm), which were chemically analysed. The results of the ICP analysis of heavy metals in various soil particle fractions are presented in Table 11.6 and Fig. 11.16 (Kostić et al. 1998b).

The greatest influx of heavy metals was identified for Cr and Fe, which are found in the coarse (sand and coarse silt) fractions, with dominant quartz, feldspars and carbonates associated with chlorite, biotite, hornblende, pyroxene and abundant accessory metallic minerals. The silt fractions contain secondary minerals: illite, vermiculite, antigorite and Fe/Mn oxide/hydroxide, and here Mn, Co, Ni and Zn are the most abundant. The fine silt and clay fractions contain Zn, Cu and Pb with dominant fine carbonates and secondary minerals (illite, smectite and vermiculite and inherited kaolinite), which are associated with wad or poorly crystallised Fe–Mn oxide/hydroxides and organo-mineral complexes. The exchangeable cations of Cu, Zn and Pb can be 'fixed' in the inter-lamellar spaces of smectite, 'open illite' or vermiculite.

Chemical speciation of heavy metals has shown that an average of 48–87% of heavy metals occur in residual forms, mainly as silicate or aluminosilicates or as heavy metals bound to Fe and Mn oxide and hydroxide, whilst the exchangeable and organic bound forms are less common.

Chromium and iron are the most resistant elements to surface weathering and are found in the soils with 80–98% in the residual fraction. The oxide fraction contains Mn/Fe oxide/hydroxide minerals, which are strong scavengers of other heavy metals: Pb > Co > Ni > Zn > Cu. The exchangeable forms are predetermined by the carbonate–calcite content present in the soils. Divalent metals with similar ionic radii and an affinity to carbonates are adsorbed by calcite in the soils, resulting in exchangeable forms with varying degrees of solubility as follows: Zn > Ni > Cu > Mn > Co > Pb, or are taken up by plants as exchangeable cations: Pb > Cu > Zn within the smectite and illite structures as in vertisol, humogley (WRB, Gleysol, humic), humic fluvisol (WRB, Fluvisol, humic) and eutric cambisol. Organic forms of the heavy metals are mainly found in ranker (WRB, Leptosol, eutric), dystric cambisol and luvisol, where they form organo-mineral complexes: Pb > Cu > Ni > Zn > Co.

The majority of heavy metals are present either as exchangeable and carbonate-bound or as organic-bound, but a few remain in residual and oxide/hydroxide forms. The only exception is the lead found in ranker (WRB, Leptosol, eutric) on peridotite, serpentinite and diabase-chert formations or humic fluvisol (WRB, Fluvisol, humic) and eutric cambisol developed on the alluvial plains of the Morava and Kolubara rivers and their tributaries in central Serbia. High lead concentrations are also found along motorways and railways, in Vojvodina and the Velika Morava valley, in clay-rich soils, where lead is found either bound to Mn/Fe oxide/hydroxide or bound to humus organo-mineral complexes, which could correlate with transportation during the time when leaded petrol was used (Jakovljević et al.1997; Kostić 1999).

11.2.2 Pesticides

Potential danger exists as pesticides are in mass use in agronomy, forestry and cattle breeding. They accumulate in soil which then acts as a store for persistent pesticides which later translocate through roots into plants and can end up in the air and water. The adverse impact of these products can be seen in their transfer to the following season's crops on which they can have potential phytotoxic effects, their negative effect on soil macro- and micro-organisms, their leaching into groundwater and the eventual impairing of soil fertility and quality.

According to reports on the state of the environment in Serbia (Reports for 2005 and 2006), approximately 10,000 t of pesticides are used annually. In our country, the problem of applying pesticides is made all the more complicated due to the inadequate technical standards of the application equipment (Sedlar et al. 2007; Bugarin et al. 2010), which is often operated by people with a lack of adequate training. Annual pesticide consumption varies depending on the cultures being grown, the technology applied and the type and characteristics of the product itself. Their use in the cultivation of field and vegetable crops is low, but for the cultivation of intensive cultures (sugar beet), the average pesticide consumption is considerably higher. The use of pesticides in the production of fruit cultures and grapevines is particularly high. Delicate fruits such as apples are treated on average 10–12 times a year, but in unfavourably wet years this can be even higher (15–16 times). For the conservation of peaches or sour cherries, the number of applications is half. The inadequate and imprecise application of even a low number of treatments (5–6 times) can have a negative impact on soil (Bugarin et al. 2010), Table 11.7.

Data from Ružić and Poznanović (2009) for 2008 show that the annual amount of pesticides used in Serbia ranges between 4 and 10 kg/ha. The products used are mainly class III poisons. In ideal conditions (optimally set sprayers and sprinklers), a minimum of 20% of liquid reaches the soil during the application of these chemicals, but levels can be twice as high (40% in cases where spraying techniques are inadequately applied), meaning that 4.8–7.2 kg/ha or 9.6–14.4 kg/ha of pesticides reaches the surface of the soil annually.

Pesticide levels in the soils of Serbia are mainly at the level of natural background values and cannot be considered contaminating. In some pastures and meadows, the occurrence of DDT and its metabolites and Lindane gHCH has been recorded, which is linked to their use in forest conservation. In some arable fields, there are somewhat elevated levels of remnants of trazinic active substances, which are used in crop production (National Strategy of Sustainable Use of Natural Resources 2012). Namely, the most recent research into concentrations of pesticides and heavy metals in agricultural soils and the cultures grown on them in the vicinity of Belgrade registered the presence of residues of herbicides (atrazine, prometryn, acetochlor, flurochloridone, chloridazon and napropamide) and insecticides (fenitrothion and chlorpyrifos) in soils and pesticide residue in potatoes, tomatoes, corn, peppers, onions and cucumbers, while the content of active ingredients in peaches and apricots was below the limit of detection (Marković et al. 2010). Concentrations of a large number of pesticides in cabbage and cucumbers were below the limit of detection (LOD), as well as below the maximum residue level (MRL). However, concentrations of metribuzin [4-amino-6-(1,1-dimethylethyl)-3-(methylthio)-1,2,4-triazin-5(4*H*)-one] and trifluralin [2,6-dinitro-*N*,*N*-dipronyl-4(trifluoromethyl) benzenamine] in tomatoes from Ovca, as well as pendimethalin [*N*-(1-ethylpropyl)-3,4-dimethyl-2,

Table 11.7 Treshold levels for pesticide concentrations and levels which may suggest significant soil contamination (Official Gazette of the Republic of Serbia, no. 88/2010)

Pesticides	Threshold level	Levels which may suggest significant contamination
DDT/DDD/DDE (total)	0.01	4
Drins	0.005	4
Aldrin	0.00006	–
Dieldrin	0.0005	–
Endrin	0.00004	–
HCH-compounds	0.01	2
α-HCH	0.003	–
β-HCH	0.009	–
γ-HCH	0.00005	–
Atrazine	0.0002	6
Carbaryl	0.00003	5
Carbofuran	0.00002	2
Chlordane	0.00003	4
Endosulfan	0.00001	4
Heptachlor	0.0007	4
Heptachlor epoxide	0.0000002	4
Maneb	0.002	35
MCPA	0.00005	4

6-dinitrobenzenamine] in peppers from Obrenovac were significantly higher than the national MRL. Likewise, concentrations of bifenthrin [2-methyl[1,10-biphenyl]-3-yl) Methyl3-(2-chloro-3,3,3-trifluoro-1-propenyl)-2,2-dimethylcyclopropanecarboxylate] and cypermethrin [cyano (3-phenoxyphenyl) methyl 3-(2,2-dichloroethenyl)-2,2-dimethylcyclopropanecarboxylate] in potatoes at several locations were several times higher than the standard national and international MRL values. Furthermore, concentrations of chlorpyrifos in onion samples at one location was over 20 times the permitted level (Table 11.7), which points to their inappropriate use.

11.2.3 Radionuclides

Radionuclides reach the surface of the soil and vegetation mainly in the form of solid particles or through rainfall. Analyses of soils from localities in the regions of Belgrade, Novi Sad, Subotica, Užice, Niš and Zaječar show that the activity of natural radionuclides in soils is within the limits for the mean values for these regions, i.e. the activity ratio of $^{238}U/^{235}U$ in samples is comparable to their ratio in natural uranium (^{214}U). Due to the long half-life of ^{137}Cs, its activity in soil is significant. ^{137}Cs activity in uncultivated soil ranged from 3.3 Bq/kg in Belgrade to 340 Bq/kg on Zlatibor Mt. in the vicinity of Užice and from 2.1 Bq/kg in Belgrade to 136.2 Bq/kg on Zlatibor in cultivated soil. The specific activity of 90Sr measured in uncultivated soil ranges from <0.037 Bq/kg in Belgrade to 3.73 Bq/kg on Zlatibor, and from <0.037 Bq/kg in Lazarevac to 4.62 Bq/kg on Zlatibor for cultivated land (National Strategy of the Sustainable Use of Natural Resources 2012). Hence, the presence of depleted uranium in the soils of Serbia in the regions of Belgrade, Niš, Užice, Zaječar, Novi Sad and Subotica has not been established. However, some localities in southern Serbia had significant levels of organic-bonded depleted uranium in surface soil. During the NATO bombing, numerous locations were under a heavy and prolonged attack, which resulted in the destruction of the terrain and disruption of the forest ecosystem (UNEP/UNCHS BTF 1999).

Namely, in the 1999 bombing of the Federal Republic of Yugoslavia, NATO forces used ammunition containing depleted uranium. Therefore, soil in contaminated areas of Serbia had a depleted uranium (DU) isotopic signature. After the military activities, most of the kinetic DU penetrators or their fragments remained buried in the ground in certain geomorphological and geochemical environments exposed to local weathering conditions. The radioactivity of ^{238}U; ^{226}Ra, ^{232}Th, ^{40}K and ^{137}Cs in sediments and soil measured by Esposito et al. (2002) revealed high ^{238}U activity and an unnatural $^{235}U/^{238}U$ activity ratio in the vicinity of DU shells. Recent radioecological research at Kopaonik Mt. revealed higher mean activity concentrations of radionuclides in the soil than the global average, with the external radiation hazard index higher than one, implying an radiation hazard in two sites, Lisina and Pančićev vrh (Mitrović et al. 2016), Table 11.8 and Fig. 11.17.

The cleaning up of the DU left over after the NATO airstrikes, along with detailed dosimetric screening and

Table 11.8 Natural and artificial radionuclides in soil samples collected in Kopaonik (Bq kg^{-1})

Sampling site	Year	^{40}K	^{238}U	^{226}Ra	^{232}Th	^{137}Cs
Gobelja	1999[a]	760 ± 110	<137	23 ± 5	52 ± 4	<0.6
	2013[b]	722 ± 21	62 ± 4	65 ± 4	63 ± 2	30 ± 1
Pančićev vrh	1999[a]	1190 ± 60	160 ± 30	115 ± 4	150 ± 4	5.3 ± 1.2
	2013[b]	1291 ± 39	160 ± 11	152 ± 11	151 ± 10	4.2 ± 0.2

[a]Esposito et al. (2002)
[b]Mitrović et al. (2016)

Fig. 11.17 Pančićev vrh (*photo* https://www.google.rs/search?q=pancicev+vrh)

decontamination, was performed by the Radiation and Environmental Protection Department of the Vinča Institute of Nuclear Sciences between 2002 and 2007 at Pljačkovica Mt., Borovac, Bratoselce and Reljan, all located in southern Serbia, in coordination with the Agency for Radiation Protection and Nuclear Safety of Serbia. Because of the loose soil at these locations, projectiles containing DU were found at depths of up to one meter. The projectiles, contaminated soil and all radioactive materials were disposed of as radioactive waste.

After the cleaning up, more dosimetric screening was performed. In 2011, the department collected and analysed samples from these locations in order to determine the activity concentrations of radionuclides and verify that the areas had been successfully cleared of DU (Sarap et al. 2014). Likewise, the results of a study of radionuclides in soil samples in Bujanovac revealed no evidence of DU contamination of the environment 5 years after the 1999 military action (Popović et al. 2008).

11.3 Climate Change

Climate change has a number of undesirable effects, from the economic damage caused by periodic climatic events (floods, drought, etc.) to decreasing agricultural yields with species preferring cold climates under threat. To a great extent, global warming and the greenhouse effect are the results of industrial development because the increase in gas emissions into the atmosphere leads to increasing average temperatures on the Earth's surface as well as an increase in average precipitation levels, mainly in the northern hemisphere. In terms of soil, the effects of climate change are above all linked to increased warming, which has an impact on the loss of soil organic matter. Namely, the increase in temperature leads to more intensive mineralization and in turn to the loss of organic matter through a reduction in the total amount of soil and disturbance of soil structure, and a global increase in CO_2 concentrations in the atmosphere. Hence, with a reduction in the amount of soil organic matter and in turn total soil fertility, what results is the accelerated and permanent loss of soil due to erosion. At the same time, the cycling of C, N, P, S and other biogenic elements is disrupted, resulting in a reduction in the productive capacity of the soil and its biogenity as organic matter is a source of C for soil microorganisms. In addition, climatic factors also have a significant effect on other chemical properties of the soil, and indirectly on its total fertility (Kadović 2013). Namely, an increase in precipitation is linked to soil acidification through the washing away of base elements, some of which are important nutrients for plants. In the same way, accelerated mineralization leads to the increased mobility of potentially toxic elements, such as heavy metals. An excess of moisture also results in the level of groundwater rising, meaning leached salts reach the plant root zone (secondary salinization), and is one more way in which the soil's productive capacity decreases.

Statistical analysis of the long-term data of annual air temperatures between 1951 and 2000 has shown that over the last two decades of the twentieth century in particular, there was an extremely high rise in annual air temperatures across the whole of Serbia. With regard to annual precipitation, results showed a negative linear trend, more distinctive in south-eastern regions of Serbia and less so in the northern half of Vojvodina. At the same time, the number of days with snow cover decreased considerably (the linear trend is negative), causing on the one hand earlier and less spring run-off, and on the other, greater potential for summer drought. Only a significant rise in summer rain can make up for this deficit, which cannot be expected in the near future (COST, Serbian National Report 2009).

Over the last few decades, the frequency of flooding in Serbia and also neighbouring regions has increased significantly. Two severe floods occurred in Serbia in 1979 and 1999, as well as in 2005 and 2006. In 2005, two municipalities in Serbia's northern province of Vojvodina were flooded by the Tamiš River due to a flood wave from Romania. This flood affected 25,000 inhabitants with more than 600 houses submerged.

In 2006, the Zapadna Morava River flooded more than 1000 ha of fertile soil in the region of Požega. More than 600 ha near the city of Smederevo were destroyed by water and thousands of hectares were flooded due to the Kolubara and Jadar rivers and some smaller rivers bursting their banks. A number of landslides occurred in central and western parts of Serbia. Later that year, another wave of floods occurred in Serbia. Namely, two weeks of flooding across wide areas of central Europe, caused by melting snow and heavy rains, swelled tributaries draining into the Danube. Out of 240,000 ha of arable land under threat in northern Serbia, more than 122,000 ha were flooded, over 112,000 ha were threatened by groundwater, with landslides threatening more than 5500 ha. Those areas under greatest threat were the districts of southern Bačka with more than 50,000 ha and central Banat with 42,000 ha under water (Serbian Environmental Protection Agency 2007).

The last severe flooding occurred in spring 2014 in Serbia and Bosnia and Herzegovina, and to a lesser extent in Croatia and Slovenia, resulting in the degradation and pollution of agriculture soil. As a consequence of this latest flooding, 486,700 ha of arable soils were flooded, or an estimated 10% of the total area of agricultural soils in Serbia. As soon as the waters receded, a clean-up operation was carried out on approximately 335,400 m^2 of land (Serbian Government Report on Floods 2014). Pollution of flooded arable soils with heavy metals such as Ni, Pb, Cr and As was proved by their concentrations found to be above the maximum allowed levels. In addition, in Western Serbia over 100,000 m^3 of tailing slurry was released into river surrounding streams and rivers (Fig. 11.18).

On the other hand, drought is quite a common natural hazard in Serbia with serious damage to the national economy, especially to agriculture and water resources. It creates unfavourable soil conditions for plant development and

Fig. 11.18 Obrenovac municipality before and after heavy floods 2014 (*photo* http://www.rgz.gov.rs/template4.asp?PageName=poplave_2014&MenuID=0040122)

growth and changes in soil structure. It triggers soil changes, particularly in conventional tillage systems, such as increased fracturing, crusting and the deterioration of soil structure and aggregation. Soil crusting has become a problem as soil surfaces have dried and hardened, especially under conventional tillage with poor crop residue cover. The changes in soil structure have a negative impact on soil–water–plant relationships, especially root system development, which reduces above-ground biomass and grain yields. Namely, poor water availability and high temperatures result in significant stress during critical phases of the development of cultures, which become prone to pathogens and reduced nutrient availability and uptake by plants.

Dry years in Serbia were particularly frequent in the last two decades of the twentieth century. Over the three decades since 1981, 12 years had lower than average annual precipitation, 8 years were droughty, 3 exceptionally droughty and one, the year 2000, was extremely droughty. Over the same period, 12 years had total precipitation exceeding the average and 4 of those can be described as rainy years (Serbian Environmental Protection Agency 2007). The Standardized Precipitation Index series indicates the increasing frequency of droughts in Serbia between 1981 and 2000, with the most extreme drought in 2000 (Spasov et al. 2002). This drought, with extremely high air temperatures and without snow cover during winter 2000/2001, was just the latest in a series of climatic extremes in the region of southern Europe. A negative trend of annual precipitation is characteristic for most of Serbia, particularly in eastern parts of the country, peaking in the region of Negotin. A positive precipitation trend is characteristic for the areas of Zlatibor Mt., Pešter plateau, parts of Kopaonik Mt. and, slightly less so, the region of Bačka (Serbian Environmental Protection Agency 2007).

References

Adriano DC (1986) Trace elements in the terrestrial environment. Springer, NY

American Coal Ash Association (ACAA) (2001) In: Proceedings of the 14th international symposium on management and use of Coal Combustion Products (CCPs). ACAA, Alexandria, VA

Antić S, Jakovljević M, Kostić N, Tomić Z, Jovšić D (2005) Distribution and content of Ni and Cr in physical fractions of soils. In: Proceedings of XI congress of soil science society of Serbia and Montenegro, Budva, Serbia and Crna Gora, pp 63–76

Banković S, Medarević M, Pantić D, Petrović N (2008) National forest inventory of the Republic of Serbia. Forest Belgrade 3:1–16 (in Serbian)

Belanović S, Knežević M, Kadović R, Danilović M (2003) Distribution of some heavy metals in the soils of beech communities in Đerdap National Park. Bull Fac Forest Belgrade 88:17–25 (in Serbian)

Belanović S, Beloica J, Perović V, Alnaass SN, Knežević M, Kadović R, Mihajlović B (2012) Current state of threats of soil resources in Rasina catchment upstream from 'Ćelije' water reservoir. In: International conference on land conservation—LANDCON 1209, 17–21 Sept 2012, Danube Region/Republic of Serbia, pp 27–28

Bugarin R, Đukić N, Sedlar A (2010) Influence of pesticide application technique on soil pollution. Contemp Agric Eng 36(2):107–116

Cao Z, Selic E, Herbell JD (2008) Utilization of fly ash from coal-fired power plants in China. J Zhejiang Univ Sci A 9(5):681–687

Central European Development Forum, CEDEF (2014) CEDEF guide. Recommendations for the management of municipal and industrial waste in Belgrade and Serbia, Central European Development Forum, Belgrade, p 112 (in Serbian)

Cooper T, Pezold T, Keenleyside C, Đorđević-Milošević S, Hart K, Ivanov S, Redman M, Vidojević D (2010) Developing a national agri environment programme for Serbia. IUCN Programme Office for South-Eastern Europe, Gland, Switzerland and Belgrade, Serbia, p 88

COST Action FP0703—ECHOES (2009) Expected climate change and options for European silviculture country report. Serbian National Report, Serbia

Dajić Z (2006) Salt stress–salinity and tolerance mechanisms in plants. In: Madhava Rao KV, Raghavendra AS, Reddy KJ (eds) Physiology and molecular biology of stress tolerance in plants. Springer, Netherlands, pp 41–99

Đalović I, Maksimović I, Kastori R, Jelić M (2010) Mechanisms of adaptation of small grains to soil acidity. Proc Nat Sci 118:107–120

Dragović S, Ćujić M, Slavković-Beškoskim L, Gajić B, Bajat B, Kilibarda M, Onjia A (2013) Trace element distribution in surface soils from a coal burning power production area: a case study from the largest power plant site in Serbia. CATENA 104:288–296

Dragović R, Gajić B, Dragović S, Đorđević M, Đorđević M, Mihailović N, Onjia A (2014) Assessment of the impact of geographical factors on the spatial distribution of heavy metals in soils around the steel production facility in Smederevo (Serbia). J Clean Prod 84:550–562

Dželetović Ž, Filipović R (1995) Grain characteristics of crops grown on power plant ash and bottom slag deposit. Resour Conserv Recy 13:105–113

Electric Power Industry of Serbia (EPS), Technical Report (2010) http://www.eps.rs/Eng/Tehnicki%20Izvestaji//TEH_Godisnjak2010_en_web.pdf

Environmental Protection Agency (2007) In: Karadžić B, Mijović A (eds) Environment in Serbia: an indicator—based review, 1st edn. Serbian Environmental Protection Agency, Belgrade, p 168

Erosion Map of Serbia (1983) Institute for Forestry and the Wood Processing Industry. Belgrade, Serbia

Esposito M, Polić P, Bartolomei P, Benzi V, Martinelli V, Cvetković O, Damjanov V, Simić M, Žunić Ž, Zivančević B, Simić S, Jovanović V (2002) Survey of natural and anthropogenic radioactivity in environmental samples from Yugoslavia. J Environ Radioact 61 (3):271–282

Francois LE, Maas EV (1994) Crop response and management of salt-affected soils. In: Pessarakli M (ed) Handbook of plant and crop stress, Marcel Dekker, New York, pp 169–201

Government of the Republic of Serbia (2014) Natural disaster report on the floods which occurred in the Republic of Serbia and the measures taken to save the local population and protect those places at risk of flooding ('Official Gazette of the Republic of Serbia', no. 52/14) (in Serbian)

Haynes RJ (2009) Reclamation and revegetation of fly ash disposal sites—challenges and research needs. J Environ Manage 90:43–53

Jakovljević M, Kostić N, Stevanović D, Blagojević S, Wilson M, Lj Martinović (1997) Factors influencing the distribution of heavy metals in the alluvial soils of the Velika Morava river valley, Serbia. App Geochem 12:637–642

Jakovljević M, Kresović M, Blagojević S, Antic-Mladenovic S (2005) Some negative chemical properties of acid soils. J Serb Chem Soc 70(5):765–774

Kabata-Pendias A, Pendias H (2001) Trace elements in soils and plants, 3rd edn. CRC Press, USA

Kadović R (2013) The potential of Serbian forests for climate change mitigation and environmental protection. In: Andjelković M (ed) Energy and environment, Scientific meetings, Book CXLIII, Serbian Academy of Science and Arts (SANU), pp 493–514

Kadović R, Belanović S (2009) Analysis of the acidification process of forest soils at 'Kopaonik' national park. Bull Fac Forest Belgrade 100:95–110 (in Serbian)

Kadović R, Knežević M (2002) Heavy metals in the forest ecosystems of Serbia. Faculty of Forestry, Belgrade University, Serbian Ministry for the Protection of Natural Resources and the Environment, Belgrade, p 278 (in Serbian)

Kadović R, Knežević M (2004) Forest health monitoring in the Republic of Serbia, Annual report, ICP forests 2003, level 1. Serbian Ministry of Science and Environmental Protection—Department for Environmental Protection, Serbian Ministry of Agriculture, Forestry and Water Management—Forestry Department, Belgrade, 1–79 (in Serbian)

Kadović R, Košanin O, Belanović S, Knežević M (2005) Heavy metals in the organic soil layer of beech forests in Serbia. Bull Fac Forest Belgrade 92:55–67 (in Serbian)

Kostić N (1999) Pedomineralogy and geochemistry of soils of Vojvodina. Mineralogija Godišnjak JAM II(2):16–28

Kostić N (2001) Pedomineralogy and geochemistry of some important soils in Serbia. In: Proceedings of 10th congress of Yugoslav soil science society (JDPZ), Vrnjačka Banja, Serbia, pp 134–148

Kostić O (2008) The influence of a monoculture of spruce and Douglas fir on pedogenesis and soil properties on Mt. Maljen. Master's thesis, Faculty of Forestry, Belgrade University, p 137 (in Serbian)

Kostić N, Hadžić V, Protić JN (1998a) Manganese geochemistry of soils of Vojvodina, Šumadija and Northern Pomoravlje. In: Proceedings of 2nd Balkan symposium of field crops. Ecology and physiology, vol 2. Novi Sad, Yugoslavia, 16–20 June 1998, pp 9–20

Kostić N, Wilson JM, Živković M, Bain D (1998b) Mineralogy and geochemical speciation of heavy metals in some serpentine soils of Serbia. In: Proceedings of the 16th world congress of soil science, vol 2. Montpellier France, pp 481–492

Kostić N, Dangić A, Jakovljević M (2002) Geochemistry of iron in soils of Vojvodina, Sumadija and Northern Pomopravlje. Zbornik Matica Srpska No 102, Novi Sad, pp 145–157

Kostić O, Mitrović M, Knežević M, Jarić S, Gajić G, Djurdjević L, Pavlović P (2012a) The potential of four woody species for the revegetation of fly ash deposits from 'Nikola Tesla-A' thermoelectric plant (Obrenovac, Serbia). Arch Biol Sci Belgrade 64(1):145–158

Kostić O, Mitrović M, Jarić S, Djurdjević L, Gajić G, Pavlović M, Pavlović P (2012b) The effects of forty years of spruce cultivation in a zone of beech forest on Mt. Maljen (Serbia). Arch Biol Sci Belgrade 64(3):1181–1195

Kostić O, Mitrović M, Vitorović G, Jarić S, Pavlović D, Pavlović M, Gajić G, Pavlović P (2015) The impact of industrial plants on potential soil contamination in peripheral settlements in the Belgrade region. In: Proceedings from the 'Sustainable soil use' conference, Rimski Šančevi, 10th Sept 2015 [Electronic resource]. Institute of Field and Vegetable Crops, Novi Sad, pp 139–146 (in Serbian)

Kresović M, Jakovljević M, Blagojević S, Maksimović S (2009) Specific transformations of mineral forms of nitrogen in acid soils. J Serb Chem Soc 77(6):93–102

Maas EV (1990) Crop salt tolerance. In: Tanji KK (ed) Agricultural salinity assessment and management, American Society of Civil Engineers, ASCE Manuals & Reports on Engineering Practice, No 71, New York, pp 262–304

Marković M, Cupać S, Đurović R, Milinović R, Kljajić P (2010) Assessment of heavy metal and pesticide levels in soil and plant products from the agricultural area of Belgrade, Serbia. Arch Environ Con Tox 58:341–351

Medarević M, Banković S, Pantić D (2003) The condition of beech forests in Serbia. Special issue: Managing beech forests. Forest Belgrade 1–2:5–23 (in Serbian)

Mitrović M, Pavlović P, Lakušić D, Stevanović B, Djurdjević L, Gajić G, Kostić O (2008) The potential of festuca rubra and calamagrostis epigejos for the revegetation of fly ash deposits. Sci Total Environ 407(1):338–347

Mitrović M, Kostić O, Jarić S, Karadžić B, Djurdjević L, Gajić G, Lj Oberan, Pavlović D, Pavlović M, Pavlović P (2012) Photosynthetic efficiency of four woody species growing on fly ash deposits of a Serbian 'Nikola Tesla A' thermoelectric plant. Polish J Environ Stud 21(5):1339–1347

Mitrović B, Ajtić J, Lazić M, Andrić V, Krtić N, Vranješ B, Vićentijević M (2016) Natural and anthropogenic radioactivity in the environment of Kopaonik mountain, Serbia. Environ Pollut 215:273–279

Mrvić V (2010) Acid soils of western Serbia and their further acidification. Geophysical research abstracts 12, EGU2010-3313, EGU General Assembly 2010

Mrvić V, Zdravković M, Sikirić B, Čakmak D, Kostić-Kravljanac Lj (2009) Harmful and hazardous elements in soil. In: Mrvić V, Antonović G, Martinović Lj (eds) The fertility and content of hazardous and harmful substances in the soils of central Serbia, Institute of Soil Science, Belgrade, Serbia, p 222 (in Serbian)

National Program of Environmental Protection (2010) Act establishing a national program for environmental protection, 'Official Gazette of the Republic of Serbia', no. 12/2010 (in Serbian)

National Strategy of the Sustainable Use of Natural Resources (2012) 'Official Gazette of the Republic of Serbia', no. 33/2012 (in Serbian)

Pavlović P, Mitrović M (2013) Thermal power plants in Serbia—the impact of ash on soil and plants. In: Anđelković M (ed) Energy and the environment, Scientific meetings, Book 4, Serbian Academy of Sciences and Arts (SANU), Belgrade, pp 429–433 (in Serbian)

Pavlović P, Mitrović M (2016) Soil contamination. In: Belanović-Simić S (ed) Degradation and protection of soil. University of Belgrade, Faculty of Forestry, pp 107–137

Pavlović P, Mitrović M, Djurdjević L (2004) An ecophysiological study of plants growing on the fly ash deposits from the 'Nikola Tesla-A' thermal power station in Serbia. Environ Manage 33:654–663

Pavlović P, Mitrović M, Djurdjević L, Gajić G, Kostić O, Bojović S (2007) Ecological potential of *Spiraea van-houttei* (Briot) Zabel for urban (the City of Belgrade) and fly ash deposit (Obrenovac) landscaping in Serbia. Polish J Environ Stud 16(3):427–431

Pavlović P, Mitrović M, Đorđević D, Sakan S, Slobodnik J, Liška I, Csany B, Jarić S, Kostić O, Pavlović D, Marinković N, Tubić B, Paunović M (2016) Assessment of the contamination of riparian soil and vegetation by trace metals—the Danube River case study. Special issue: River conservation under multiple stressors: integration of ecological status, pollution and hydrological variability. Sci Tot Environ 540:396–409

Perović V, Beloica J, Kadović R, Todosijević M, Čakmak D, Milčanović V, Belanović S (2012) The effect of acid deposition on soil solution reaction in the drainage basin of the Rasina river. Erosion Belgrade 38:45–61 (in Serbian)

Pessarakli M, Szabolcs I (1994) Soil salinity and sodicity as particular plant/crop stress factors. In: Pessarakli M (ed) Handbook of plant and crop stress, Marcel Dekker, New York, pp 1–15

Popović D, Todorović D, Frontasyeva M, Ajtić J, Tasić M, Rajšić S (2008) Radionuclides and heavy metals in Borovac, southern Serbia. Environ Sci Pollut Res 15:509–520

Popović AR, Djordjević DS, Relić DJ, Djinović-Stojanović JM (2013) Thermal power plants in Serbia as possible sources of contamination of surface and groundwater macro- and microelements. Energy and the environment, Serbian Academy of Sciences and Arts, Scientific meetings, Book CXVIII, Department of Chemical and Biological Sciences, Book 4, Belgrade, pp 373–401 (in Serbian)

Project Report (2001) 'Solving the problems of ash emission and deposition at 'Nikola Tesla' thermal power plant (TENT) in Obrenovac', Subproject 3: 'Assessment of the current situation', Activity 3.5: 'Evaluation of soil quality in the vicinity of TENT', Subproject 9: 'Reducing the harmful effects of ash deposits', Activity 9.4.: 'Recultivation of fly ash deposit sites at the end of exploitation', Ministry for the Protection of Natural Resources and the Environment of the Republic of Serbia, Project implementers: Institute for Physical Chemistry, Institute for Biological Research 'Siniša Stanković', Faculty of Mining and Geology, Faculty of Chemistry, Belgrade University. User: Electric Power Industry of Serbia (EPS) (in Serbian)

Project Report (2002) 'Solving the ecological problems resulting from operations at the 'Nicola Tesla A and B' thermal power plants', Subproject: 'Proposal for possible solutions to the biological recultivation of ash and slag deposits at TENT-A', Ministry for the Protection of Natural Resources and the Environment of the Republic of Serbia, Project implementers: Institute for Physical Chemistry, Institute for Biological Research 'Siniša Stanković', Faculty of Mining and Geology, Faculty of Biology, Faculty of Chemistry, Belgrade University. User: Electric Power Industry of Serbia (EPS) (in Serbian)

Protić R, Rajković S, Janković S, Protić N (2004) Grain yield of stubble crops in the Republic of Serbia and some more important agricultural measures. Sci Res PKB INI Agroeconomic 10(1):79–90

Quevauville P, Rauret G, Muntau H, Ure AM, Rubio R, López-Sánchez JF, Fiedler HD, Griepink B (1994) Evaluation of sequential extraction procedure for determination of extractable trace metal contents in sediments. Fresenius J Anal Chem 348:808–814

Rademacher P (2001) Atmospheric heavy metals and forest ecosystems. Federal Research Centre for Forestry and Forest Products (BFH), UN/ECE, Geneve

Regulation on a programme for the systematic monitoring of soil quality, soil degradation risk assessment indicators, and methodology for the development of remediation programmes, Official Gazette of the Republic of Serbia, no. 88/2010 (in Serbian)

Regulations on permitted levels of hazardous and harmful substances in soil and irrigation water and methods for their testing ('Official Gazette of the Republic of Serbia', no. 23/94) (in Serbian)

Relić D, Đorđević D, Sakan S, Anđelković I, Miletić S, Đuričić J (2011) Aqua regia extracted metals in sediments from the industrial area and surroundings of Pančevo, Serbia. J Hazard Mat 186:1893–1901

Ristić R, Kašanin-Grubin M, Radić B, Nikić Z, Vasiljević N (2012) Land degradation at the Stara Planina Ski resort. Environ Manage 49:580–592

Ružić D, Poznanović N (2009) Hazardous agricultural materials in the autonomous province of Vojvodina. Contemp Agric Eng 35 (3):191–200 (in Serbian)

Sakan S, Đorđević D, Dević G, Relić D, Anđelković I, Đuričić J (2011) A study of trace element contamination in river sediments in Serbia using microwave-assisted *aqua regia* digestion and multivariate statistical analysis. Microchem J 99:492–502

Sarap NB, Janković MM, Todorović DJ, Nikolić JD, Kovačević MS (2014) Environmental radioactivity in southern Serbia at locations where depleted uranium was used. Arh Hig Rada Toksikol 65:189–197 (in Serbian)

Sedlar A, Đukić N, Bugarin R (2007) Initial inspections of sprayers in Serbia. Contemp Agric Eng 1–2:12–19 (in Serbian)

Spasov P, Spasov D, Petrović P (2002) Changes in drought occurrences in Serbia. In: Proceedings of international conference on drought mitigation and prevention of land desertification, Bled, Slovenia, 21–25 April 2002. CD-ROM, International Commission on Irrigation and Drainage (ICID), Ljubljana, Slovenia

Statistical Office of the Republic of Serbia (2008) Statistical pocketbook of the Republic of Serbia. Belgrade

Statistical Office of the Republic of Serbia (2012) Waste statistics and waste management in the Republic of Serbia 2008–2010. Statistical Office of the Republic of Serbia, Belgrade (in Serbian)

The Privatization Agency—Republic of Serbia (2006) Environmental assessment of the damage resulting from the past operations of the Bor mine and smelting combine—final report. Environmental Resources Management, the trading name of ERM Italia S.r.l., p 518 (in Serbian)

UNEP/UNCHS Balkans Task Force (BTF) (1999) Assessment of the damage to biodiversity in protected areas of the Federal Republic of Yugoslavia. http://www.grid.unep.ch/btf/missions/september/biodiversity.pdf

United Nations Environment Programme (UNEP) (2002) Depleted uranium in Serbia and Montenegro. Post-conflict environmental assessment in the Federal Republic of Yugoslavia. http://postconflict.unep.ch/publications/duserbiamont.pdf

Vanmechelen L, Groenemans R, Van Ranst E (1997) Forest soil condition in Europe, results of a large-scale soil survey. EC, UN/ECE, ICP Forests

Vidojević D, Manojlović M (2006) Overview of the soil information and policies in the Republic of Serbia. ESBN workshop, 28–30 Sept 2006, Joint Research Center EC, Zagreb, p 23

Waste management strategy for 2010–2019 (2010) Official Gazette of the Republic of Serbia no. 29/10 (in Serbian)

Životić MM, Stojiljković DD, Jovović MA, Čudić VV (2012) Potential usage of fly ash and bottom ash from the deposit sites at the 'Nikola Tesla' power plant. Chemical Industry Belgrade 66(3):403–412 (in Serbian)

Zlatić M, Dragović N, Todosijević M, Tomićević J (2008) Contribution of the World Overview of Conservation Approaches and Technologies—WOCAT programme to the Serbian practice and education, 15th ISCO congress 'Soil and water conservation, climate change and environmental sensitivity', Book of abstracts, 18–23 May 2008, Budapest

Index

A
Abrasion terraces, 66
Acrohistosol, 171
Aeolian deposits, 102, 104
Aeroprecipitates, 94
Alluvial deposits, 25, 42, 52, 92, 103, 117, 157, 160, 182
Alluvial plains, 2, 61, 62, 79
Alluvial terrace, 62, 164, 212, 213
Amphibole, 73
Amphibole schist, 58, 74
Amphibolite facie, 58, 72, 74
Andesite, 57, 76, 78, 84, 109, 111
Andosol, 83, 148, 150–154
Anthrosol, 145
Anthrosol hortic, 145
Arenosol, 25, 27, 79–81, 91, 104–106, 183, 184, 212
Automorphic soils, 90, 101, 161
Azonal vegetation, 41, 184

B
Basalt, 76, 101, 109
Beech forest, 45
Beidellite, 66, 81, 84
Bentonite, 66, 78
Biodiversity, 51, 181, 184, 196
Biotite, 79
Boreal forest, 41
Boreal grassland, 42, 50
Brunipodzol, 26, 94, 143, 144

C
Calcite, 62, 63, 79–81, 104, 213, 214
Cambisol dystric, 26, 28, 45, 81, 82, 84, 93, 101, 109, 131–135, 195, 203, 211, 213, 214
Cambisol eutric, 25, 27, 28, 43, 48, 82, 84, 93, 101, 118, 128–131, 157, 163, 195, 213, 214
Carboniferous, 57–59, 71, 73, 74
Chasmophytic vegetation, 42, 50, 51
Chernozem, 25, 28, 50, 80–82, 92, 117–121, 123, 130, 184, 195, 196, 211, 213
Chlorite, 58, 73, 74, 81, 82, 84
Clay, 59, 61–64, 66, 71, 82–85, 91, 93, 103–106, 110, 112–114, 120, 127, 134, 158, 163, 167, 175, 178, 212
Clay fraction, 79, 80, 82–84
Clay minerals, 79
Climate, 31, 52
Climate change, 184, 217
Colluvial soils, 25, 91, 105, 181
Conglomerates, 66, 68
Coniferous forests, 39, 41, 45–47, 51, 131, 145, 183
Correnite, 82
Cretaceous, 58, 59, 68, 71, 84
Crystalline schists, 58, 59, 72, 76, 93

D
Dacite, 58, 72, 74, 76, 78, 84, 109, 111, 151
Deciduous forests, 41
Deposol, 28, 94, 145, 146, 197
Depression, 59, 61, 63, 66
Devonian, 57, 58, 68
Diabase, 56, 60, 61, 75, 76, 101, 109
Diorite, 101
Dolomite, 55, 57–59, 61–64, 71, 79, 80, 91, 101, 104, 113, 116
Drained gyttja, 97
Drained sapropel, 97
Dunite, 75, 76
Dy, 97

E
Endemic plants, 113, 171, 172
Eolian deposits, 61, 64, 93, 157
Erosion, 4, 55, 66, 101–105, 107, 109, 113, 192
Eugley, 28, 79–81, 96, 165, 170, 171, 182, 212
Exchangeable cations, 214
Extrazonal vegetation, 41

F
Feldspars, 71, 79–81, 104
Flotisol, 94
Fluvisol, 25, 26, 28, 29, 80–82, 95, 146, 147, 160–164, 166, 181, 184, 211, 213
Fly ash, 26, 145–148, 197, 199, 200, 202, 208
Flysch, 55–59, 61, 67–69, 71, 76, 79, 81, 109, 111
Forest fringe vegetation, 49

G
Gabbro, 58, 59, 75, 76, 101, 109
Glacial deposits, 103, 114
Gleysol, 24, 26, 95, 96, 165, 167, 182
Gleysol humic, 25, 28, 29, 79–81, 95, 164, 182, 203, 211–214

P. Pavlović et al., *The Soils of Serbia*, World Soils Book Series,
DOI 10.1007/978-94-017-8660-7

Gneiss, 58, 59, 71, 73, 74, 102, 109–111
Goethite, 79, 83, 84
Granite, 58, 66, 71, 72, 74, 75, 101, 102, 109, 111
Granitoid, 58, 59, 72–74, 76
Gravel, 42, 59, 61, 62, 64, 92, 104, 113, 114
Greenschist, 58, 57–59, 73, 74
Greenschist facie, 73
Greywacke, 61, 67, 68
Gyttja, 97, 170, 180

H
Halloysite, 66, 82–84
Halomorphic soils, 25, 41, 50, 90, 175, 184
Halophytic vegetation, 42, 50
Harzburgite, 75, 76, 83
Heath vegetation, 42, 50
Heavy metals, 191, 197, 201, 206–208, 210–213, 217
Histosol, 96, 169, 172, 182
Holocene, 59
Hornblende, 76
Horsts, 58, 59
Hortisol, 94, 145
Humofluvisol, 79–81, 95, 163, 165, 167, 181
Humogley, 25, 28, 29, 79–81, 95, 164, 165, 168, 182, 211–214
Hydromorphic soils, 80, 90, 157, 168, 172, 196
Hydromeliorated soil, 165, 173

I
Illite, 66, 71, 79, 81–84, 119
Intercalations, 67, 71, 72, 74, 78

J
Jurassic, 55

K
Kalkocambisol, 26, 28, 84, 93, 101, 127, 136–138
Kalkomelanosol, 26, 27, 92, 101, 125–128, 135–137, 184, 203
Kaolinite, 66, 71, 79, 81–84, 119
Karst, 94, 101

L
Lacustrine vegetation, 158
Lake terrace, 66
Landforms, 60, 62
Land use, 28, 102, 104, 106, 113, 117, 121, 124, 127, 130, 134, 136, 139, 142, 143, 145, 154, 159, 163, 164, 167, 171, 172, 176, 178, 181, 182, 187, 195
Lepthynolithe, 72, 74
Leptosol, 83
Leptosol dystric, 51, 203
Leptosol eutric, 25, 28, 46, 48, 51, 83, 92, 107, 161, 201, 203, 211–213
Limestone, 55, 57–59, 61, 64, 67, 71, 74, 76, 78, 79, 84, 91, 92, 94, 101–103, 113–115, 117
Lithosol, 27, 46, 51, 91, 101, 102, 104, 109, 110, 115, 184, 203
Loess, 25, 57, 59, 62, 65, 79, 91–93, 103, 104, 113, 114, 116, 118–121, 178
Loess plateau, 2, 4, 61, 62, 65, 66, 79, 81, 117, 213
Loess terrace, 25, 50, 62, 117, 120, 213
Luvisol, 25–27, 45, 79, 81, 82, 94, 130, 140–142, 157, 211–214

M
Marl, 59, 61, 64, 66, 92, 103, 113, 114, 116
Marlstone, 59, 61, 67, 68, 70, 71, 78, 81, 82, 102–104
Mesozoic, 55, 58, 59, 71, 74, 77, 79, 80
Micas, 71, 79, 80
Mica-schist, 58, 72, 74, 109
Microcline, 80
Migmatite, 71–73
Miocene, 57–59, 63, 64, 66, 74, 76, 82
Mixed-layer-silicate (M-L-S), 79, 84
Montmorillonite, 66, 81, 119, 123, 175
Moraine, 92, 113, 116
Muscovite, 79

N
Neogene, 55–59, 64, 68, 76, 79, 81, 213

O
Oak forests, 42, 49
Olivine, 76
Ophiolite, 55–57, 76
Orthoclase, 79, 80

P
Palaeozoic, 55, 57
Parent material, 16, 31, 52, 59, 71, 79, 82, 150
Pesticides, 155, 191, 214, 215
Phyllite, 57, 74, 75, 84, 94, 109
Pillow lavas, 75, 76
Plagioclase, 76, 79, 80
Planohistosol, 169
Plano-acrohistosol, 172
Podzol, 26, 94, 109, 132, 143, 157, 171
Porphyrite, 56, 74, 77, 78
Protopedon, 97, 180
Psamophytic vegetation, 42, 184
Pseudogley, 25, 28, 81, 82, 93–95, 157–160, 164, 195, 203, 211–213
Pyroclastic material, 68, 148
Pyroxene, 76

Q
Quartz, 71, 73, 74, 79–81, 84, 91, 93, 94, 104, 107, 109
Quartz-latite, 74, 76, 83, 84, 111

R
Radionuclides, 215–217
Ranker, 25–28, 48, 51, 80–84, 92, 107–114, 130, 132, 184, 195, 201, 203, 206, 211, 213, 214
Regosol, 25, 46, 48, 51, 91, 102–104, 106, 109–111, 114, 115, 118, 157, 161, 203
Rendzina, 25, 43, 48, 51, 81, 82, 92, 113–117, 130, 184, 203, 211, 213
Rigosol, 94, 145
Riparian forests, 41, 42
River terrace, 2, 59, 61, 65, 158, 161, 163, 164

S
Sand plains, 61–63, 79
Sanidine, 76
Sapropel, 97, 170, 180

Scree vegetation, 51
Scrub vegetation, 41, 42, 48
Serpentine flora, 48
Serpentinite, 57, 58, 76, 93, 101, 102, 108, 109, 111–113
Shale, 56, 59, 61, 64, 66–68, 70, 71, 78, 93, 109
Silici-clastic rocks, 67
Siltstone, 69
Slates, 59, 74, 77
Smectite, 66, 71, 79, 81–84
Soil acidification, 193, 195
Soil erosion, 183, 191, 192
Soil mineralogy, 18
Soil pollution, 25, 26, 184, 191, 207, 208, 210
Soil salinization, 52, 89, 140, 194
Solonchak, 80, 97, 121, 175–178
Solonetz, 25, 27, 29, 80, 97, 121, 175–179, 196
Solontchak, 27, 29, 196
Spilite, 75, 76
Stagnosol, 28, 81, 213
Steppe vegetation, 42, 48, 50, 118, 121
Subalpine herbaceous vegetation, 42, 50
Subaquatic soils, 179, 180
Subaqueous (subhydric) soils, 90
Swampy forests, 41, 42

T
Talc, 79–81, 83
Tall-herb grasslands, 50
Terra rossa, 93, 138
Tertiary, 56, 57
Triassic, 56, 57
Tuff, 56, 78, 109, 111
Tuffite, 67, 78

U
Ultramafites, 76

V
Vermiculite, 79, 81, 82, 84, 85
Vertisol, 25, 27, 28, 43, 80–82, 92, 122–125, 130, 184, 211, 213, 214

W
Waste, 26, 191, 197
Water erosion, 192
Wind erosion, 191, 201
World Reference Base (WRB), 25–27, 46, 48, 79–84, 91, 95, 101, 104, 105, 107, 113, 117, 122, 125, 128, 131, 135, 136, 140, 143, 145, 150, 157, 160, 163–165, 169, 171, 172, 175, 177, 181, 184, 201, 203, 206, 211–214

Z
Zonal vegetation, 41